BERR

Department for Business
Enterprise & Regulatory Reform

Digest of United Kingdom Energy Statistics 2008

Production team: Iain MacLeay
Kevin Harris
Chris Michaels
and chapter authors

A National Statistics publication

London: TSO

ISBN 9780115155222

Digest of United Kingdom Energy Statistics

Enquiries about statistics in this publication should be made to the contact named at the end of the relevant chapter. Brief extracts from this publication may be reproduced provided that the source is fully acknowledged. General enquiries about the publication, and proposals for reproduction of larger extracts, should be addressed to Kevin Harris, at the address given in paragraph XXIX of the Introduction.

The Department for Business, Enterprise and Regulatory Reform reserves the right to revise or discontinue the text or any table contained in this Digest without prior notice.

About TSO's Standing Order Service

The Standing Order Service, open to all TSO account holders, allows customers to automatically receive the publications they require in a specified subject area, thereby saving them the time, trouble and expense of placing individual orders, also without handling charges normally incurred when placing ad-hoc orders.

Customers may choose from over 4,000 classifications arranged in 250 sub groups under 30 major subject areas. These classifications enable customers to choose from a wide variety of subjects, those publications that are of special interest to them. This is a particularly valuable service for the specialist library or research body. All publications will be dispatched immediately after publication date. A Standing Orders Handbook describing the service in detail and a complete list of classifications may be obtained on request. Write to TSO, Standing Order Department, PO Box 29, St Crispins, Duke Street, Norwich, NR3 1GN, quoting reference 12.01.013. Alternatively telephone 0870 600 5522 and select the Standing Order Department (option 2); fax us on 0870 600 5533; or finally e-mail us at book.standing.orders@tso.co.uk

---

National Statistics

National Statistics are produced to high professional standards set out in the National Statistics Code of Practice. They undergo regular quality assurance reviews to ensure that they meet customer needs. They are produced free from any political interference.

You can find a range of National Statistics on the Internet – www.statisticsauthority.gov.uk

---

Contents

Monthly and quarterly data are also available for Energy, Solid fuels and derived gases, Petroleum, Gas and Electricity at:

www.berr.gov.uk/energy/statistics/source/index.html

Information on Energy Prices is available at:

www.berr.gov.uk/energy/statistics/publications/prices/index.html

A list of tables

Chapter 7   Renewable sources of energy

Annex A   Energy and commodity balances, conversion factors and calorific values

Introduction

I   This issue of the Digest of United Kingdom Energy Statistics continues a series which commenced with the Ministry of Fuel and Power Statistical Digest for the years 1948 and 1949, published in 1950. The Ministry of Fuel and Power Statistical Digest was previously published as a Command Paper, the first being that for the years 1938 to 1943, published in July 1944 (Cmd. 6538).

II   The current issue updates the figures given in the Department for Business, Enterprise & Regulatory Reform's (BERR) *Digest of United Kingdom Energy Statistics 2007*, published in July 2007.

III   This printed and bound issue consists of seven chapters and four annexes. The first chapter deals with overall energy. The other chapters cover the specific fuels, combined heat and power and renewable sources of energy. The annexes cover conversion factors and calorific values, a glossary of terms, further sources of information and major events in the energy industries.

IV   This Digest is also available on the internet. Some additional information appears on the internet only. The tables on the internet are provided in Microsoft Excel format. Most internet versions of the tables include data for earlier years, which are not provided in the printed copy publication. For example commodity and energy balances (see VII and VIII, below) for 1998 to 2004 are included on the internet, and tables that show five years in this printed version show ten years in their internet form because page sizes are not a limiting factor. In addition, the following appear on the internet version only:

> Long term trends text and tables
> Major events from 1990 to 2008 - Annex D
> (only Major events for 2006 to 2008 appear in the printed and bound version)
> Energy and the environment – Annex E
> UK oil and gas resources - Annex F
> Foreign trade – Annex G
> Flow charts – Annex H
> Energy balance: net calorific values – Annex I
> Heat reconciliation – Annex J

V   Annual information on prices is included in the publication *Quarterly Energy Prices*. This is available together with *Energy Trends* on subscription from the Department for Business, Enterprise and Regulatory Reform. Further information on these publications can be found in Annex C.

VI   Where necessary, data have been converted or adjusted to provide consistent series. However, in some cases changes in methods of data collection have affected the continuity of the series. The presence of remaining discontinuities is indicated in the chapter text or in footnotes to the tables.

VII   Chapters 2, 3, 4, 5 and 7 contain production and consumption of individual fuels and are presented using *commodity balances*. A commodity balance illustrates the flows of an individual fuel through from production to final consumption, showing its use in transformation (including heat generation) and energy industry own use. Further details of commodity balances and their use are given in Annex A, paragraphs A.7 to A.42.

VIII   The individual commodity balances are combined in an *energy balance,* presented in Chapter 1, *Energy*. The energy balance differs from a commodity balance in that it shows the interactions between different fuels in addition to illustrating their consumption. The energy balance thus gives a fuller picture of the production, transformation and use of energy showing all the flows. Expenditure on energy is also presented in energy balance format in Chapter 1. Further details of the energy balance and its use, including the methodology introduced in the 2003 Digest for heat, are given in Annex A, paragraphs A.43 to A.58.

IX   Chapter 1 also covers general energy statistics and includes tables showing energy consumption by final users and an analysis of energy consumption by main industrial groups. Fuel production and consumption statistics are derived mainly from the records of fuel producers and suppliers.

X   Chapters 6 and 7 summarise the results of surveys conducted by AEA Energy & Environment on behalf of BERR. These chapters estimate the contribution made by combined heat and power (CHP) and renewable energy sources to energy production and consumption in the United Kingdom.

XI   Some of the data shown in this Digest may contain unpublished revisions and estimates of trade from additional sources.

Definitions

XII   The text at the beginning of each chapter explains the main features of the tables. Technical notes and definitions, given at the end of this text, provide detailed explanations of the figures in the tables and how they are derived. Explanations of the logic behind an energy balance and for commodity balances are given in Annex A.

XIII   Most chapters contain some information on 'oil' or 'petroleum'; these terms are used in a general sense and vary according to usage in the field examined. In their widest sense they are used to include all mineral oil and related hydrocarbons (except methane) and any derived products.

XIV   An explanation of the terms used to describe electricity generating companies is given in Chapter 5, paragraphs 5.50 to 5.51.

XV   Data in this issue have been prepared on the basis of the Standard Industrial Classification (SIC 2003) as far as is practicable. For further details of classification of consumers see Chapter 1, paragraphs 1.54 to 1.58.

XVI   Where appropriate, further explanations and qualifications are given in footnotes to the tables.

Proposed change to use net calorific values when producing energy statistics

XVII   A consultation was launched in the 2005 edition of DUKES seeking views of users as to whether Net Calorific Values (NCVs) should be used in place of Gross Calorific Values (GCVs). As a result of this consultation, BERR recognised that there are good arguments both for and against moving from GCV to NCV. However at present it has been concluded that there would be no demonstrable advantage to changing the method of presenting UK Energy statistics, and so GCVs continue to be used in this edition and will be used in future editions of the Digest. The fuel specific NCVs will continue to be published, and are shown in Annex A. The total energy balances on a net calorifc basis are now being produced as part of the internet version of the Digest, Annex I.

Geographical coverage

XVIII   The geographical coverage of the statistics is the United Kingdom. Shipments to the Channel Islands and the Isle of Man from the United Kingdom are not classed as exports. Supplies of solid fuel and petroleum to these islands are therefore included as part of United Kingdom inland consumption or deliveries.

Periods

XIX   Data in this Digest are for calendar years or periods of 52 weeks, depending on the reporting procedures within the fuel industry concerned. Actual periods covered are given in the notes to the individual fuel chapters

Revisions

XX   The tables contain revisions to some of the previously published figures, and where practicable the revised data have been indicated by an 'r'. The 'r' marker is used whenever the figure has been revised from that published in the printed copy of the 2007 Digest, even though some figures may have been amended on the internet version of the tables. Statistics on energy in this Digest are classified as National Statistics. This means that they are produced to the professional standards set out in the National Statistics Code of Practice and relevant protocols. The National Statistics protocol on revisions requires that "Each organisation responsible for producing National Statistics will publish and maintain a general statement describing its practice on revisions". The following statement outlines the policy on revisions for energy statistics.

Revisions to data published in the *Digest of UK Energy Statistics*.
It is intended that any revisions should be made to previous years' data only at the time of the publication of the Digest (ie in July 2008 when this Digest is published, revisions can be made to 2006 and earlier years). In exceptional circumstances previous years' data can be amended between Digest publication dates, but this will only take place when quarterly *Energy Trends* is published. The reasons for substantial revisions will be explained in the 'Highlights' sheet of the internet version of the table concerned. Valid reasons for revisions of Digest data include:
- revised and validated data received from a data supplier;
- the figure in the Digest was wrong because of a typographical or similar error.

In addition, when provisional annual data for a new calendar year (eg 2008) are published in *Energy Trends* in March of the following year (eg March 2009), percentage growth rates are liable to be distorted if the prior year (ie 2007) data are constrained to the Digest total, when revisions are known to have been made. In these circumstances the prior year (ie 2007) data will be amended for all affected tables in *Energy Trends* and internet versions of all affected Digest tables will be clearly annotated to show that the data has been up-dated in *Energy Trends*.

Revisions to current years data published in *Energy Trends* but not in the *Digest of UK Energy Statistics*.
- All validated amendments from data suppliers will be updated when received and published in the next statistical release.
- All errors will be amended as soon as identified and published in the next statistical release.
- Data in energy and commodity balances format will be revised on a quarterly basis, to coincide with the publication of *Energy Trends*.

Further details on National Statistics Code of Practice and related protocols can be found at: www.statistics.gov.uk/about_ns/cop/default.asp.

Energy data on the internet

XXI   Energy data are held on the energy area of the BERR web site, under "statistics". The Digest is available at www.berr.gov.uk/energy/statistics/publications/dukes/page45537.html. Information on further BERR energy publications available both in printed copy format and on the internet is given in Annex C.

XXII   The Department for Business, Enterprise and Regulatory Reform (BERR) was created on 28 June 2007. This Department took over the energy responsibilities of the Department of Trade & Industry (DTI). Within this publication references to BERR's predecessor Department refer to DTI.

XXIII   Short term statistics are published:
- monthly, by BERR on the internet at www.berr.gov.uk/energy/statistics/source/index.html.

- quarterly, by BERR in paper and on the internet in *Energy Trends*, and *Quarterly Energy Prices*: www.berr.gov.uk/energy/statistics/publications.

- quarterly, by BERR in Statistical Press Release which provides a summary of information published in *Energy Trends* and *Quarterly Energy Prices* publications: www.gnn.gov.uk/

- monthly, by the Office for National Statistics in the Monthly Digest of Statistics (Palgrave Macmillan).

To subscribe to *Energy Trends* and *Quarterly Energy Prices,* please contact Clive Sarjantson at the address given at paragraph XXIX. Single copies are available from the BERR Publications Orderline, as given in Annex C, priced £6 for Energy Trends and £8 for Quarterly Energy Prices.

Table numbering

XXIV   Page 10 contains a list showing the tables in the order in which they appear in this issue, and their corresponding numbers in previous issues.

Symbols used

XXV   The following symbols are used in this Digest:

..	not available
-	nil or negligible (less than half the final digit shown)
r	Revised since the previous edition

Rounding convention

XXVI   Individual entries in the tables are rounded independently and this can result in totals, which are different from the sum of their constituent items.

Acknowledgements

XXVII   Acknowledgement is made to the main coal producing companies, the electricity companies, the oil companies, the gas pipeline operators, the gas suppliers, National Grid, the Institute of Petroleum, the Coal Authority, the United Kingdom Iron and Steel Statistics Bureau, AEA Energy & Environment, the Department for Environment, Food and Rural Affairs, the Department for Transport, OFGEM, Building Research Establishment, HM Revenue and Customs, the Office for National Statistics, and other contributors to the enquiries used in producing this publication.

Cover photograph

XXVIII   The cover illustration used for this Digest and other BERR energy statistics publications is from a photograph by Peter Askew. It was a winning entry in the DTI News Photographic Competition in 2002.

Contacts

XXIX   For general enquiries on energy statistics contact:

Clive Sarjantson on 020 7215 2698   Kevin Harris on 020 7215 6049
(E-mail:clive.sarjantson@berr.gsi.gov.uk)   **or**   (E-mail:kevin.harris@berr.gsi.gov.uk)

Department for Business, Enterprise and Regulatory Reform
Energy Markets Unit
Bay 299
1 Victoria Street
London SW1H 0ET
Fax: 020 7215 2723

Enquirers with hearing difficulties can contact the Department on the BERR Textphone: 020 7215 6740.

XXX   For enquiries concerning particular data series or chapters contact those named on page 9 or at the end of the relevant chapter.

Kevin Harris, Production Team
July 2008

Contact List

The following people in the Department for Business, Enterprise and Regulatory Reform may be contacted for further information about the topics listed:

Chapter	Contact	Telephone 020 7215	E-mail
Total energy statistics	Iain MacLeay	6898	Iain.MacLeay@berr.gsi.gov.uk
Solid fuels and derived gases	James Hemingway Mita Kerai	2717 3839	James.Hemingway@berr.gsi.gov.uk Mita.Kerai@berr.gsi.gov.uk
Oil and upstream gas resources	Clive Evans Martin Young	5189 5184	Clive.Evans@berr.gsi.gov.uk Martin.Young@berr.gsi.gov.uk
North Sea profits, operating costs and investments	Suhail Siddiqui	5262	Suhail.Siddiqui@berr.gsi.gov.uk
Petroleum (downstream)	Lisa Vine Martin Young	6072 5184	Lisa.Vine@berr.gsi.gov.uk Martin.Young@berr.gsi.gov.uk
Gas supply (downstream)	James Hemingway Joe Ewins	2717 5190	James.Hemingway@berr.gsi.gov.uk Joe.Ewins@berr.gsi.gov.uk
Electricity	Mike Janes Joe Ewins	5186 5190	Mike.Janes@berr.gsi.gov.uk Joe.Ewins@berr.gsi.gov.uk
Combined heat and power	Mike Janes	5186	Mike.Janes@berr.gsi.gov.uk
Renewable sources of energy	Mike Janes	5186	Mike.Janes@berr.gsi.gov.uk
Prices and values Industrial, international and oil prices	Iain MacLeay Jo Marvin	6898 6935	Iain.MacLeay@berr.gsi.gov.uk Jo.Marvin@berr.gsi.gov.uk
Regional and Local Authority Energy Statistics	Julian Prime Jennifer Knight	6178 6490	Julian.Prime@berr.gsi.gov.uk Jennifer.Knight@berr.gsi.gov.uk
Calorific values and conversion factors	Iain MacLeay Lisa Vine Julian Prime	6898 6072 6178	Iain.MacLeay@berr.gsi.gov.uk Lisa.Vine@berr.gsi.gov.uk Julian.Prime@berr.gsi.gov.uk
General enquiries (energy helpdesk)	Clive Sarjantson	2698	Clive.Sarjantson@berr.gsi.gov.uk

All the above can be contacted by fax on 020 7215 2723

Tables as they appear in this issue and their corresponding numbers in the previous three issues

Chapter	2005	2006	2007	2008
ENERGY	-	-	-	1.1
	-	-	1.1	1.2
	-	1.1	1.2	1.3
	1.1	1.2	1.3	-
	1.2	1.3	-	-
	1.3	-	-	-
	-	-	-	1.4
	-	-	1.4	1.5
	-	1.4	1.5	1.6
	1.4	1.5	1.6	-
	1.5	1.6	-	-
	1.6	-	-	-
	1.7	1.7	1.7	1.7
	1.8	1.8	1.8	1.8
	1.9	1.9	1.9	1.9
SOLID FUELS & DERIVED GASES	-	-	-	2.1
	-	-	2.1	2.2
	-	2.1	2.2	2.3
	2.1	2.2	2.3	-
	2.2	2.3	-	-
	2.3	-	-	-
	-	-	-	2.4
	-	-	2.4	2.5
	-	2.4	2.5	2.6
	2.4	2.5	2.6	-
	2.5	2.6	-	-
	2.6	-	-	-
	2.7	2.7	2.7	2.7
	2.8	2.8	2.8	2.8
	2.9	2.9	2.9	2.9
	2.10	2.10	2.10	2.10
	2.11	2.11	2.11	2.11
PETROLEUM	-	-	-	3.1
	-	-	3.1	3.1
	-	3.1	3.1	3.1
	3.1	3.2	3.1	-
	3.2	3.3	-	-
	3.3	-	-	-
	-	-	-	3.2
	-	-	3.2	3.3
	-	3.4	3.3	3.4
	3.4	3.5	3.4	-
	3.5	3.6	-	-
	3.6	-	-	-
	3.7	3.7	3.5	3.5
	3.8	3.8	3.6	3.6
	3.9	3.9	3.7	-
	3.10	3.10	3.8	3.7

Chapter	2005	2006	2007	2008
GAS	4.1	4.1	4.1	4.1
	4.2	4.2	4.2	4.2
	4.3	4.3	4.3	4.3
	-	-	4.4	4.4
	-	-	-	4.5
ELECTRICITY	5.1	5.1	5.1	5.1
	5.2	5.2	5.2	5.2
	5.3	5.3	5.3	5.3
	5.4	5.4	5.4	5.4
	5.5	5.5	5.5	5.5
	5.6	5.6	5.6	5.6
	5.7	5.7	5.7	5.7
	5.8	5.8	5.8	5.8
	5.9	5.9	5.9	5.9
	5.10	5.10	5.10	5.10
	5.11	5.11	5.11	5.11
	5.12	5.12	5.12	5.12
COMBINED HEAT AND POWER	6.1	6.1	6.1	6.1
	6.2	6.2	6.2	6.2
	6.3	6.3	6.3	6.3
	6.4	6.4	6.4	6.4
	6.5	6.5	6.5	6.5
	6.6	6.6	6.6	6.6
	6.7	6.7	6.7	6.7
	6.8	6.8	6.8	6.8
	6.9	6.9	6.9	6.9
RENEWABLE SOURCES OF ENERGY	-	-	-	7.1
	-	-	7.1	7.2
	-	7.1	7.2	7.3
	7.1	7.2	7.3	-
	7.2	7.3	-	-
	7.3	-	-	-
	7.4	7.4	7.4	7.4
	7.5	7.5	7.5	7.5
	7.6	7.6	7.6	-
	7.7	7.7	7.7	7.6
ANNEX A CALORIFIC VALUES	A.1	A.1	A.1	A.1
	A.2	A.2	A.2	A.2
	A.3	A.3	A.3	A.3

Chapter 1
Energy

Introduction

1.1   This chapter presents figures on overall energy production and consumption. Figures showing the flow of energy from production, transformation and energy industry use through to final consumption are presented in the format of an energy balance based on the individual commodity balances presented in Chapters 2 to 5 and 7.

1.2   The chapter begins with aggregate energy balances covering the last three years (Tables 1.1 to 1.3) starting with the latest year, 2007. Energy value balances then follow this for the same years (Tables 1.4 to 1.6) and Table 1.7 shows sales of electricity and gas by sector in value terms. Table 1.8 covers final energy consumption by the main industrial sectors over the last five years followed by Table 1.9, which shows the fuels used for electricity generation by these industrial sectors. The explanation of the principles behind the energy balance and commodity balance presentations is set out in Annex A. Information on long term trends (Tables 1.1.1 to 1.1.8) for production, consumption, and expenditure on energy; as well as long term temperature data and analyses such as the relationship between energy consumption and the economy of the UK are available on BERR's energy statistics web site at: www.berr.gov.uk/energy/statistics/publications/dukes/page45537.html

Calorific values when producing energy statistics

1.3   A consultation was launched in the 2005 edition of the Digest seeking views of users as to whether Net Calorific Values (NCVs) should be used in place of Gross Calorific Values (GCVs). In conclusion BERR recognised that there are good arguments both for and against moving from GCV to NCV. However, it was concluded that there was no demonstrable advantage to changing the method of presenting UK Energy statistics, and so GCVs continue to be used in the main DUKES publication. The fuel specific NCVs are shown at Annex A. However, as the new EU renewables target is calculated on data converted using net calorific values, aggregate energy balances for the most recent years have been calculated using NCVs and are available on the internet version, Annex I, of this publication at: www.berr.gov.uk/energy/statistics/publications/dukes/page45537.html

The energy industries

1.4   The energy industries in the UK play a central role in the economy by producing, transforming and supplying energy in its various forms to all sectors. They are also major contributors to the UK's Balance of Payments through the exports of crude oil and oil products. The box below summarises the energy industries' contribution to the economy:

- 4.8 per cent of GDP;

- 8.6 per cent of total investment;

- 44.3 per cent of industrial investment;

- 137,800 people directly employed (5 per cent of industrial employment);

- Many others indirectly employed (eg an estimated 260,000 in support of UK Continental Shelf activities).

Aggregate energy balance (Tables 1.1, 1.2 and 1.3)

1.5   These tables show the flows of energy in the United Kingdom from production to final consumption through conversion into secondary fuels such as coke, petroleum products, secondary electricity and heat sold. The principles behind the presentation used in the Digest and how this links with the figures presented in other chapters are explained in Annex A. The figures are presented on an energy supplied basis, in tonnes of oil equivalent.

1.6　In 2007, the primary supply of fuels was 235.9 million tonnes of oil equivalent, a 3.4 per cent decrease compared to 2006. Indigenous production in 2007 was 5.6 per cent lower than in 2006. Chart 1.1 illustrates the figures for the production and consumption of individual primary fuels in 2007. In 2007, overall primary fuel consumption was not met by indigenous production; this continues the trend since 2004 when the UK became a net importer of fuel. The UK imported more coal, manufactured fuels, crude oil, electricity and gas than it exported; however, we remained a net exporter of petroleum products.

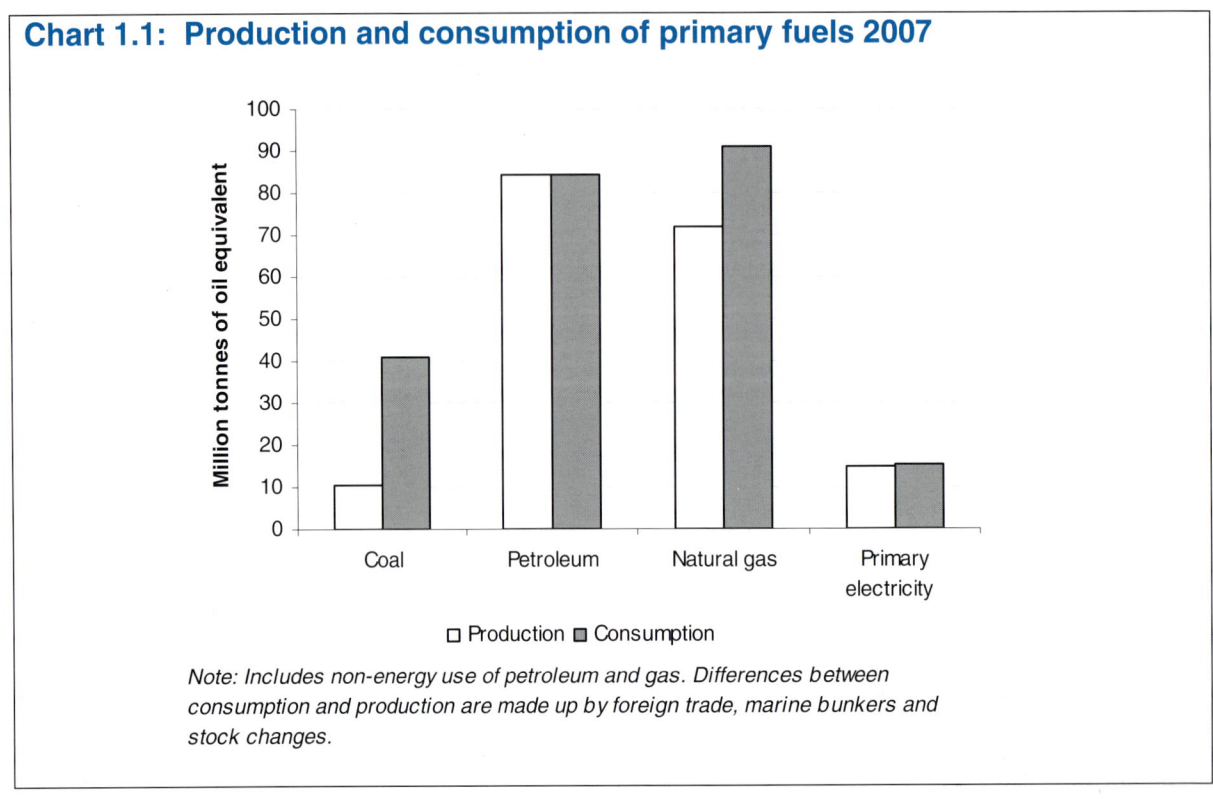

Chart 1.1: Production and consumption of primary fuels 2007

Note: Includes non-energy use of petroleum and gas. Differences between consumption and production are made up by foreign trade, marine bunkers and stock changes.

1.7　Total primary energy demand was 3.5 per cent lower in 2007 than in 2006 at 235.7 million tonnes of oil equivalent. Chart 1.2 shows the composition of primary demand in 2007.

1.8　The transfers row in Tables 1.1 to 1.3 should ideally sum to zero with transfers from primary oils to petroleum products amounting to a net figure of zero. Similarly the manufactured gases and natural gas transfers should sum to zero. However differences in calorific values between the transferred fuels can result in non-zero values.

1.9　The transformation section of the energy balance shows, for each fuel, the net inputs for transformation uses. For example, Table 1.1 shows, 4,319 thousand tonnes of oil equivalent of coal feeds into the production of 4,170 thousand tonnes of oil equivalent of coke, representing a loss of 149 thousand tonnes of oil equivalent in the manufacture of coke in 2007. In 2007, energy losses during the production of electricity and other secondary fuels amounted to just over 53 million tonnes of oil equivalent, shown in the transformation row in Table 1.1.

1.10　In 2007 coal prices rose, whilst gas prices to generators fell. As a result a number of electricity generators switched some production from using coal to gas fired stations. Generation from coal-fired stations was 9.2 per cent lower in 2007 than in 2006, with generation from gas in 2007 at a record level, 16 per cent higher than in 2006. Generation from nuclear sources fell by 16 per cent as the nuclear sector was affected by a high level of outages for repairs and maintenance and the closure of two Magnox stations.

1.11   This switch from coal to gas fired electricity generation contributed to the fall in carbon dioxide emissions between 2006 and 2007.   More details of carbon dioxide emissions are available in Annex E to the Digest, which is available on the BERR website at:

www.berr.gov.uk/energy/statistics/publications/dukes/page45537.html

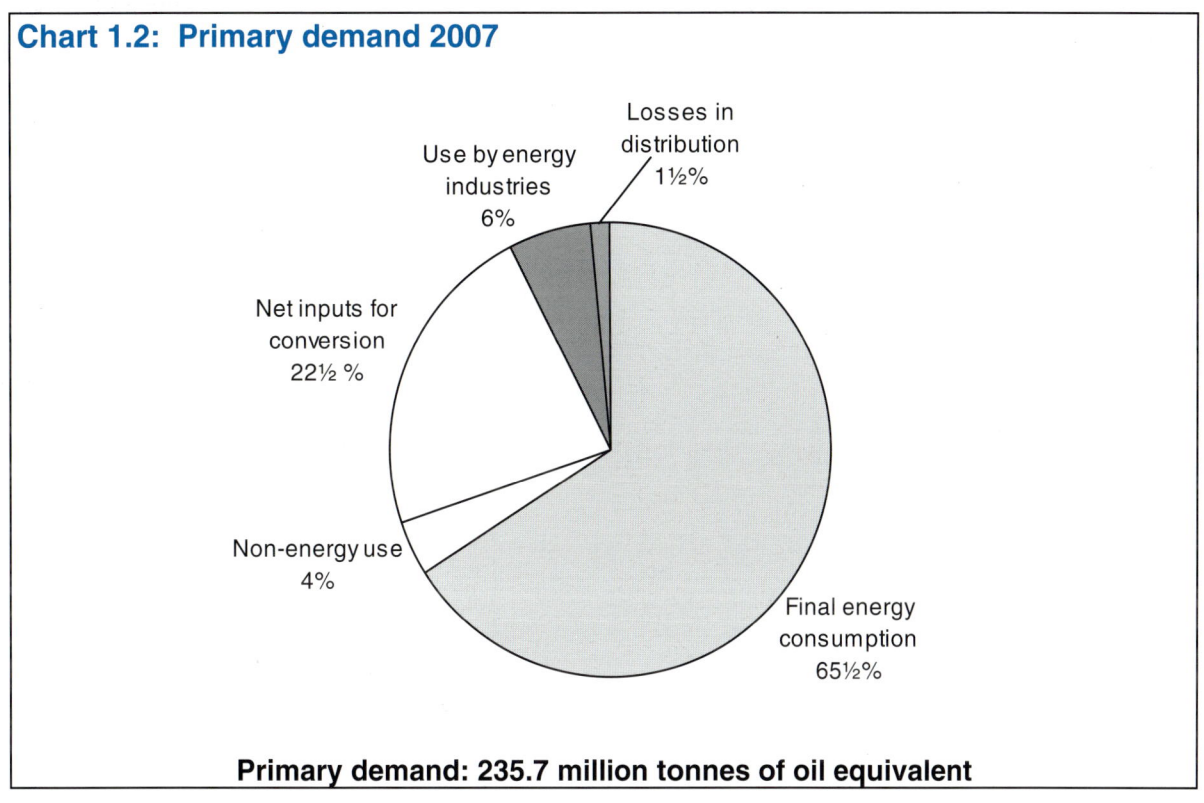

Chart 1.2: Primary demand 2007

Losses in distribution 1½%

Use by energy industries 6%

Net inputs for conversion 22½ %

Non-energy use 4%

Final energy consumption 65½%

Primary demand: 235.7 million tonnes of oil equivalent

1.12   The energy industry use section of the table represents use of fuels by the energy industries themselves.   This section also includes consumption by those parts of the iron and steel industry which behave like an energy industry i.e. they are involved in the transformation processes (see paragraph A.29 of Annex A).   In 2007, energy industry use amounted to 14.3 million tonnes of oil equivalent of energy, a decrease of 6.4 per cent on 2006.   The main reason for the decline was reduced production combined with an increase in efficiency driven by a rise in costs.

1.13   Losses presented in the energy balance include distribution and transmission losses in the supply of manufactured gases, natural gas, and electricity.   Recorded losses decreased marginally by 1.3 per cent between 2006 and 2007, to remain at broadly similar levels to those of the last six years. Losses in North Sea gas production are no longer separately identified in the simplified Petroleum Product Reporting System, which was introduced in January 2001. This has improved the quality of production data and reduced reported losses.   Further details can be found in paragraph 4.30 in Chapter 4.

1.14   Total final consumption, which includes non-energy use of fuels, in 2007 was 164.6 million tonnes of oil equivalent; this is a 5.4 million tones of oil equivalent reduction on the consumption in 2006.   Final energy consumption in 2007 was mainly accounted for by the transport sector (36 per cent), the domestic sector (27 per cent), the industrial sector (19 per cent), the commercial sector (6 per cent) and non-energy use (6 per cent).   These figures are illustrated in Chart 1.3. Recent trends in industrial consumption are shown in Table 1.8 and are discussed in paragraphs 1.22 to 1.24.

1.15   The main fuels used by final consumers in 2007 were petroleum products (48 per cent), natural gas (31 per cent) and electricity (18 per cent). Of the petroleum products consumed by final users 11 per cent was for non-energy purposes; for natural gas 1.8 per cent was consumed for non-energy purposes. The amount of heat that was bought for final consumption accounted for 0.7 per cent of the total final energy consumption.

1.16 Non-energy use of fuels includes use as chemical feedstocks and other uses such as lubricants. Non-energy use of fuels for 2007 are shown in Table 1A. Further details of non-energy use are given in Chapter 3, paragraphs 3.60 to 3.66 and Chapter 4, paragraphs 4.38.

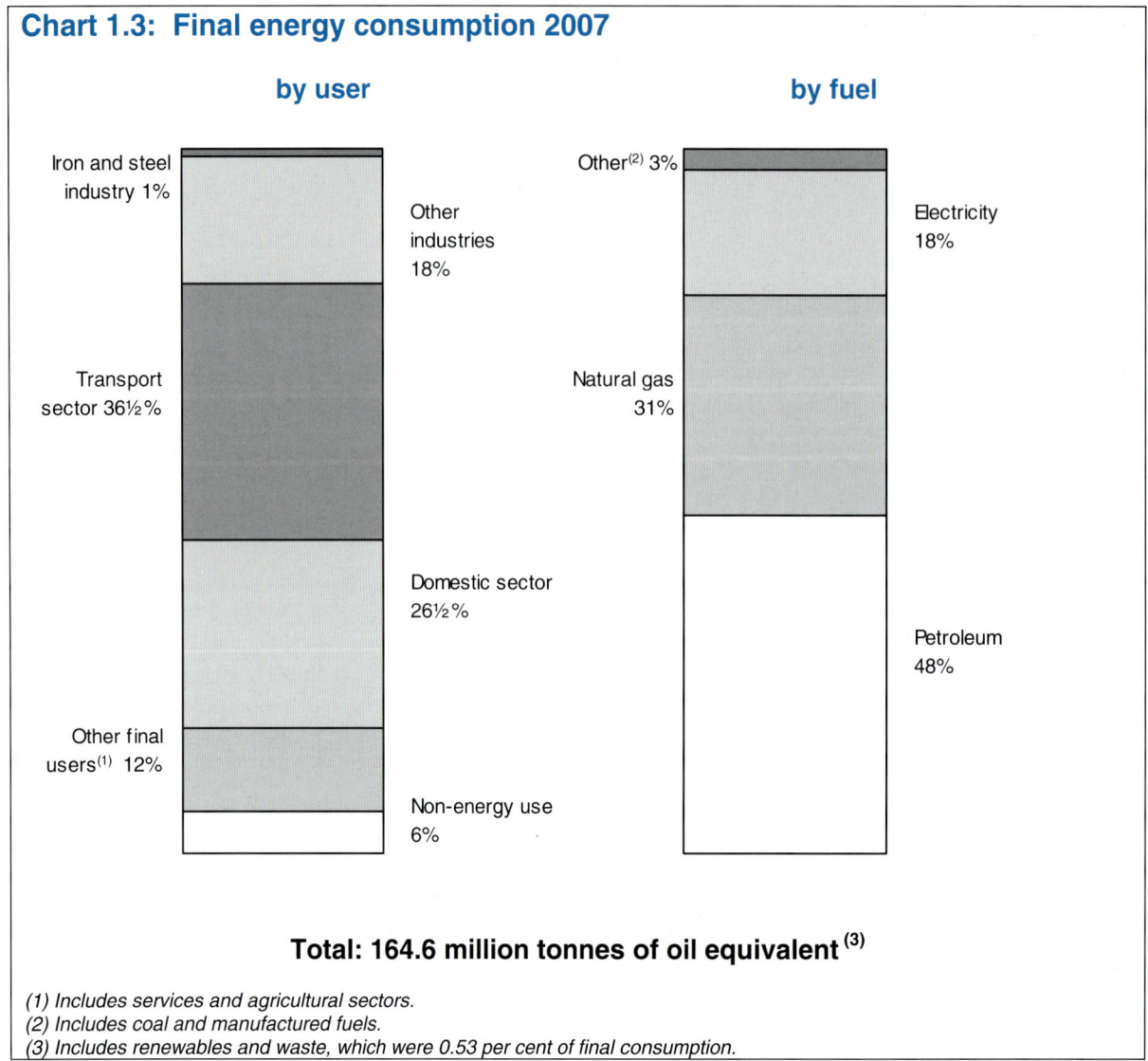

Chart 1.3: Final energy consumption 2007

by user

by fuel

Iron and steel industry 1%

Transport sector 36½%

Other final users[1] 12%

Other industries 18%

Domestic sector 26½%

Non-energy use 6%

Other[2] 3%

Natural gas 31%

Electricity 18%

Petroleum 48%

Total: 164.6 million tonnes of oil equivalent [3]

(1) Includes services and agricultural sectors.
(2) Includes coal and manufactured fuels.
(3) Includes renewables and waste, which were 0.53 per cent of final consumption.

Table 1A: Non-energy use of fuels 2007

	Thousand tonnes of oil equivalent	
	Petroleum	Natural gas
Petrochemical feedstocks	5,217	899
Other	3,613	-
Total	**8,830**	**899**

Value balance of traded energy (Tables 1.4, 1.5 and 1.6)

1.17 Tables 1.4 to 1.6 present the value of traded energy in a similar format to the energy balances. The balance shows how the value of inland energy supply is made up from the value of indigenous production, trade, tax and margins (profit and distribution costs). The lower half of the table shows how this value is generated from the final expenditure on energy through transformation processes and other energy sector users as well as from the industrial and domestic sectors. The balances only contain values of energy which is traded ie where a transparent market price is applicable. Further technical notes are given in paragraphs 1.26 to 1.59. In keeping with the energy balances, the value balances since 2000 have included data on heat generation and heat sold. Additionally, an estimate of the amount of Climate Change Levy paid is included in Tables 1.4, 1.5 and 1.6. This levy was introduced in April 2001 and is payable by non-domestic final consumers of gas, electricity, coal, coke and LPG.

1.18   Total expenditure by final consumers in 2007 is estimated at £100,260 million, (£99,825 million shown as actual final consumption and £435 million of coal consumed by the iron and steel sector in producing coke for their own consumption). This is up by 2.4 per cent on 2006, reflecting small rises in energy prices. Chart 1.4 below shows energy consumption and expenditure by final users.

Chart 1.4: Energy consumption and estimated expenditure on energy by final users

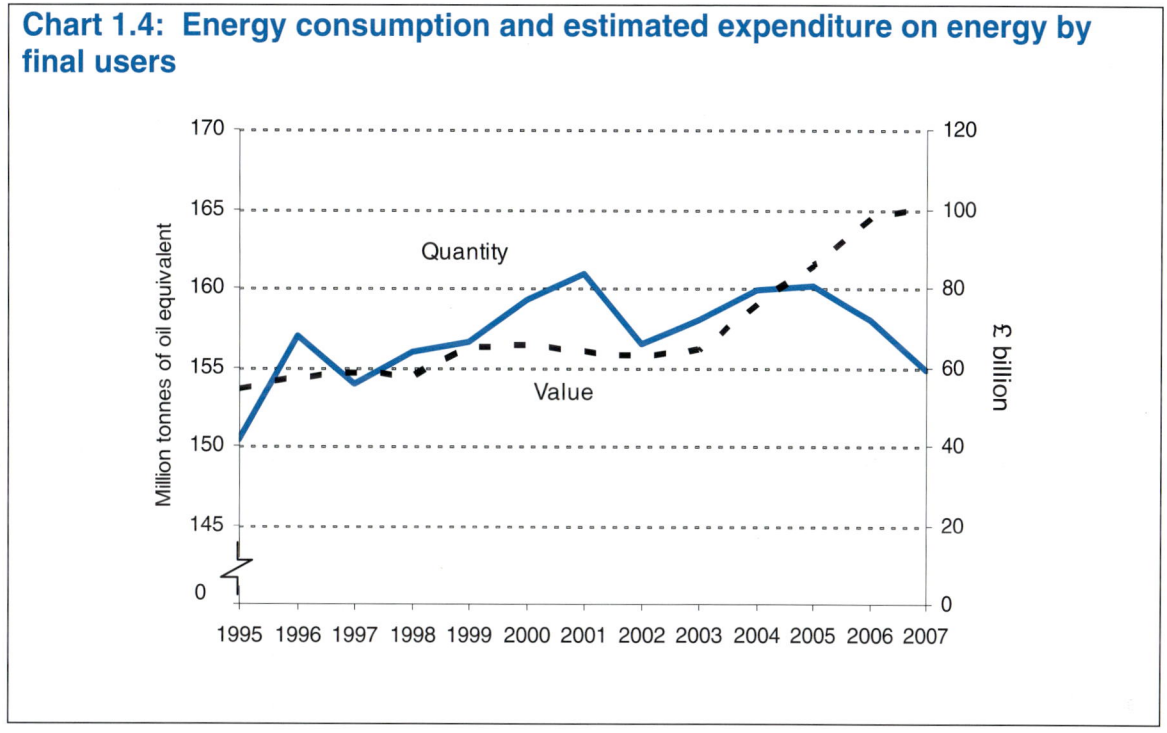

1.19   The value balance provides a guide on how the value chain works in the production and consumption of energy.   For example in 2007, £21,000 million of crude oil were indigenously produced, of which £12,935 million were exported, and £13,290 million of crude oil were imported. Allowing for stock changes this provides a total value of inland crude oil supply of £21,535 million. This fuel was then completely consumed within the petroleum industry in the process of producing £27,375 million of petroleum products.   Again some external trade and stock changes took place before arriving at a basic value of petroleum products of £26,430 million. In supplying the fuel to final consumers distribution costs were incurred and some profit was made amounting to £2,590 million, whilst duty and tax meant a further £31,985 million was added to the basic price to arrive at the final market value of £61,010 million.   This was the value of petroleum products purchased of which industry purchased £2,420 million, domestic consumers for heating purposes purchased £1,130 million, with the vast majority purchased within the transportation sectors, £54,600 million.

1.20   Of the total final expenditure on energy in 2007 (£100,260 million) the biggest share, 55 per cent, fell to the transport sector. Of the remaining 45 per cent, industry purchased around a quarter or £12,260 million, with the domestic sector purchasing over a half or £23,445 million.

Sales of electricity and gas by sector (Table 1.7)

1.21   Table 1.7 shows broad estimates for the total value of electricity and gas to final consumption. Net selling values provide some indication of typical prices paid in broad sectors and can be of use to supplement more detailed and accurate information contained in the rest of this chapter.

Energy consumption by main industrial groups (Table 1.8)

1.22   This table presents final energy consumption for the main industrial sub-sectors over the last 5 years.

1.23 So far as is practicable, the user categories have been grouped on the basis of the 2003 Standard Industrial Classification (see paragraphs 1.54 to 1.58). However, some data suppliers have difficulty in classifying consumers to this level of detail and the breakdown presented in these tables must therefore be treated with caution. The groupings used are consistent with those used in Table 1.9 which shows industrial sectors' use of fuels for generation of electricity (autogeneration).

1.24 In 2007, 31.7 million tonnes of oil equivalent were consumed by the main industrial groups. The largest consuming groups were chemicals (17.6 per cent), metal products, machinery and equipment (12.5 per cent), food, beverages and tobacco (11.5 per cent), iron and steel and non-ferrous metals (8.6 per cent), mineral products (8.1 per cent), and paper, printing and publishing (7.2 per cent). The figures are illustrated in Chart 1.5.

Chart 1.5: Energy consumption by main industrial groups 2007

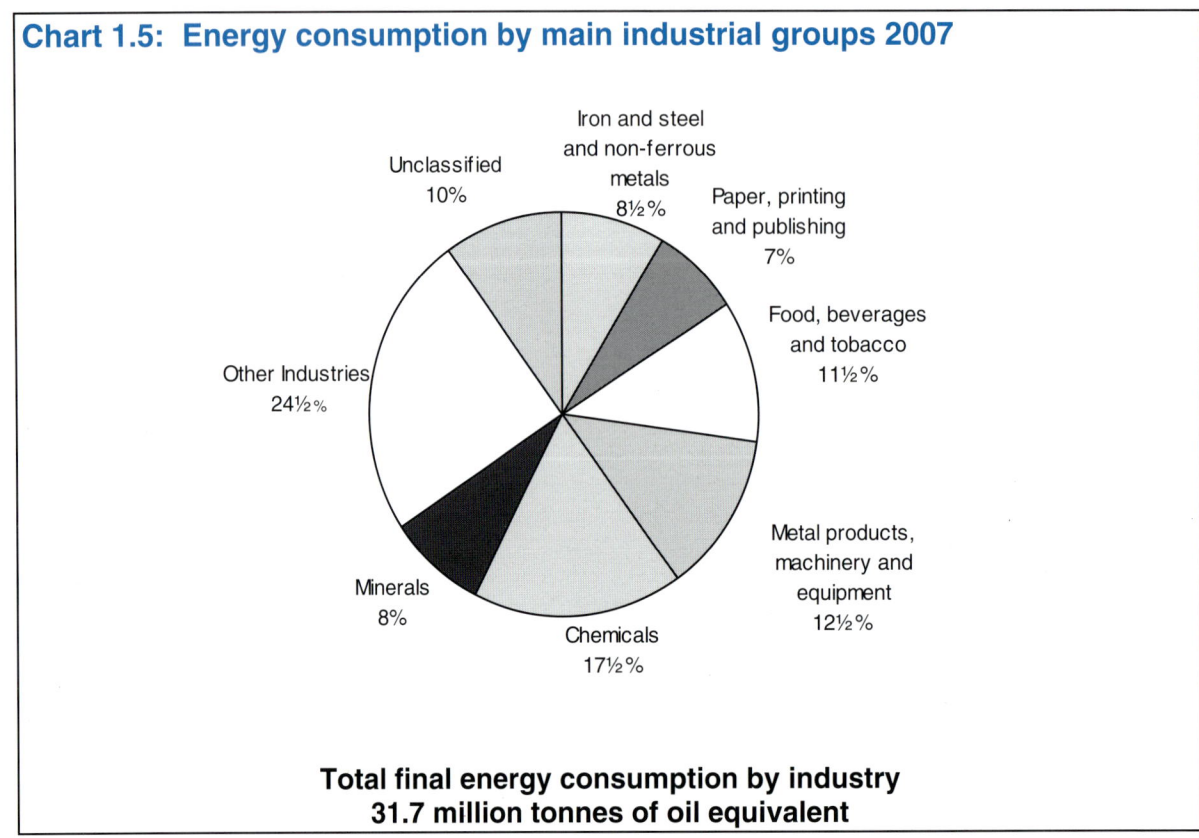

**Total final energy consumption by industry
31.7 million tonnes of oil equivalent**

Fuels consumed for electricity generation by main industrial groups (autogeneration) (Table 1.9)

1.25 This table gives details of the amount of each fuel consumed by industries in order to generate electricity for their own use. Fuel consumption is consistent with the figures given for "other generators" in Table 5.4 of Chapter 5. The term autogeneration is explained further in paragraphs 1.31 and 1.32. Electricity produced via autogeneration is included within the figures for electricity consumed by industrial sectors in Table 1.8. Table 1.9 has been produced using the information currently available and shows the same sector detail as Table 1.8, data cannot be given in as much detail as in the individual commodity balances and the energy balance because it could disclose information about individual companies. Table 1.9 allows users to allocate the fuel used for autogeneration to individual industry groups in place of the electricity consumed. Further information on the way Table 1.9 links with the other tables is given in paragraph 1.32. Because CHP schemes have now been classified according to the sector that receives the majority of the heat there are some changes to the allocation between sectors this year. 2005, 2006 and 2007 are all on the new basis but there are some discontinuities between 2004 and 2005 (see Chapter 6 paragraphs 6.19, 6.33 and Table 6G for more information).

Technical notes and definitions

I Units and measurement of energy

Units of measurement

1.26 The original units of measurement appropriate to each fuel are used in the individual fuel chapters. A common unit of measurement, the tonne of oil equivalent (toe), which enables different fuels to be compared and aggregated, is used in Chapter 1. In common with the International Energy Agency and with the Statistical Office of the European Communities, the tonne of oil equivalent is defined as follows:

1 tonne of oil equivalent
$= 10^7$ kilocalories
$= 396.83$ therms
$= 41.868$ Gigajoules (GJ)
$= 11,630$ Kilowatt hours (kWh)

1.27 This unit should be regarded as a measure of energy content rather than a physical quantity. There is no intention to represent an actual physical tonne of oil, and indeed actual tonnes of oil will normally have measurements in tonnes of oil equivalent which differ from units.

Thermal content - energy supplied basis of measurement

1.28 Tables 1.1 to 1.3, 1.8 and 1.1.1 to 1.1.5 (available on BERR's energy statistics site at www.berr.gov.uk/energy/statistics/source/total/page18424.html) are compiled on an energy-supplied basis. Detailed data for individual fuels are converted from original units to tonnes of oil equivalent using gross calorific values and conversion factors appropriate to each category of fuel. The results are then aggregated according to the categories used in the tables. Gross calorific values represent the total energy content of the fuel, including the energy needed to evaporate the water present in the fuel (see also paragraph 1.52).

1.29 Estimated gross and net calorific values for 2007 are given on page 211. Calorific values are reviewed each year in collaboration with the fuel industries, and figures for earlier years can be found in Table A.2 and A.3 on pages 212 and 213. To construct energy balances on an energy supplied basis calorific values are required for production, trade, and stocks, as follows:

Coal The weighted average gross calorific value of all indigenous coal consumed is used to derive the thermal content of coal production and undistributed stocks. Thermal contents of imports and exports allow for the quality of coal. Thermal contents of changes in coal stocks at secondary fuel producers are the average calorific values of indigenous coal consumed.

Petroleum Work carried out in 1997 to revise calorific values for petroleum products did not find any recent work on the subject.   In the absence of such work, the gross calorific values, included in Annex A, and used in the construction of these energy balances from 1990 onwards have been calculated using a formula derived by the US Bureau of Standards. This formula estimates the gross calorific value of products according to their density as follows:

$Gj = 51.83 - 8.78 \times d^2$, where d is the density of the product in terms of kilograms per litre.

For crude petroleum and refinery losses, the weighted average calorific value for all petroleum products from UK refineries is used. A notional figure of 42.9 GJ per tonne is used for non-energy petroleum products (industrial and white spirits, lubricants, bitumen, petroleum coke, waxes and miscellaneous products).

Gases Although the original unit for gases is the cubic metre, figures for gases are generally presented in the fuel sections of this Digest in gigawatt hours (GWh), having been converted from cubic metres using gross calorific values provided by the industries concerned. Conversion factors between units of energy are given on the flap inside the back cover and on page 209.

Electricity and heat Unlike other fuels, the original unit used to measure electricity and heat is a measure of energy. The figures for electricity and heat can therefore be converted directly to toe using the conversion factors on the flap inside the back cover and on page 209.

Primary electricity   Hydro electricity and net imports of electricity are presented in terms of the energy content of the electricity produced (the energy supplied basis).   This is consistent with international practice.   Primary inputs for nuclear electricity assume the thermal efficiencies at nuclear stations given in Chapter 5, Table 5.10 (38.6 per cent in 2007). (See Chapter 5, paragraphs 5.26 and 5.58.)

Non-energy uses of fuel

1.30   Energy use of fuel mainly comprises use for lighting, heating, motive power and power for appliances.   Non-energy use includes use as chemical feedstocks, solvents, lubricants and road making material.   It should be noted that the amounts of non-energy use of natural gas included in the Digest are approximate.   Further discussion of non-energy uses of lubricating oils and petroleum coke appears in Chapter 3, paragraphs 3.63 to 3.70.

Autogeneration of electricity

1.31   Autogeneration is defined as the generation of electricity by companies whose main business is not electricity generation, the electricity being produced mainly for that company's own use.   Estimated amounts of fuel used for thermal generation of electricity by such companies, the output of electricity and the thermal losses incurred in generation are included within the Transformation sector in the energy balances shown in Tables 1.1 to 1.3.   Electricity used in the power generation process by autogenerators is shown within the Energy Industry Use section.   Electricity consumed by industry and commerce from its own generation is included as part of Final consumption.   This treatment is in line with the practice in international energy statistics.

1.32   Figures on total amount of fuel used and electricity generated by autogenerators, and the amount of electricity for own consumption is shown in Tables 1.9, 5.1, 5.3 to 5.6.   Table 1.9 summarises the figures by broad industrial groups.   Much of the power generated is from combined heat and power (CHP) plants and data from Chapter 6 are included within Table 1.9.   Differences will occur where CHP plants are classified to major power producers, and this mainly affects the chemicals sector.   The method of allocating fuel used in CHP plants between electricity production and heat production is described in Chapter 6 paragraphs 6.35 to 6.37.   This method can give rise to high implied conversion efficiencies in some sectors, most notably in the iron and steel sector.

Final consumption, deliveries, stock changes

1.33   Figures for final consumption relate to deliveries, if fuels can be stored by users and data on actual consumption are not available.   Final consumption of petroleum and solid fuels is on a deliveries basis throughout, except for the use of solid fuels by the iron and steel industry.   Figures for domestic use of coal are based on deliveries to merchants.   Figures for stock changes in Tables 1.1 to 1.3 cover stocks held by primary and secondary fuel producers, major distributors of petroleum products, and stocks of coke and breeze held by the iron and steel industry; for coal they also include an estimate of volumes in transit.   Figures for stock changes in natural gas represent the net amount put into storage by gas companies operating pipelines.

1.34   Figures for final consumption of electricity include sales by the public distribution system and consumption of electricity produced by generators other than the major electricity producing companies.   Thus electricity consumption includes that produced by industry and figures for deliveries of other fuels to industry exclude amounts used to generate electricity (except for years prior to 1987, shown in tables giving long term trends).

Heat sold

1.35   Heat sold is defined as heat that is produced and sold under the provision of a contract.   The heat sold figures have been derived from two sources covering CHP plants and community heating schemes without CHP plants.   Data for heat sold were supplied by CHP plants to the Combined Heat and Power Quality Assurance Programme and were processed by AEA.   Data for heat consumption from community heating schemes were derived from the Building Research Establishment's (BRE) 'Nationwide Survey of Community Heating' that was carried out in 1997, a database of community heating schemes in social housing in 2000, and Community Heating Sales Surveys undertaken between 2003 and 2005.   The estimates from these sources have been used to derive heat sold figures since 1999. When information about where the heat was generated was not available from the BRE sources, it was assumed that domestic sector heat consumption was provided by the commercial sector, public sector heat consumption was provided by the public administration and industrial sectors (using proportions derived from CHP statistics) and that industrial sector heat consumption was provided by the industrial sector. The introduction of heat sold into the energy balances has not affected the individual fuel totals, since the energy used to generate the heat has been deducted from the final consumption section of the energy balance and transferred to the transformation section. The figures that are included in the balances should be treated as indicative of the amount of heat sold. Annex J of the Digest, at www.berr.gov.uk/energy/statistics/publications/dukes/page45537.html, shows the quantity of fuel by consuming sector used to produce heat that is subsequently sold.

II Energy balances (Tables 1.1, 1.2 and 1.3)

1.36   Tables 1.1, 1.2 and 1.3 show the energy flows as the primary fuels are processed (or used) and as the consequent secondary fuels are used. The net inputs to transformation are shown in the transformation rows and hence outputs from transformation processes into which primary fuels are input (such as electricity generation, heat generation or petroleum refining) appear as positive figures under the secondary product's heading in the tables. Similarly the net inputs are shown as negative figures under the primary fuel headings.

III Value balances (Tables 1.4, 1.5 and 1.6)

Valuation of energy purchases

1.37   In common with the rest of the chapter, these tables covering energy expenditure follow a balance format. While a user may derive data on a similar basis as that previously published, the balance table allows for more varied use and interpretation of traded energy value data. That said, the table continues to only show values for energy that has to be purchased and therefore does not include estimated values of a sector's internal consumption, such as coal used in the process of coal extraction.

The value balance

1.38   The table balances around **market value of inland consumption** with the lower half of the table showing the total value of consumption by end users, sub divided into energy sector users and final users both for energy and non-energy use. The top half of the table shows the supply components that go to make up the final market value of inland consumption, namely upstream cost of production, imports, taxes and the margins and costs of delivering and packaging the fuel for the final consumer. The total final consumers' value of energy consumption is represented by the lines 'total non energy sector use' and iron and steel sectors purchases of coal for use in solid fuel manufacture.

1.39   All figures are estimates and have been rounded to the nearest £5 million.

Fuel definitions in value balances

1.40   **Crude oil** includes NGLs and refinery feedstocks. **Natural gas** does not include colliery methane. **Electricity** only includes electricity delivered via the public distribution system and therefore does not value electricity produced and consumed by autogenerators, however the fuels used by autogenerators are included under Transformation. **Manufactured solid fuels** includes coke, breeze and other solid manufactured fuels, mainly products from patent fuel and carbonisation plants. **Other fuels** includes all other fuels not separately listed, where they can be clearly considered as traded and some reasonable valuation can be made. Fuels mainly contributing to this year's values are wood, coke oven and colliery methane gases sold on to other industrial users and some use of waste products such as poultry litter.

Energy end use

1.41   Values represent the cost to the final user including transportation of the fuel. They are derived, except where actual values are available, from the traded element of the volumes presented in aggregate energy balance and end user prices collected from information supplied by users or energy suppliers. The **energy sector** consists of those industries engaged in the production and sale of energy products, but values are not given for consumption of self-generated fuels eg coke oven gas used by coke producers. Many of the processes in the **iron and steel** industry are considered to be part of the energy sector in the energy balances, but for the purposes of this economic balance their genuine purchases are treated as those of final consumers, except for purchases of coal directly used in coke manufacture, which is shown separately as part of manufacture of solid fuel. Coal used directly in or to heat blast furnaces is shown as iron and steel final use. **Transformation** includes those fuels used directly in producing other fuels eg crude oil in petroleum products. **Electricity generators** keep and use significant stocks of coal and the stocks used in consumption each year are shown separately. The value and margins for these being assumed to be the same as other coal purchased in the year. **Road transport** includes all motor spirit and DERV use. **Commercial and other users** includes public administration and miscellaneous uses not classified to the industrial sector.

Supply

1.42 The supply side money chain is derived using various methods. **Indigenous production** represents the estimated basic value of in-year sales by the upstream producers. This value is gross of any taxes or cost they must meet. The valuation problems in attributing network losses in gas and electricity between upstream and downstream within this value chain means any costs borne are included in the production value. **Imports and exports** are valued in accordance with data published by HM Revenue and Customs, contained in Annex G (which can be found on the Internet at www.berr.gov.uk/energy/statistics/publications/dukes/page45537.html). However crude oil is treated differently where the value is formed from price data taken from a census survey of refiners and volume data taken from Tables 3.1 to 3.3. These values are considered to reflect the complete money chain more accurately than Tables G.1 to G.4. **Stock changes** are those for undistributed stocks except for coal where coke oven and generators stocks are included. A stock increase takes money out of the money chain and is therefore represented as a negative. **Distribution costs** are arrived at by removing an estimate of producers value along with any taxes from the end user values shown. For most fuel the estimate of producer value is derived from the consumption used for end use and the producer price taken from survey of producers. No sector breakdown is given for gas and electricity margins because it is not possible to accurately measure delivery costs for each sector. **Taxes** include VAT where not refundable and duties paid on downstream sales. Excluded are the gas and fossil fuel levies, petroleum revenue tax and production royalties and licence fees. The proceeds from the fossil fuel levy are redistributed across the electricity industry, whilst the rest are treated as part of the production costs.

Sales of electricity and gas by sector (Table 1.7)

1.43 This table provides data on the total value of gas and electricity sold to final consumers. The data are collected from the energy supply companies. The data are useful in indicating relative total expenditure between sectors, but the quality of data provided in terms of industrial classification has been worsening in recent years. Net selling values provide an indication of typical prices paid in broad sectors.

IV Measurement of energy consumption

Primary fuel input basis

1.44 Energy consumption is usually measured in one of three different ways. The first, known as the primary fuel input basis, assesses the total input of primary fuels and their equivalents. This measure includes energy used or lost in the conversion of primary fuels to secondary fuels (for example in power stations and oil refineries), energy lost in the distribution of fuels (for example in transmission lines) and energy conversion losses by final users. Primary demands as in Table 1.1, 1.2 and 1.3 are on this basis.

Final consumption - energy supplied basis

1.45 The second method, known as the energy supplied basis, measures the energy content of the fuels, both primary and secondary, supplied to final users. Thus it is net of fuel industry own use and conversion, transmission and distribution losses, but it includes conversion losses by final users. Table 1B presents shares of final consumption on this basis. The final consumption figures are presented on this basis throughout Chapter 1.

1.46   Although this is the usual and most direct way to measure final energy consumption, it is also possible to present final consumption on a primary fuel input basis. This can be done by allocating the conversion losses, distribution losses and energy industry use to final users. This approach can be used to compare the total primary fuel use which each sector of the economy accounts for. Table 1C presents shares of final consumption on this basis.

Final consumption - useful energy basis

1.47   Thirdly, final consumption may be expressed in the form of useful energy available after deduction of the losses incurred when final users convert energy supplied into space or process heat, motive power or light. Such losses depend on the type and quality of fuel and the equipment used and on the purpose, conditions, duration and intensity of use.   Statistics on useful energy are not sufficiently reliable to be given in this Digest; there is a lack of data on utilisation efficiencies and on the purposes for which fuels are used.

Shares of each fuel in energy supply and demand

1.48   The relative importance of the energy consumption of each sector of the economy depends on the method used to measure consumption. Shares of final consumption on an energy supplied basis (that is in terms of the primary and secondary fuels directly consumed) in 2007 are presented in Table 1B. For comparison, Table 1C presents shares of final consumption on a primary fuel input basis.

Table 1B: Primary and secondary fuels consumed by final users in 2007 – energy supplied basis

	Percentage of each fuel						Percentage of each sector				
	Industry	Transport	Domestic	Others	Total		Solid fuels	Petrol- eum	Gas	Secondary electricity	Total
Solid fuels	75	-	25	1	100	Industry	7	22	38	33	100
Petroleum	10	84	4	2	100	Transport	-	99	-	1	100
Gas	23	-	60	17	100	Domestic	2	7	69	23	100
Electricity	34	2	34	30	100	Others	-	8	46	46	100
All fuels	**20**	**39**	**28**	**12**	**100**	**All users**	**2**	**46**	**33**	**19**	**100**

Table 1C: Total primary fuel consumption by final users in 2007 - primary input basis

	Percentage of each fuel						Percentage of each sector				
	Industry	Transport	Domestic	Others	Total		Coal	Petrol- eum	Gas	Primary electricity	Total
Coal	38	2	33	27	100	Industry	28	15	47	10	100
Petroleum	10	83	4	2	100	Transport	1	97	1	1	100
Gas	27	1	50	22	100	Domestic	20	5	67	8	100
Primary electricity	34	2	34	30	100	Others	29	5	53	13	100
All fuels	**24**	**30**	**30**	**16**	**100**	**All users**	**18**	**35**	**40**	**7**	**100**

1.49   In 2007, every 1 toe of secondary electricity consumed by final users required, on average, 1.0 toe of coal, 0.9 toe of natural gas, 0.4 toe of primary electricity (nuclear, natural flow hydro and imports) and 0.2 toe of oil and renewables combined.   The extent of this primary consumption is hidden in Table 1B, which presents final consumption only in terms of the fuels directly consumed. When all such primary consumption is allocated to final users, as in Table 1C, the relative importance of fuels and sectors changes; the transport sector, which uses very little electricity, declines in importance, whilst the true cost of final consumption in terms of coal use can now be seen.

1.50   Another view comes from shares of users' expenditure on each fuel (Table 1D based on Table 1.4). In this case the importance of fuels which require most handling by the user (solids and liquid fuels) is slightly understated, and the importance of uses taxed at higher rates (transport) is overstated in the "All users" line.

Table 1D: Value of fuels purchased by final users in 2007

	Solid fuels	Petroleum	Gas	Secondary electricity	Heat	Total
					Percentage of each sector	
Industry	6	20	17	57	-	100
Transport	-	99	-	1	-	100
Domestic	1	5	42	52	-	100
Others	-	6	22	71	1	100
All users	**1**	**59**	**14**	**26**	**0**	**100**

Systems of measurement - international statistics

1.51   The systems of energy measurement used in various international statistics differ from the methods of the Digest as follows:

Net calorific values

1.52   Calorific values (thermal contents) used internationally are net rather than gross. The difference between the net and gross thermal content is the amount of energy necessary to evaporate the water present in the fuel or formed during the combustion process. The differences between gross and net values are taken to be 5 per cent for liquid and solid fuels (except for coke and coke breeze where there is no difference), 10 per cent for gases (except for blast furnace gas, 1 per cent), 15 per cent for straw, and 16 per cent for poultry litter. The calorific value of wood is highly dependent on its moisture content. In Annex A, the gross calorific value is given as 10 GJ per tonne at 50 per cent moisture content and this rises to 14.5 GJ at 25 per cent moisture content and 19 GJ for dry wood (equivalent to a net calorific value). Both gross and net calorific values are shown in Annex A. BERR and the Iron and Steel Statistics Bureau are currently reviewing the relationship between net and gross calorific values for fuels used by the Iron and Steel industry.

V Definitions of fuels

1.53   The following paragraphs explain what is covered under the terms "primary" and "secondary" fuels.

Primary fuels

Coal - Production comprises all grades of coal, including slurry.

Primary oils - This includes crude oil, natural gas liquids (NGLs) and feedstock.

Natural gas liquids - Natural gas liquids (NGLs) consist of condensates (C_5 or heavier) and petroleum gases other than methane C_1, that is ethane C_2, propane C_3 and butane C_4, obtained from the onshore processing of associated and non-associated gas. These are treated as primary fuels when looking at primary supply but in the consumption data presented in this chapter these fuels are treated as secondary fuels, being transferred from the primary oils column in Tables 1.1, 1.2 and 1.3.

Natural gas - Production relates to associated or non-associated methane C_1 from land and the United Kingdom sector of the Continental Shelf. It includes that used for drilling production and pumping operations, but excludes gas flared or re-injected. It also includes colliery methane piped to the surface and consumed by collieries or others.

Nuclear electricity - Electricity generated by nuclear power stations belonging to the major power producers. See Chapter 5, paragraphs 5.50 and 5.52.

Natural flow hydro-electricity - Electricity generated by natural flow hydroelectric power stations, whether they belong to major power producers or other generators. Pumped storage stations are not included (see under secondary electricity below).

Renewable energy sources - In this chapter figures are presented for renewables and waste in total. Further details, including a detailed breakdown of the commodities and technologies covered are in Chapter 7.

Secondary fuels

Manufactured fuel - This heading includes manufactured solid fuels such as coke and breeze, other manufactured solid fuels, liquids such as benzole and tars and gases such as coke oven gas and blast furnace gas. Further details are given in Chapter 2, Tables 2.4, 2.5 and 2.6.

Coke and breeze – Coke, oven coke and hard coke breeze. Further details are given in Chapter 2, Tables 2.4, 2.5 and 2.6.

Other manufactured solid fuels – Manufactured solid fuels produced at low temperature carbonisation plants and other manufactured fuel and briquetting plants. Further details are given in Chapter 2, Tables 2.4, 2.5 and 2.6.

Coke oven gas - Gas produced at coke ovens, excluding low temperature carbonisation plants. Gas bled or burnt to waste is included in production and losses. Further details are given in Chapter 2, Tables 2.4, 2.5 and 2.6.

Blast furnace gas - Blast furnace gas is mainly produced and consumed within the iron and steel industry. Further details are given in Chapter 2, Tables 2.4, 2.5 and 2.6.

Petroleum products - Petroleum products produced mainly at refineries, together with inland deliveries of natural gas liquids.

Secondary electricity - Secondary electricity is that generated by the combustion of another fuel, usually coal, natural gas, biofuels or oil. The figure for outputs from transformation in the electricity column of Tables 1.1, 1.2 and 1.3 is the total of primary and secondary electricity, and the subsequent analysis of consumption is based on this total.

Heat sold – Heat sold is heat that is produced and sold under the provision of a contract.

VI Classification of consumers

1.54 The Digest has been prepared, as far as is practicable, on the basis of the *Standard Industrial Classification (SIC)2003* (www.statistics.gov.uk/about/data/classifications/default.asp). SIC(2003) replaced SIC(1992) on 1 January 2003. SIC(1992) had been the basis of the industrial classification of energy statistics since 1995. Between 1986 and 1994 data in the Digest were prepared on the basis of the previous classification, SIC(1980). The changes in classification between SIC(1992) and SIC(2003) are mainly in the very detailed classifications at the four or five digit level. As such the classifications used for energy statistics are unaffected by these changes. However, not all consumption/disposals data are on this new basis, and where they are, there are sometimes constraints on the detail available. In particular the sectoral breakdown in the petroleum chapter is based on data that continue to be classified according to SIC (1968) by the oil industry. The main differences between the 1968 SIC (which was used as the basis for most data published for years prior to 1984) and the 1980 SIC were described in the 1986 and 1987 issues of the Digest. The differences between SIC 1980 and SIC 1992 are relatively minor. At the time of the change from the 1980 SIC to the 1992 SIC the main difference was that under the former showrooms belonging to the fuel supply industries were classified to the energy sector, whilst in the latter they are in the commercial sector. Since privatisation few gas, coal and electricity companies have retained showrooms and the difference is therefore minimal.

1.55 Table 1E shows the categories of consumers together with their codes in SIC 2003. The coverage varies between tables (eg in some instances the 'other' category is split into major constituents, whereas elsewhere it may include transport). This is because the coverage is dictated by what data suppliers can provide. The table also shows the disaggregation available within industry. This disaggregation forms the basis of virtually all the tables that show a disaggregated industrial breakdown.

Table 1E: SIC 2003 classifications

Fuel producers	10-12, 23, 40

Final consumers:

Industrial

Unclassified	See paragraph 1.56 below
Iron and steel	27, *excluding 27.4*, 27.53, 27.54
Non-ferrous metals	27.4, 27.53, 27.54
Mineral products	14, 26
Chemicals	24
Mechanical engineering and metal products	28, 29
Electrical and instrument engineering	30-33
Vehicles	34, 35
Food, beverages & tobacco	15, 16
Textiles, clothing, leather, & footwear	17-19
Paper, printing & publishing	21, 22
Other industries	13, 20, 25, 36, 37, 41
Construction	45

Transport, storage and communications 60-63

Other final users

Domestic	Not covered by SIC 2003.
Public administration	75, 80, 85
Commercial	50-52, 55, 64-67, 70-74
Agriculture	01, 02, 05
Miscellaneous	90-93, 99

1.56   There is also an 'unclassified' category in the industry sector (see Table 1E). Wherever the data supplier is unable to allocate an amount between categories, but the Department for Business, Enterprise and Regulatory Reform has additional information, from other data sources, with which to allocate between categories, then this has been done. Where such additional information is not available the data are included in the 'unclassified' category, enabling the reader to decide whether to accept a residual, pro-rate, or otherwise adjust the figures. The 'miscellaneous' category also contains some unallocated figures for the services sector.

1.57   In Tables 6.8 and 6.9 of Chapter 6 the following abbreviated grouping of industries, based on SIC 2003, is used in order to prevent disclosure of information about individual companies.

Table 1F: Abbreviated grouping of Industry

Iron and steel and non-ferrous metal	27
Chemicals	24
Oil refineries	23.2
Paper, printing and publishing	21, 22
Food, beverages and tobacco	15, 16
Metal products, machinery and equipment	28, 29, 30, 31, 32, 34, 35
Mineral products, extraction, mining and agglomeration of solid fuels	10, 11, 14, 26
Sewage Treatment	(parts of 41 and 90)
Electricity supply	40.1
Other industrial branches	12, 13, 17, 18, 19, 20, 23.1, 23.3, 25, 33, 36, 37, 40.2, 41 (remainder) 45
Transport, commerce, and administration	1, 2, 5, 50 to 99 (except 90 and 92)
Other	40.3, 90 (remainder), 92

1.58 In Tables 1.8 and 1.9 the list above is further condensed and includes only manufacturing industry and construction as follows.

Table 1G: Abbreviated grouping of Industry for Tables 1.8 and 1.9

Iron and steel and non-ferrous metals	27
Chemicals	24
Paper, printing and publishing	21, 22
Food, beverages and tobacco	15, 16
Metal products, machinery and equipment	28, 29, 30, 31, 32, 34, 35
Other (including construction)	12, 13, 14, 17, 18, 19, 20, 23.1, 23.3, 25, 26, 33, 36, 37, 45

VII Monthly and quarterly data

1.59   Monthly and quarterly data on energy production and consumption (including on a seasonally adjusted and temperature corrected basis) split by fuel type are provided on the BERR website at www.berr.gov.uk/energy/statistics/source/total/page18424.html. Quarterly figures are also published in BERR's quarterly statistical bulletin *Energy Trends* and *Quarterly Energy Prices*. See Annex C for more information about these bulletins.

Contact: *Iain MacLeay (Statistician)*
          Energy Markets Unit
          iain.macleay@berr.gsi.gov.uk
          020 7215 6898

          Chris Michaels
          Energy Markets Unit
          chris.michaels@berr.gsi.gov.uk
          020 7215 2710

1.1 Aggregate energy balance 2007

Thousand tonnes of oil equivalent

	Coal	Manufact ured fuel(1)	Primary oils	Petroleum products	Natural gas(2)	Renewable & waste(3)	Primary electricity	Electric ity	Heat	Total
Supply										
Indigenous production	10,668	-	84,211	-	72,125	3,975	14,928	-	-	185,906
Imports	28,200	728	62,311	26,090	29,065	378	-	741	-	147,512
Exports	-401	-182	-55,754	-32,108	-10,590	-	-	-292	-	-99,326
Marine bunkers	-	-	-	-2,513	-	-	-	-	-	-2,513
Stock change(4)	1,896	-22	856	1,084	471	-	-	-	-	4,286
Primary supply	**40,362**	**524**	**91,625**	**-7,446**	**91,071**	**4,354**	**14,928**	**448**	**-**	**235,865**
Statistical difference(5)	**+107**	**-28**	**-117**	**-95**	**+131**	-	-	**+130**	**-**	**+127**
Primary demand	**40,255**	**552**	**91,741**	**-7,351**	**90,941**	**4,354**	**14,928**	**318**	**-**	**235,738**
Transfers	-	-126	-3,211	+3,234	-7	-	-892	+892	-	-110
Transformation	**-38,578**	**1,725**	**-88,530**	**87,808**	**-32,155**	**-3,487**	**-14,036**	**32,865**	**1,190**	**-53,197**
Electricity generation	-32,897	-937	-	-677	-30,397	-3,487	-14,036	32,865	-	-49,566
Major power producers	-31,964	-	-	-211	-27,501	-683	-14,036	30,116	-	-44,279
Autogenerators	-933	-937	-	-466	-2,896	-2,805	-	2,749	-	-5,287
Heat generation	-286	-51	-	-60	-1,758	-	-	-	1,190	-966
Petroleum refineries	-	-	-88,530	88,755	-	-	-	-	-	225
Coke manufacture	-4,319	4,170	-	-	-	-	-	-	-	-149
Blast furnaces	-904	-1,633	-	-210	-	-	-	-	-	-2,747
Patent fuel manufacture	-172	176	-	-	-	-	-	-	-	5
Other	-	-	-	-	-	-	-	-	-	-
Energy industry use	**4**	**881**	**-**	**4,553**	**6,406**	**-**	**-**	**2,404**	**60**	**14,308**
Electricity generation	-	-	-	-	-	-	-	1,555	-	1,555
Oil and gas extraction	-	-	-	-	5,523	-	-	48	-	5,571
Petroleum refineries	-	-	-	4,553	291	-	-	401	60	5,304
Coal extraction	4	-	-	-	8	-	-	85	-	96
Coke manufacture	-	424	-	-	26	-	-	8	-	458
Blast furnaces	-	458	-	-	62	-	-	41	-	561
Patent fuel manufacture	-	-	-	-	-	-	-	-	-	-
Pumped storage	-	-	-	-	-	-	-	104	-	104
Other	-	-	-	-	497	-	-	162	-	658
Losses	**-**	**216**	**-**	**-**	**1,038**	**-**	**-**	**2,270**	**-**	**3,524**
Final consumption	**1,673**	**1,053**	**-**	**79,139**	**51,335**	**866**	**-**	**29,402**	**1,130**	**164,599**
Industry	**1,173**	**861**	**-**	**6,827**	**11,760**	**256**	**-**	**10,123**	**692**	**31,693**
Unclassified	-	237	-	2,640	3	256	-	-	-	3,137
Iron and steel	1	624	-	20	631	-	-	423	-	1,699
Non-ferrous metals	22	-	-	48	292	-	-	673	-	1,035
Mineral products	691	-	-	197	964	-	-	705	-	2,558
Chemicals	91	-	-	192	3,201	-	-	1,829	278	5,592
Mechanical engineering etc	7	-	-	107	673	-	-	750	2	1,539
Electrical engineering etc	4	-	-	36	342	-	-	625	-	1,007
Vehicles	35	-	-	123	759	-	-	510	-	1,427
Food, beverages etc	30	-	-	268	2,289	-	-	1,058	1	3,645
Textiles, leather etc	52	-	-	119	549	-	-	288	-	1,008
Paper, printing etc	101	-	-	65	945	-	-	1,178	1	2,291
Other industries	140	-	-	2,843	881	-	-	1,945	411	6,220
Construction	-	-	-	169	231	-	-	137	-	537
Transport (6)	**-**	**-**	**-**	**59,104**	**-**	**-**	**-**	**710**	**-**	**59,814**
Air	-	-	-	13,971	-	-	-	-	-	13,971
Rail	-	-	-	700	-	-	-	-	-	700
Road	-	-	-	42,815	-	-	-	-	-	42,815
National navigation	-	-	-	1,618	-	-	-	-	-	1,618
Pipelines	-	-	-	-	-	-	-	-	-	-
Other	**501**	**192**	**-**	**4,377**	**38,676**	**610**	**-**	**18,569**	**437**	**63,362**
Domestic	487	192	-	2,877	30,090	426	-	9,893	52	44,015
Public administration	5	-	-	487	3,834	91	-	1,879	376	6,673
Commercial	4	-	-	409	3,091	-	-	6,469	9	9,982
Agriculture	3	-	-	294	172	74	-	329	-	872
Miscellaneous	1	-	-	310	1,490	20	-	-	-	1,821
Non energy use	**-**	**-**	**-**	**8,830**	**899**	**-**	**-**	**-**	**-**	**9,730**

(1) Includes all manufactured solid fuels, benzole, tars, coke oven gas and blast furnace gas.
(2) Includes colliery methane.
(3) Includes geothermal and solar heat.
(4) Stock fall (+), stock rise (-).
(5) Primary supply minus primary demand.
(6) See paragraph 5.14 regarding electricity use in transport

1.2 Aggregate energy balance 2006

Thousand tonnes of oil equivalent

	Coal	Manufact ured fuel(1)	Primary oils	Petroleum products	Natural gas(2)	Renewable & waste(3)	Primary electricity	Electric ity	Heat	Total
Supply										
Indigenous production	11,370r	-	83,958	-	80,012r	3,731r	17,889r	-	-	196,960r
Imports	32,668r	695r	64,872	29,335	20,983	497	-	884	-	149,933r
Exports	-342r	-120	-54,875	-31,474	-10,369	-	-	-238	-	-97,418r
Marine bunkers	-	-	-	-2,486	-	-	-	-	-	-2,486
Stock change(4)	-807r	-153r	-391	-917	-553	-	-	-	-	-2,820r
Primary supply	**42,889r**	**422r**	**93,564**	**-5,543**	**90,072r**	**4,228r**	**17,889r**	**646**	**-**	**244,169r**
Statistical difference(5)	-100r	-19r	-127	+66r	-58r	-	-	+129r	-	-108r
Primary demand	**42,989r**	**441r**	**93,691**	**-5,609r**	**90,130r**	**4,228r**	**17,889r**	**517r**	**-**	**244,277r**
Transfers	-	-110r	-2,835	+2,869	-5	-	-759r	+759r	-	-80r
Transformation	**-41,410r**	**1,796r**	**-90,856**	**89,770r**	**-28,582r**	**-3,471r**	**-17,130r**	**33,202r**	**1,305r**	**-55,377r**
Electricity generation	-35,799r	-967	-	-692	-26,689r	-3,471r	-17,130r	33,202r	-	-51,545r
Major power producers	-34,869	-	-	-277	-23,915	-731r	-17,130r	30,500r	-	-46,422r
Autogenerators	-930r	-967	-	-415	-2,774r	-2,740r	-	2,702r	-	-5,123r
Heat generation	-286r	-51	-	-60r	-1,894r	-	-	-	1,305r	-987r
Petroleum refineries	-	-	-90,856	90,760	-	-	-	-	-	-96
Coke manufacture	-4,315	4,272r	-	-	-	-	-	-	-	-44r
Blast furnaces	-816	-1,659r	-	-238	-	-	-	-	-	-2,713r
Patent fuel manufacture	-194	202	-	-	-	-	-	-	-	8
Other	-	-	-	-	-	-	-	-	-	-
Energy industry use	**3**	**871**	**-**	**4,986**	**6,895r**	**-**	**-**	**2,478r**	**60r**	**15,293r**
Electricity generation	-	-	-	-	-	-	-	1,652r	-	1,652r
Oil and gas extraction	-	-	-	-	5,954	-	-	47	-	6,001
Petroleum refineries	-	-	-	4,986	277r	-	-	392r	60r	5,716r
Coal extraction	3	-	-	-	10	-	-	89r	-	101r
Coke manufacture	-	414	-	-	23	-	-	8	-	446
Blast furnaces	-	457	-	-	53	-	-	43r	-	552r
Patent fuel manufacture	-	-	-	-	-	-	-	-	-	-
Pumped storage	-	-	-	-	-	-	-	92	-	92
Other	-	-	-	-	578	-	-	156r	-	734r
Losses	-	177	-	-	1,033	-	-	2,362r	-	3,572r
Final consumption	**1,576r**	**1,079r**	**-**	**82,044**	**53,616r**	**757r**	**-**	**29,638r**	**1,245r**	**169,956r**
Industry	**1,130r**	**871r**	**-**	**7,220**	**12,366r**	**198r**	**-**	**10,172r**	**809r**	**32,766r**
Unclassified	-	253r	-	3,019	4r	198r	-	-	-	3,474r
Iron and steel	-	618	-	20	723r	-	-	504r	-	1,864r
Non-ferrous metals	36r	-	-	53	276r	-	-	661	-	1,025r
Mineral products	691	-	-	200	1,058r	-	-	685r	-	2,634r
Chemicals	54r	-	-	193	3,289r	-	-	1,970r	371r	5,876r
Mechanical engineering etc	9	-	-	106	746r	-	-	730r	2r	1,593r
Electrical engineering etc	4	-	-	85	358	-	-	630	-	1,076r
Vehicles	37	-	-	124	797	-	-	503r	-	1,461r
Food, beverages etc	18r	-	-	283	2,346r	-	-	1,059r	1	3,706r
Textiles, leather etc	49r	-	-	131	571r	-	-	291r	-	1,042r
Paper, printing etc	99	-	-	59	1,033r	-	-	1,154r	22	2,367r
Other industries	133r	-	-	2,774	948r	-	-	1,845r	414r	6,114r
Construction	-	-	-	174	220	-	-	141	-	535
Transport (6)	-	-	-	59,047	-	-	-	706r	-	**59,753r**
Air	-	-	-	13,999	-	-	-	-	-	13,999
Rail	-	-	-	726	-	-	-	-	-	726
Road	-	-	-	42,509	-	-	-	-	-	42,509
National navigation	-	-	-	1,812	-	-	-	-	-	1,812
Pipelines	-	-	-	-	-	-	-	-	-	-
Other	**446r**	**208r**	**-**	**4,779**	**40,239r**	**560r**	**-**	**18,760r**	**436r**	**65,429r**
Domestic	426r	208r	-	3,251	31,371r	382r	-	10,013	52	45,703r
Public administration	10r	-	-	489	4,197r	85r	-	1,911r	376r	7,068r
Commercial	4	-	-	393	2,947	-	-	6,481r	8	9,834
Agriculture	3	-	-	306	173	74	-	355	-	912
Miscellaneous	3	-	-	340	1,550r	20	-	-	-	1,913r
Non energy use	-	-	-	10,997	1,011r	-	-	-	-	12,008r

(1) Includes all manufactured solid fuels, benzole, tars, coke oven gas and blast furnace gas.
(2) Includes colliery methane.
(3) Includes geothermal and solar heat.
(4) Stock fall (+), stock rise (-).
(5) Primary supply minus primary demand.
(6) See paragraph 5.14 regarding electricity use in transport

1.3 Aggregate energy balance 2005

Thousand tonnes of oil equivalent

	Coal	Manufactured fuel(1)	Primary oils	Petroleum products	Natural gas(2)	Renewable & waste(3)	Primary electricity	Electricity	Heat	Total
Supply										
Indigenous production	12,714r	-	92,883	-	88,219r	3,674r	19,044	-	-	216,534r
Imports	28,534r	623r	64,255	24,577	14,904r	421	-	960r	-	134,273r
Exports	-420r	-89r	-59,177	-32,321	-8,270r	-	-	-244r	-	-100,521r
Marine bunkers	-	-	-	-2,181		-	-	-	-	-2,181r
Stock change(4)	-1,406r	-97r	-416	+1,157	+114r	-	-	-	-	-648r
Primary supply	**39,422r**	**437r**	**97,546**	**-8,768**	**94,966r**	**4,095r**	**19,044**	**715r**	**-**	**247,457r**
Statistical difference(5)	+31r	-7r	-121	-176	+271r	+0	-	+205r	-	+203r
Primary demand	**39,392r**	**443r**	**97,667**	**-8,592**	**94,695r**	**4,094r**	**19,044**	**511r**	**-**	**247,254r**
Transfers	-	-113r	-3,643	+3,649r	-4r	-	-	-674	+674r	-112r
Transformation	**-37,700r**	**1,737r**	**-94,023**	**93,063**	**-30,219r**	**-3,368r**	**-18,370**	**33,323r**	**1,366r**	**-54,192r**
Electricity generation	-32,408r	-990r	-	-704	-28,285r	-3,368r	-18,370	33,323r	-	-50,803r
Major power producers	-31,528r	-	-	-263	-25,421r	-831r	-18,370	30,564r	-	-45,850r
Autogenerators	-880r	-990r	-	-441	-2,865r	-2,537r	-	2,759r	-	-4,953r
Heat generation	-286r	-51	-	-61	-1,934r	-	-	-	1,366r	-966r
Petroleum refineries	-	-	-94,023	94,107		-	-	-	-	84r
Coke manufacture	-4,053r	4,024r	-		-	-	-	-	-	-29r
Blast furnaces	-756r	-1,446r	-	-280	-	-	-	-	-	-2,482r
Patent fuel manufacture	-197r	200r	-	-	-	-	-	-	-	3r
Other	-	-	-	-	-	-	-	-	-	-
Energy industry use	**4r**	**820r**	**-**	**5,974**	**7,418r**	**-**	**-**	**2,336r**	**98**	**16,650r**
Electricity generation	-	-	-	-	-	-	-	1,537r	26	1,563r
Oil and gas extraction	-	-	-	-	6,309r	-	-	43r	-	6,352r
Petroleum refineries	-	-	-	5,974	368r	-	-	383r	71	6,796r
Coal extraction	4r	-	-	-	10	-	-	92r	-	106r
Coke manufacture	-	396r	-	-	-	-	-	8	-	404r
Blast furnaces	-	424r	-	-	81r	-	-	44	-	549r
Patent fuel manufacture	-	-	-	-	-	-	-	-	-	-
Pumped storage	-	-	-	-	-	-	-	67r	-	67r
Other	-	-	-	-	651	-	-	162r	-	813r
Losses		**211r**	**-**	**-**	**943r**	**-**	**-**	**2,380r**	**-**	**3,533r**
Final consumption	**1,687r**	**1,035r**	**-**	**82,146**	**56,111r**	**727r**	**-**	**29,791r**	**1,268**	**172,766r**
Industry	**1,180r**	**812r**	**-**	**7,227**	**13,017r**	**189r**	**-**	**10,363r**	**831r**	**33,618r**
Unclassified	-	226r	-	2,665	5	189r	-	-	-	3,085r
Iron and steel	-	586r	-	15	728r	-	-	432r	-	1,760r
Non-ferrous metals	24	-	-	53	277r	-	-	661	-	1,015r
Mineral products	739r	-	-	216	1,144r	-	-	686	-	2,784r
Chemicals	84r	-	-	194	3,532r	-	-	2,074r	392r	6,276r
Mechanical engineering etc	10r	-	-	118	746r	-	-	742r	3	1,618r
Electrical engineering etc	3r	-	-	35	373r	-	-	638r	-	1,049r
Vehicles	38r	-	-	139	882r	-	-	502	-	1,561r
Food, beverages etc	19r	-	-	323	2,400r	-	-	1,098r	1	3,841r
Textiles, leather etc	50r	-	-	110	613r	-	-	292r	-	1,065r
Paper, printing etc	98r	-	-	86	1,177r	-	-	1,180r	31	2,572r
Other industries	116r	-	-	3,083	911r	-	-	1,891r	405	6,406r
Construction	-	-	-	190	230r	-	-	166r	-	586r
Transport (6)	**-**	**-**	**-**	**58,325**	**-**	**-**	**-**	**758r**	**-**	**59,083r**
Air	-	-	-	13,856		-	-	-	-	13,856r
Rail	-	-	-	707		-	-	-	-	707r
Road	-	-	-	42,390		-	-	-	-	42,390r
National navigation	-	-	-	1,372		-	-	-	-	1,372r
Pipelines	-	-	-	-		-	-	-	-	-
Other	**508r**	**223r**	**-**	**4,847**	**42,261r**	**538r**	**-**	**18,670r**	**437r**	**67,485r**
Domestic	474r	223r	-	3,093	33,019r	340r	-	10,044r	52	47,245r
Public administration	22r	-	-	541	4,327r	105r	-	1,795r	376	7,166r
Commercial	4	-	-	388	3,018r	-	-	6,474r	10	9,894r
Agriculture	6r	-	-	364	194r	74	-	357r	-	995r
Miscellaneous	1r	-	-	461	1,704r	19r	-	-	-	2,185r
Non energy use	**-**	**-**	**-**	**11,748**	**832r**	**-**	**-**	**-**	**-**	**12,580r**

(1) Includes all manufactured solid fuels, benzole, tars, coke oven gas and blast furnace gas.
(2) Includes colliery methane.
(3) Includes geothermal and solar heat.
(4) Stock fall (+), stock rise (-).
(5) Primary supply minus primary demand.
(6) See paragraph 5.14 regarding electricity use in transport

1.4 Value balance of traded energy in 2007[1]

£million

	Coal	Manufactured solid fuels	Crude oil	Petroleum products	Natural gas	Electricity	Heat sold	Other fuels	Total
Supply									
Indigenous production	580	55	21,000	27,375	7,780	8,470	230	120	65,605
Imports	1,960	140	13,290	9,205	2,885	240	-	-	27,715
Exports	-40	-30	-12,935	-9,900	-995	-110	-	-	-24,005
Marine bunkers	-	-	-	-555	-	-	-	-	-555
Stock change	100	-10	175	305	5	-	-	-	575
Basic value of inland consumption	**2,600**	**160**	**21,535**	**26,430**	**9,670**	**8,600**	**230**	**120**	**69,340**
Tax and margins									
Distribution costs and margins	**355**	**50**	**-**	**2,590**	**8,325**	**17,000**	**-**	**-**	**28,325**
Electricity generation	110	-	-	5	-	-	-	-	115
Solid fuel manufacture	110	-	-	-	-	-	-	-	110
of which iron & steel sector	100	-	-	-	-	-	-	-	100
Iron & steel final use	20	35	-	20	-	-	-	-	75
Other industry	10	10	-	450	-	-	-	-	465
Air transport	-	-	-	25	-	-	-	-	25
Rail and national navigation	-	-	-	20	-	-	-	-	20
Road transport	-	-	-	1,740	-	-	-	-	1,740
Domestic	100	10	-	20	-	-	-	-	130
Agriculture	-	-	-	10	-	-	-	-	10
Commercial and other services	-	-	-	40	-	-	-	-	40
Non energy use	-	-	-	260	155	-	-	-	415
VAT and duties	**10**	**5**	**-**	**31,985**	**470**	**580**	**-**	**-**	**33,050**
Electricity generation	-	-	-	30	-	-	-	-	30
Iron & steel final use	-	-	-	-	-	-	-	-	-
Other industry	-	-	-	350	-	-	-	-	350
Air transport	-	-	-	15	-	-	-	-	15
Rail and national navigation	-	-	-	195	-	-	-	-	195
Road transport	-	-	-	31,210	-	-	-	-	31,210
Domestic	10	5	-	70	470	580	-	-	1,130
Agriculture	-	-	-	15	-	-	-	-	15
Commercial and other services	-	-	-	105	-	-	-	-	105
Climate Change Levy	**5**	**-**	**-**	**-**	**165**	**515**	**-**	**-**	**685**
Total tax and margins	**365**	**55**	**-**	**34,580**	**8,965**	**18,095**	**-**	**-**	**62,060**
Market value of inland consumption	**2,965**	**215**	**21,535**	**61,010**	**18,635**	**26,695**	**135**	**120**	**131,310**
Energy end use									
Total energy sector	**2,625**	**-**	**21,535**	**190**	**4,505**	**325**	**-**	**50**	**29,225**
Transformation	**2,625**	**-**	**21,535**	**190**	**4,385**	**-**	**-**	**50**	**28,785**
Electricity generation	2,130	-	-	175	4,360	-	-	50	6,720
of which from stocks	60	-	-	-	-	-	-	-	60
Heat Generation	20	-	-	15	25	-	-	-	60
Petroleum refineries	-	-	21,535	-	-	-	-	-	21,535
Solid fuel manufacture	480	-	-	-	-	-	-	-	480
of which iron & steel sector	435	-	-	-	-	-	-	-	435
Other energy sector use	**-**	**-**	**-**	**-**	**115**	**325**	**-**	**-**	**440**
Oil & gas extraction	-	-	-	-	-	40	-	-	40
Petroleum refineries	-	-	-	-	40	220	-	-	260
Coal extraction	-	-	-	-	-	65	-	-	65
Other energy sector	-	-	-	-	75	-	-	-	75
Total non energy sector use	**340**	**215**	**-**	**58,720**	**13,975**	**26,370**	**135**	**70**	**99,825**
Industry	**180**	**145**	**-**	**2,420**	**2,025**	**6,985**	**55**	**20**	**11,825**
Iron & steel final use	95	125	-	55	110	125	-	-	515
Other industry	80	20	-	2,360	1,915	6,855	55	20	11,310
Transport	**-**	**-**	**-**	**54,600**	**-**	**460**	**-**	**-**	**55,060**
Air	-	-	-	4,675	-	-	-	-	4,675
Rail and national navigation	-	-	-	790	-	460	-	-	1,245
Road	-	-	-	49,135	-	-	-	-	49,135
Other final users	**160**	**70**	**-**	**1,700**	**11,950**	**18,930**	**80**	**50**	**32,945**
Domestic	160	70	-	1,130	9,865	12,160	10	50	23,445
Agriculture	-	-	-	105	50	340	-	-	495
Commercial and other services	-	-	-	465	2,035	6,430	70	-	9,000
Total value of energy end use	**2,965**	**215**	**21,535**	**58,905**	**18,480**	**26,695**	**135**	**120**	**129,050**
Value of non energy end use	**-**	**-**	**-**	**2,100**	**155**	**-**	**-**	**-**	**2,255**
Market value of inland consumption	**2,965**	**215**	**21,535**	**61,010**	**18,635**	**26,695**	**135**	**120**	**131,310**

(1) For further information see paragraphs 1.37 to 1.42.

1.5 Value balance of traded energy in 2006[1]

£million

	Coal	Manufactured solid fuels	Crude oil	Petroleum products	Natural gas	Electricity	Heat sold	Other fuels	Total
Supply									
Indigenous production	545	100	20,495r	26,295r	10,270r	8,190r	250r	110r	66,255r
Imports	2,135r	90	13,600r	9,790r	2,510	420r	-	-	28,545r
Exports	-30	-20	-12,830	-9,625r	-1,315	-105r	-	-	-23,925r
Marine bunkers	-	-	-	-595	-	-	-	-	-595
Stock change	-45	-	-95	-260	-10	-	-	-	-405r
Basic value of inland consumption	**2,605r**	**175**	**21,175r**	**25,600r**	**11,460r**	**8,505r**	**250r**	**110r**	**69,875r**
Tax and margins									
Distribution costs and margins	**415**	**35r**	**-**	**3,515r**	**7,050r**	**15,990**	**-**	**-**	**27,000r**
Electricity generation	175	-	-	20	-	-	-	-	195
Solid fuel manufacture	125	-	-	-	-	-	-	-	125
of which iron & steel sector	115	-	-	-	-	-	-	-	115
Iron & steel final use	25	15	-	25	-	-	-	-	65
Other industry	10	10r	-	440	-	-	-	-	460r
Air transport	-	-	-	115r	-	-	-	-	115r
Rail and national navigation	-	-	-	50r	-	-	-	-	50r
Road transport	-	-	-	2,325	-	-	-	-	2,325
Domestic	80	10	-	140r	-	-	-	-	230r
Agriculture	-	-	-	15	-	-	-	-	15
Commercial and other services	-	-	-	60	-	-	-	-	60
Non energy use	-	-	-	320r	210r	-	-	-	530r
VAT and duties	**5**	**5**		**30,190r**	**480r**	**560r**			**31,245r**
Electricity generation	-	-	-	30	-	-	-	-	30
Iron & steel final use	-	-	-	-	-	-	-	-	-
Other industry	-	-	-	285r	-	-	-	-	285r
Air transport	-	-	-	20	-	-	-	-	20
Rail and national navigation	-	-	-	170r	-	-	-	-	170r
Road transport	-	-	-	29,520	-	-	-	-	29,520
Domestic	5	5	-	70r	480r	560r	-	-	1,125r
Agriculture	-	-	-	10	-	-	-	-	10
Commercial and other services	-	-	-	85	-	-	-	-	85
Climate Change Levy	**5**	**-**	**-**	**-**	**185r**	**525**			**720r**
Total tax and margins	**425**	**40r**	**-**	**33,705r**	**7,715r**	**17,075r**	**-**	**-**	**58,960r**
Market value of inland consumption	**3,030r**	**215r**	**21,175r**	**59,305r**	**19,175r**	**25,580r**	**155r**	**110r**	**128,740r**
Energy end use									
Total energy sector	**2,720**	**-**	**21,175r**	**190r**	**4,135r**	**295**	**-**	**40**	**28,555r**
Transformation	**2,720**	**-**	**21,175r**	**190r**	**4,005**	**-**	**-**	**40**	**28,130r**
Electricity generation	2,145	-	-	180	3,980r	-	-	40	6,340
of which from stocks	55	-	-	-	-	-	-	-	55
Heat Generation	15	-	-	15	30	-	-	-	60
Petroleum refineries	-	-	21,175r	-	-	-	-	-	21,175r
Solid fuel manufacture	555	-	-	-	-	-	-	-	555
of which iron & steel sector	505	-	-	-	-	-	-	-	505
Other energy sector use	**-**	**-**	**-**	**-**	**130r**	**295**	**-**	**-**	**425r**
Oil & gas extraction	-	-	-	-	-	35	-	-	35
Petroleum refineries	-	-	-	-	40r	195	-	-	240r
Coal extraction	-	-	-	-	-	65	-	-	65
Other energy sector	-	-	-	-	90	-	-	-	90
Total non energy sector use	**310r**	**215r**	**-**	**56,555r**	**14,825r**	**25,285r**	**155r**	**70r**	**97,410r**
Industry	**175**	**140r**	**-**	**2,420r**	**2,600**	**6,775**	**70r**	**30**	**12,215r**
Iron & steel final use	100	120	-	65	155r	210	-	-	650r
Other industry	75	20r	-	2,355	2,445r	6,565	70r	30	11,565r
Transport	**-**	**-**	**-**	**52,330r**	**-**	**460**	**-**	**-**	**52,790r**
Air	-	-	-	4,595r	-	-	-	-	4,595r
Rail and national navigation	-	-	-	855	-	460	-	-	1,315
Road	-	-	-	46,880	-	-	-	-	46,880
Other final users	**135r**	**70r**	**-**	**1,800r**	**12,225r**	**18,050r**	**80r**	**45r**	**32,410r**
Domestic	135r	70r	-	1,225r	10,070r	11,790r	10	45r	23,345r
Agriculture	-	-	-	105	45	340	-	-	490
Commercial and other services	-	-	-	470	2,105r	5,920	70r	-	8,570r
Total value of energy end use	**3,030r**	**215r**	**21,175r**	**56,745**	**18,965r**	**25,580r**	**155r**	**110r**	**125,970r**
Value of non energy end use	**-**	**-**	**-**	**2,560r**	**210r**	**-**	**-**	**-**	**2,770r**
Market value of inland consumption	**3,030r**	**215r**	**21,175r**	**59,305r**	**19,175r**	**25,580r**	**155r**	**110r**	**128,740r**

(1) For further information see paragraphs 1.37 to 1.42.

1.6 Value balance of traded energy in 2005[1]

£million

	Coal	Manufactured solid fuels	Crude oil	Petroleum products	Natural gas	Electricity	Heat sold	Other fuels	Total
Supply									
Indigenous production	560r	95r	18,930	22,485r	8,260r	5,145r	260	95r	55,825r
Imports	1,870r	110r	11,565	7,850	1,730	440	-	-	23,570r
Exports	-40	-15	-11,290	-8,305	-735	-100	-	-	-20,490
Marine bunkers	-	-	-	-415	-	-	-	-	-415
Stock change	-80	-5	-60	260	-	-	-	-	115r
Basic value of inland consumption	**2,315r**	**175r**	**19,145**	**21,880r**	**9,255r**	**5,485r**	**260**	**95r**	**58,610r**
Tax and margins									
Distribution costs and margins	**280r**	**70r**	**-**	**3,340r**	**5,750r**	**13,885**	**-**	**-**	**23,325r**
Electricity generation	50r	-	-	20r	-	-	-	-	70r
Solid fuel manufacture	105	-	-	-	-	-	-	-	105
of which iron & steel sector	90	-	-	-	-	-	-	-	90
Iron & steel final use	20	50	-	15	-	-	-	-	85
Other industry	15	10	-	340	-	-	-	-	365
Air transport	-	-	-	25	-	-	-	-	25
Rail and national navigation	-	-	-	20	-	-	-	-	20
Road transport	-	-	-	2,370	-	-	-	-	2,370
Domestic	95	10r	-	170r	-	-	-	-	275r
Agriculture	-	-	-	15	-	-	-	-	15
Commercial and other services	-	-	-	55	-	-	-	-	55
Non energy use	-	-	-	315r	135	-	-	-	450r
VAT and duties	**5**	**5**	**-**	**29,840r**	**390r**	**460r**	**-**	**-**	**30,705r**
Electricity generation	-	-	-	25	-	-	-	-	25
Iron & steel final use	-	-	-	-	-	-	-	-	-
Other industry	-	-	-	250	-	-	-	-	250
Air transport	-	-	-	20	-	-	-	-	20
Rail and national navigation	-	-	-	115	-	-	-	-	115
Road transport	-	-	-	29,285	-	-	-	-	29,285
Domestic	5	5	-	55r	390r	460r	-	-	920r
Agriculture	-	-	-	10	-	-	-	-	10
Commercial and other services	-	-	-	75	-	-	-	-	75
Climate Change Levy	**5**	**-**	**-**	**-**	**195r**	**535r**	**-**	**-**	**735r**
Total tax and margins	**295r**	**70**	**-**	**33,185r**	**6,335r**	**14,880r**	**-**	**-**	**54,765r**
Market value of inland consumption	**2,610r**	**250**	**19,145**	**55,065r**	**15,590r**	**20,365r**	**160**	**95r**	**113,275r**
Energy end use									
Total energy sector	**2,300**	**-**	**19,145**	**175**	**3,475**	**230**	**5**	**35**	**25,365r**
Transformation	**2,300**	**-**	**19,145**	**175**	**3,355**	**-**	**5**	**35**	**25,015r**
Electricity generation	1,830	-	-	160r	3,335	-	5	35	5,365r
of which from stocks	60	-	-	-	-	-	-	-	60
Heat Generation	15	-	-	15	25	-	-	-	50
Petroleum refineries	-	-	19,145	-	-	-	-	-	19,145
Solid fuel manufacture	455	-	-	-	-	-	-	-	455
of which iron & steel sector	405			-	-	-	-	-	405
Other energy sector use	**-**	**-**	**-**	**-**	**120**	**230**	**-**	**-**	**350**
Oil & gas extraction	-	-	-	-	-	25	-	-	25
Petroleum refineries	-	-	-	-	45	155	-	-	200
Coal extraction	-	-	-	-	-	50	-	-	50
Other energy sector	-	-	-	-	75	-	-	-	75
Total non energy sector use	**305**	**250**	**-**	**52,505r**	**11,980r**	**20,135r**	**155**	**60r**	**85,390r**
Industry	**160**	**175**	**-**	**2,015**	**2,110r**	**5,060**	**75r**	**20**	**9,615r**
Iron & steel final use	80	155	-	55	125	120	-	-	535
Other industry	80	20	-	1,965	1,985r	4,940	75r	20	9,085r
Transport	**-**	**-**	**-**	**48,925**	**-**	**345**	**-**	**-**	**49,270**
Air	-	-	-	3,810	-	-	-	-	3,810
Rail and national navigation	-	-	-	580	-	345	-	-	925
Road	-	-	-	44,535	-	-	-	-	44,535
Other final users	**145**	**70**	**-**	**1,565r**	**9,870r**	**14,725r**	**80**	**40r**	**26,500r**
Domestic	145	70	-	1,010r	8,215r	9,665r	10	40r	19,155r
Agriculture	-	-	-	110	35	275	-	-	420
Commercial and other services	5r	-	-	445	1,620r	4,785	70	-	6,925r
Total value of energy end use	**2,610**	**250**	**19,145**	**52,680r**	**15,455r**	**20,365r**	**160**	**95r**	**110,755r**
Value of non energy end use	**-**	**-**	**-**	**2,385r**	**135**	**-**	**-**	**-**	**2,520r**
Market value of inland consumption	**2,610**	**250**	**19,145**	**55,065r**	**15,590r**	**20,365r**	**160**	**95r**	**113,275r**

(1) For further information see paragraphs 1.37 to 1.42.

1.7 Sales of electricity and gas by sector

United Kingdom

	2003	2004	2005	2006	2007
Total selling value (£ million)[1]					
Electricity generation - Gas	2,209	2,590r	3,333r	3,978r	4,362
Industrial - Gas	1,567	1,492	2,223r	2,699r	2,049
- Electricity	3,071	3,428	5,292	7,071	7,306
of which:					
Fuel industries	144	171	232	296	323
Industrial sector	2,927	3,257	5,060	6,775	6,983
Domestic sector - Gas	5,964	7,889	7,822r	9,592r	9,396
- Electricity	7,295	8,688	9,205r	11,227r	11,582
Other - Gas	1,325	1,449	1,797r	2,335r	2,259
- Electricity	3,827	4,354	5,408	6,720	7,228
of which:					
Agricultural sector	193	228	276	340	342
Commercial sector	2,748	3,115	3,857	4,776	5,184
Transport sector	216	260	346	459	459
Public lighting	75	82	106	134	151
Public admin. and other services	595	669	823	1,011	1,092
Total, all consumers	**25,258**	**29,889**	**35,080r**	**43,621r**	**44,182**
of which gas	**11,065**	**13,419**	**15,176r**	**18,604r**	**18,066**
of which electricity	**14,193**	**16,470**	**19,905r**	**25,018r**	**26,116**
Average net selling value per kWh sold (pence)[1]					
Electricity generation - Gas	0.682	0.761	1.015	1.284	1.236
Industrial - Gas	0.872	0.969	1.469	1.877	1.498
- Electricity	2.971	3.320	4.998	6.707	6.895
of which:					
Fuel industries	2.974	3.672	4.791	6.305	6.778
Industrial sector	2.971	3.304	5.008	6.726	6.901
Domestic sector - Gas	1.543	1.658r	2.037r	2.629r	2.685
- Electricity	6.302	7.333r	7.880r	9.641r	10.067
Other - Gas	1.133	1.277	1.672	2.264r	2.262
- Electricity	3.637	4.149	5.084	6.314	6.823
of which:					
Agricultural sector	4.804	5.447	6.648	8.224	8.944
Commercial sector	3.701	4.197	5.122	6.336	6.891
Transport sector	2.872	3.546	4.642	6.109	6.103
Public lighting	3.651	4.140	5.052	6.249	6.797
Public admin. and other services	3.423	3.881	4.737	5.860	6.373
Average, all consumers	**1.897**	**2.251**	**2.698r**	**3.490r**	**3.489**
of which gas	**1.099**	**1.337**	**1.563r**	**2.019r**	**1.923**
of which electricity	**4.376**	**5.088**	**6.049r**	**7.620r**	**7.988**

(1) Excludes VAT where payable - see paragraph 1.43 for a definition of average net selling value.

1.8 Final energy consumption by main industrial groups[1]

Thousand tonnes of oil equivalent

	2003	2004	2005	2006	2007
Iron and steel and non-ferrous metals					
Coal	8	7	24	36r	22
Manufactured solid fuels [2]	504	482	479r	434	451
Blast furnace gas	36	32	28	78	59
Coke oven gas	49	67	79	106r	114
Natural gas	1,299	1,110	1,005r	998r	922
Petroleum	67	87	68	73	68
Electricity	1,094	1,112	1,093	1,164r	1,096
Total iron and steel and non-ferrous metals	**3,057**	**2,898**	**2,775r**	**2,889r**	**2,733**
Chemicals					
Coal	46	94	84r	54r	91
Natural gas	3,873	3,611	3,532r	3,289r	3,201
Petroleum	197	203	194	193	192
Electricity	1,801	1,817	2,074r	1,970r	1,829
Heat purchased from other sectors [3]	1,097	394	392r	371r	278
Total chemicals	**7,014**	**6,119**	**6,276r**	**5,876r**	**5,592**
Metal products, machinery and equipment					
Coal	61	70	51	50	45
Natural gas	2,162	1,977	2,001r	1,901r	1,774
Petroleum	279	264	292	315	266
Electricity	1,764	1,806	1,883r	1,863r	1,885
Heat purchased from other sectors [3]	26	2	3	2r	2
Total metal products, machinery and equipment	**4,292**	**4,119**	**4,229r**	**4,130r**	**3,972**
Food, beverages and tobacco					
Coal	36	26	19r	18r	30
Natural gas	2,476	2,428	2,400r	2,346r	2,289
Petroleum	222	345	323	283	268
Electricity	1,027	1,062	1,098r	1,059r	1,058
Heat purchased from other sectors [3]	5	2	1	1	1
Total food, beverages and tobacco	**3,767**	**3,863**	**3,841r**	**3,706r**	**3,645**

(1) Industrial categories used are described in Table 1G. Data excludes energy used to generate heat for all fuels except manufactured solid fuels and electricity.

(2) Includes tars, benzole, coke and breeze and other manufactured solid fuels.

(3) Data equates to heat sold information in the energy balances.

1.8 Final energy consumption by main industrial groups[1] (continued)

Thousand tonnes of oil equivalent

	2003	2004	2005	2006	2007
Paper, printing and publishing					
Coal	88	96	98	99	101
Natural gas	1,367	1,193	1,177r	1,033r	945
Petroleum	56	59	86	59	65
Electricity	1,096	1,175	1,180r	1,154r	1,178
Heat purchased from other sectors (3)	-	27	31	22	1
Total paper, printing and publishing	**2,607**	**2,551**	**2,572r**	**2,367r**	**2,291**
Other industries					
Coal	1,009	941	905r	873r	883
Natural gas	3,108	2,912	2,898r	2,796r	2,625
Petroleum	3,414	3,329	3,599	3,280	3,328
Electricity	2,965	2,989	3,035r	2,962r	3,076
Heat purchased from other sectors (3)	-	407	405	414r	411
Total other industries	**10,496**	**10,577**	**10,841r**	**10,324r**	**10,322**
Unclassified					
Manufactured solid fuels (2)	135	145	226	253r	237
Coke oven gas	5	-	-	-	-
Natural gas	6	6	5	4r	3
Petroleum	2,505	2,632	2,665	3,019	2,640
Renewables & waste	267	265	189r	198r	256
Total unclassified	**2,917**	**3,048**	**3,085r**	**3,474r**	**3,137**
Total					
Coal	1,248	1,234	1,180r	1,130r	1,173
Manufactured solid fuels (2)	639	627	705r	687r	688
Blast furnace gas	36	32	28	78	59
Coke oven gas	53	67	79	106r	114
Natural gas	14,292	13,238	13,017r	12,366r	11,760
Petroleum	6,740	6,918	7,227	7,220	6,827
Renewables & waste	267	265	189r	198r	256
Electricity	9,747	9,961	10,363r	10,172r	10,123
Heat purchased from other sectors (3)	1,128	832	831r	809r	692
Total	**34,151**	**33,175**	**33,618r**	**32,766r**	**31,693**

1.9 Fuels consumed for electricity generation (autogeneration) by main industrial groups[1]

Thousand tonnes of oil equivalent
(except where shown otherwise)

	2003	2004	2005	2006	2007
Iron and steel and non-ferrous metals					
Coal	766	764	767	768	766
Blast furnace gas	774	790	801	780	756
Coke oven gas	136	107	162	161r	157
Natural gas	57	61	44	39r	39
Petroleum	14	32	19	20	20
Other (including renewables) (2)	47	64	70r	55	56
Total fuel input (3)	**1,795**	**1,817**	**1,863r**	**1,823r**	**1,793**
Electricity generated by iron & steel and non-ferrous	**474**	**481**	**488**	**481r**	**475**
metals (4) (in GWh)	5,511	5,592	5,670	5,592r	5,527
Electricity consumed by iron and steel and non-ferrous	**381**	**417**	**427**	**418r**	**398**
metals from own generation (5) (in GWh)	4,435	4,852	4,969	4,862r	4,629
Chemicals					
Coal	160	113	109	140r	139
Natural gas	938	884	900r	812r	799
Petroleum	7	14	20r	15r	8
Other (including renewables) (2)	353	147	138r	116r	148
Total fuel input (3)	**1,458**	**1,159**	**1,167r**	**1,082r**	**1,094**
Electricity generated by chemicals (4)	**791**	**737**	**875r**	**910r**	**871**
(in GWh)	9,204	8,572	10,177r	10,588r	10,134
Electricity consumed by chemicals from own generation (5)	**470**	**560**	**768r**	**671r**	**538**
(in GWh)	5,460	6,517	8,935r	7,807r	6,254
Metal products, machinery and equipment					
Coal	-	-	-	-	-
Natural gas	94	70	57r	29r	74
Petroleum	6	6	6	6	6
Other (including renewables) (2)	-	-	-	-	-
Total fuel input (3)	**100**	**76**	**63r**	**35r**	**79**
Electricity generated by metal products, machinery	**43**	**31**	**18r**	**15r**	**35**
and equipment (4) (in GWh)	505	364	213r	172r	407
Electricity consumed by metal products, machinery	**41**	**30**	**18r**	**14r**	**34**
and equipment from own generation (5) (in GWh)	476	351	205r	165r	391
Food, beverages and tobacco					
Coal	18	12	11r	7r	5
Natural gas	326	277	351r	330r	335
Petroleum	5	47	9	8	6
Other (including renewables) (2)	-	-	-	-	-
Total fuel input (3)	**349**	**335**	**371r**	**345r**	**346**
Electricity generated by food, beverages and tobacco (4)	**170**	**164**	**182r**	**171r**	**168**
(in GWh)	1,978	1,904	2,115r	1,983r	1,951
Electricity consumed by food, beverages and tobacco	**96**	**145**	**162r**	**129r**	**128**
from own generation (5) (in GWh)	1,112	1,682	1,886r	1,498r	1,491

(1) Industrial categories used are described in Table 1G.
(2) Includes hydro electricity, solid and gaseous renewables and waste.
(3) Total fuels used for generation of electricity. Consistent with figures for fuels used by other generators in Table 5.4.

1.9 Fuels consumed for electricity generation (autogeneration) by main industrial groups[1] (continued)

Thousand tonnes of oil equivalent
(except where shown otherwise)

	2003	2004	2005	2006	2007
Paper, printing and publishing					
Coal	32	25	25	54	50
Natural gas	842	887	827r	782r	812
Petroleum	7	7	11	7	2
Other (including renewables) (2)	-	5	7	8	7
Total fuel input (3)	**881**	**925**	**869r**	**851r**	**871**
Electricity generated by paper, printing and publishing (4)	**411**	**439**	**408r**	**378r**	**385**
(in GWh)	4,774	5,100	4,749r	4,400r	4,473
Electricity consumed by paper, printing and publishing	**272**	**314**	**300r**	**280r**	**321**
from own generation (5) (in GWh)	3,159	3,652	3,490r	3,260r	3,728
Other industries					
Coal	-	-	-	-	-
Coke oven gas	7	24	28	26r	24
Natural gas	201	180	84r	79r	115
Petroleum	4	6	5r	6r	3
Other (including renewables) (2)	1,187	1,446	1,556r	1,613r	1,693
Total fuel input (3)	**1,399**	**1,657**	**1,673r**	**1,724r**	**1,836**
Electricity generated by other industries (4)	**147**	**139**	**93r**	**90r**	**118**
(in GWh)	1,714	1,614	1,085r	1,043r	1,373
Electricity consumed by other industries from own	**93**	**83**	**73r**	**68r**	**71**
generation (5) (in GWh)	1,083	962	851r	796r	827
Total					
Coal	977	914	911r	969r	959
Blast furnace gas	774	790	801	780	756
Coke oven gas	143	131	190	187	181
Natural gas	2,458	2,359	2,263r	2,071r	2,174
Petroleum	44	111	69r	61r	45
Other (including renewables) (2)	1,587	1,663	1,770r	1,791r	1,904
Total fuel input (3)	**5,982**	**5,968**	**6,005r**	**5,859r**	**6,020**
Electricity generated (4)	**2,037**	**1,990**	**2,064r**	**2,045r**	**2,052**
(in GWh)	23,685	23,147	24,009r	23,778r	23,866
Electricity consumed from own generation (5)	**1,352**	**1,549**	**1,749r**	**1,581r**	**1,489**
(in GWh)	15,725	18,015	20,335r	18,387r	17,320

(4) Combined heat and power (CHP) generation (ie electrical output from Table 6.8) plus non-chp generation, so that the total electricity generated is consistent with the "other generators" figures in Table 5.6.

(5) This is the electricity consumed by the industrial sector from its own generation and is consistent with the other generators final users figures used within the electricity balances (Tables 5.1 and 5.2). These figures are less than the total generated because some of the electricity is sold to the public distribution system and other users.

(6) The figures presented here are consistent with other figures presented elsewhere in this publication as detailed at (3), (4), and (5) above but are further dissaggregated. Overall totals covering all autogenerators can be derived by adding in figures for transport, services and the fuel industries. These can be summarised as follows:

Thousand tonnes of oil equivalent

Fuel input	2003	2004	2005	2006	2007
All industry	5,982	5,968	6,005r	5,859r	6,020
Fuel industries	1,449	1,203	1,465r	1,442r	1,433
Transport, Commerce and Administration	251	303	275r	274r	273
Services	977	920	1,172r	1,320r	1,344
Total fuel input	**8,659**	**8,394**	**8,917r**	**8,894r**	**9,070**
Electricity generated	**3,062**	**3,057**	**3,104r**	**3,144r**	**3,285**
Electricity consumed	**1,918**	**2,066**	**2,232r**	**2,145r**	**2,042**
					GWh
Electricity generated	**35,609**	**35,548**	**36,101r**	**36,563r**	**38,205**
Electricity consumed	**23,303**	**24,029**	**25,958r**	**24,947r**	**23,752**

Chapter 2
Solid fuels and derived gases

Introduction

2.1   This chapter presents figures on the supply and demand for coal and solid fuels derived from coal, and on the production and consumption of gases derived from the processing of solid fuels. An energy flow chart for 2007, showing the flows of coal from production and imports through to consumption, is included for the first time, overleaf. This is a way of simplifying the figures that can be found in the commodity balance for coal in Table 2.1. It illustrates the flow of coal from the point at which it become available from home production or imports (on the left) to the eventual final use of coal (on the right).

2.2   Balances for coal and manufactured fuels, covering each of the last three years, form the first six tables of this chapter (Tables 2.1 to 2.6). These are followed by a five year table showing the supply and consumption of coal as a time series (Table 2.7). Comparable five year tables bring together data for coke oven coke, coke breeze and manufactured solid fuels (Table 2.8) and coke oven gas, blast furnace gas, benzole and tars (Table 2.9). As in previous years, tables showing deep mines in production (Table 2.10) and opencast sites in production (Table 2.11) complete the chapter. The long term trends commentary and tables on coal production and stocks, and on coal consumption are on the BERR energy statistics web site at:
www.berr.gov.uk/energy/statistics/publications/dukes/page45537.html

2.3   Detailed statistics of imports and exports of solid fuels are in Annex G, also available on the BERR energy statistics web site at:
www.berr.gov.uk/energy/statistics/publications/dukes/page45537.html

2.4   Figures for actual consumption of coal are available for all fuel and power producers and for final use by the iron and steel industry. The remaining final users consumption figures are based on information on disposals to consumers by producers and on imports. For further details see the technical notes and definitions section which begins at paragraph 2.32 of this chapter.

Commodity balances for coal (Tables 2.1, 2.2 and 2.3)

2.5   These balance tables separately identify the three main types of coal: steam coal, coking coal and anthracite. They show the variation both in the sources of supply and where the various types of coal are mainly used.

2.6   In 2007, 86.5 per cent of coal demand was for steam coal, 11.5 per cent was for coking coal and 2 per cent was for anthracite. Electricity generation accounted for 95 per cent of demand for steam coal and 62 per cent of demand for anthracite. Coking coal was nearly all used in coke ovens (83 per cent) but 17 per cent was directly injected into blast furnaces.

2.7   Only 4 per cent of the total demand for coal was for final consumption, where it was used for steam raising, space or hot water heating, or heat for processing. Steam coal accounted for 89 per cent of this final consumption. 77.5 per cent of this was consumed by industry, where mineral products (e.g. cement, glass and bricks) and paper and printing were the largest users. The domestic sector accounted for 27 per cent of the final demand for coal, with 71 per cent of this demand being for steam coal and the remainder for anthracite.

2.8   Chart 2.1, compares the sources of coal supplies in the UK in 2007, along with a breakdown of consumption by user, and serves to illustrate some of the features brought out below.

2.9   In 2007, 13 per cent of supply was from deep-mined production, 15 per cent from opencast operations, 71.5 per cent from net imports and 0.5 per cent from other sources such as slurry. Between the end of 2006 and the end of 2007, total stock levels fell by 14 per cent.

Coal flow chart 2007 (million tonnes of coal)

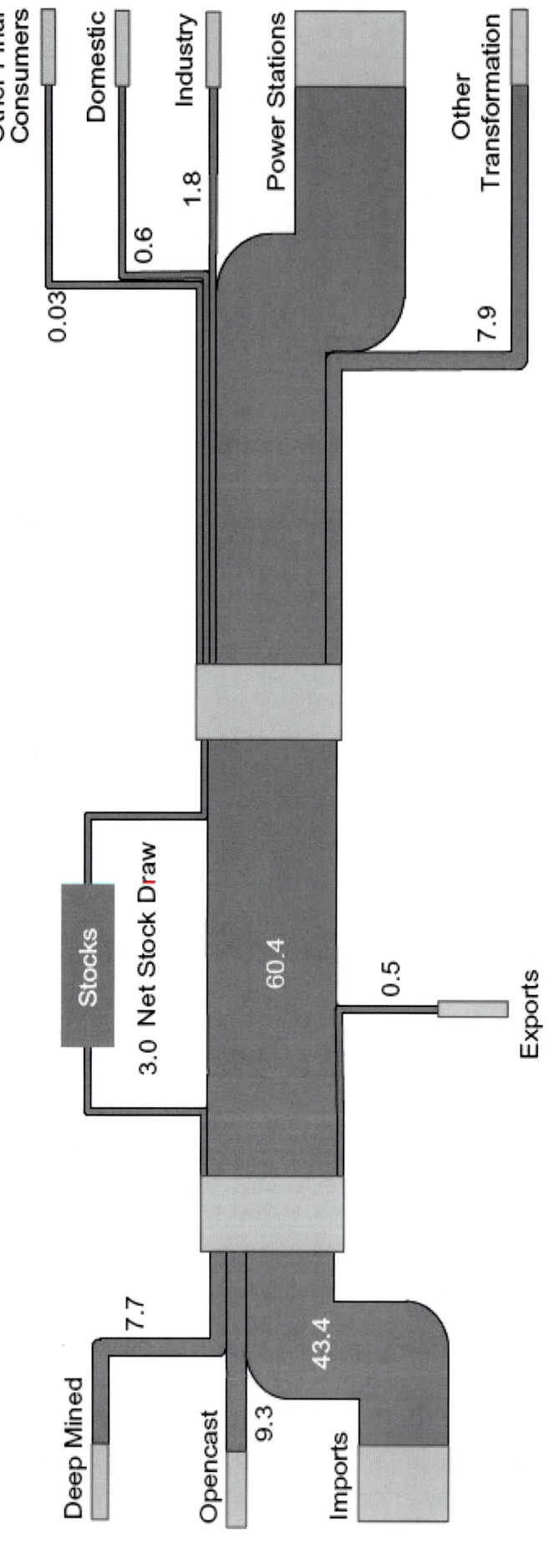

Notes:
This flow chart is based on the data that appear in Tables 2.1 and 2.7.
Opencast includes slurry and recovered coal.

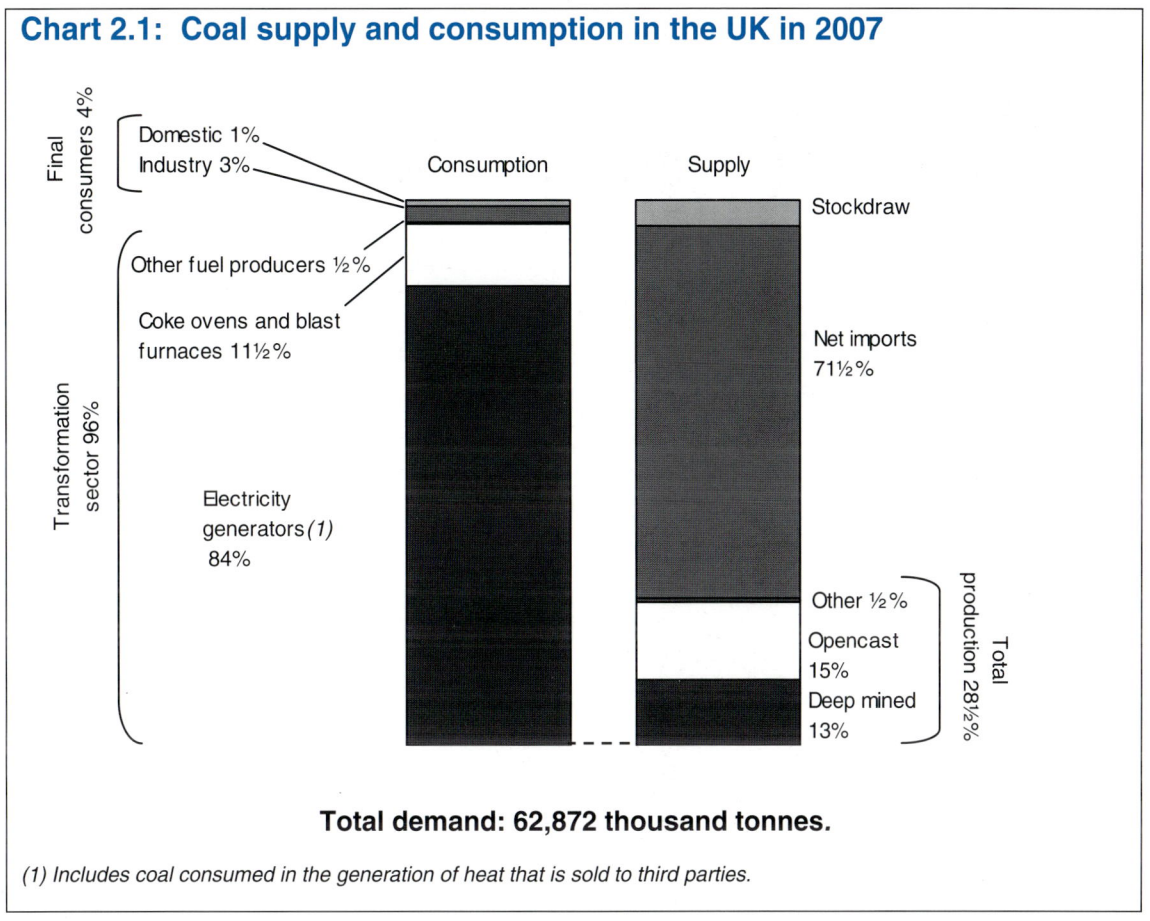

Chart 2.1: Coal supply and consumption in the UK in 2007

Final consumers 4%

Domestic 1%
Industry 3%

Consumption

Supply

Stockdraw

Other fuel producers ½%

Net imports
71½%

Coke ovens and blast
furnaces 11½%

Transformation sector 96%

Electricity
generators(1)
84%

Other ½%

Opencast
15%

Deep mined
13%

Total production 28½%

Total demand: 62,872 thousand tonnes.

(1) Includes coal consumed in the generation of heat that is sold to third parties.

2.10   Recent trends in coal production and consumption are described in paragraphs 2.18 to 2.24.

Commodity balances for manufactured fuels (Table 2.4, 2.5 and 2.6)

2.11   These tables cover fuels manufactured from coal and gases produced when coal is used in coke ovens and blast furnaces. Definitions of terms associated with coke, breeze, other manufactured solid fuels and manufactured gases are set out in paragraphs 2.44 to 2.48.

2.12   Around 85 per cent of coke oven coke and coke breeze is home produced with the rest imported. In 2007 the volume of imports was up 5.9 per cent from 2006. About 5 per cent of home production was exported. The coke screened out by producers as breeze and fines amounted to about a fifth of production plus imports in 2007; this appears as transfers in the coke breeze column of the balances. Transfers out of coke oven coke have not always been equal to transfers into coke oven breeze. This was due to differences arising from the timing, location of measurement and the practice adopted by the Iron and Steel works. But since 2000, the Iron and Steel Statistics Bureau have been able to reconcile these data. In 2007, 97 per cent of the demand for coke oven coke was at blast furnaces (part of the transformation sector) with most of the remainder going into final consumption in the unclassified sector (e.g. foundry coke).

2.13   Most of the supply of **coke breeze** is from re-screened coke oven coke, with direct production accounting for only 2 per cent of total supply. Some breeze is re-used in coke manufacture or in blast furnaces, but the majority is boiler fuel. In 2007, most of the supply of coke breeze was reclassified to coke oven coke following better information received by the Iron and Steel Statistics Bureau.

2.14   Patent fuels are manufactured smokeless fuels, produced mainly for the domestic market, as the balances show. A small amount of these fuels (only 4 per cent of total supply in 2007) is imported, but exports generally exceed this. Imports and exports of manufactured smokeless fuels can contain small quantities of non-smokeless fuels.

2.15   Chart 2.2 shows the sources of coke, breeze and other manufactured solid fuels, and a breakdown of their consumption.

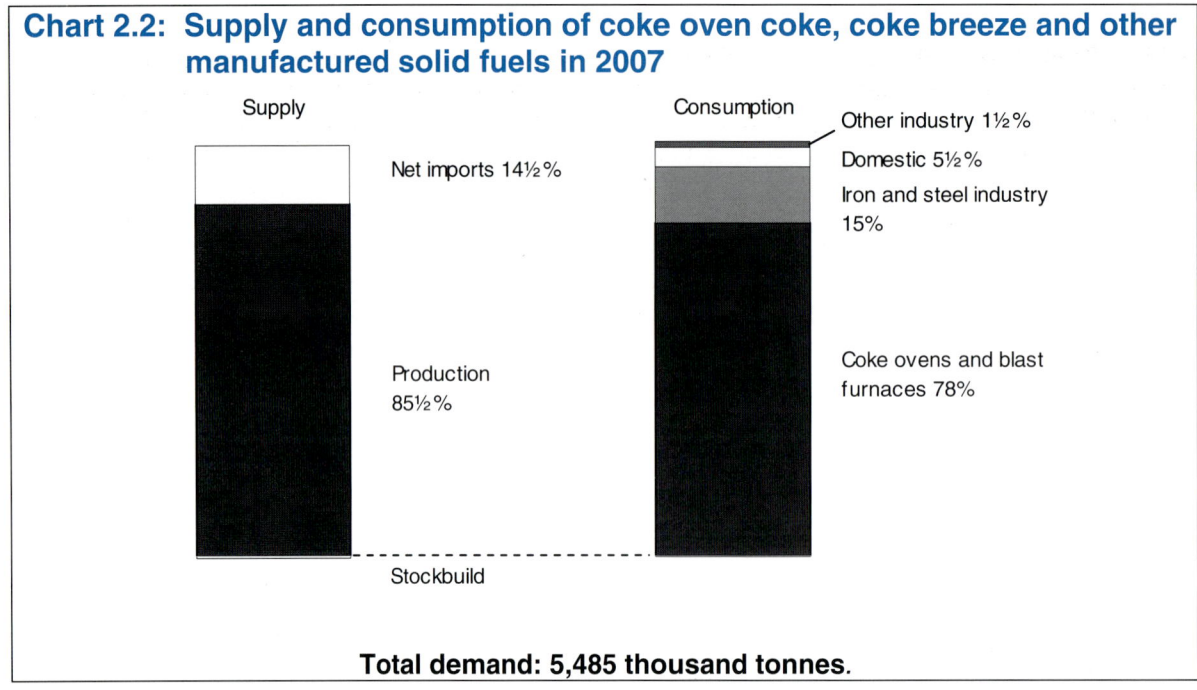

Chart 2.2: Supply and consumption of coke oven coke, coke breeze and other manufactured solid fuels in 2007

Supply

Consumption

Net imports 14½%

Production 85½%

Stockbuild

Other industry 1½%

Domestic 5½%

Iron and steel industry 15%

Coke ovens and blast furnaces 78%

Total demand: 5,485 thousand tonnes.

2.16 The carbonisation and gasification of solid fuels at coke ovens produces **coke oven gas** as a by-product. Some of this (43.5 per cent in 2007) is used to fuel the coke ovens themselves, while some is piped to blast furnaces and used in the production of steel (13.5 per cent in 2007). Elsewhere at steel works, the gas is used for electricity generation (22 per cent), or for heat production and for other iron and steel making processes (16.5 per cent). The remaining 4.5 per cent is lost.

2.17 **Blast furnace gas** is a by-product of iron smelting in a blast furnace. A similar product is obtained when steel is made in basic oxygen steel converters and "BOS" gas is included in this category. Most of these gases are used in other parts of integrated steel works. The generation of electricity in 2007 used 52 per cent of blast furnace gas and BOS gas, while 30 per cent was used in coke ovens and blast furnaces themselves. One per cent was used for general heat production. The remaining 16 per cent was lost or burned as waste.

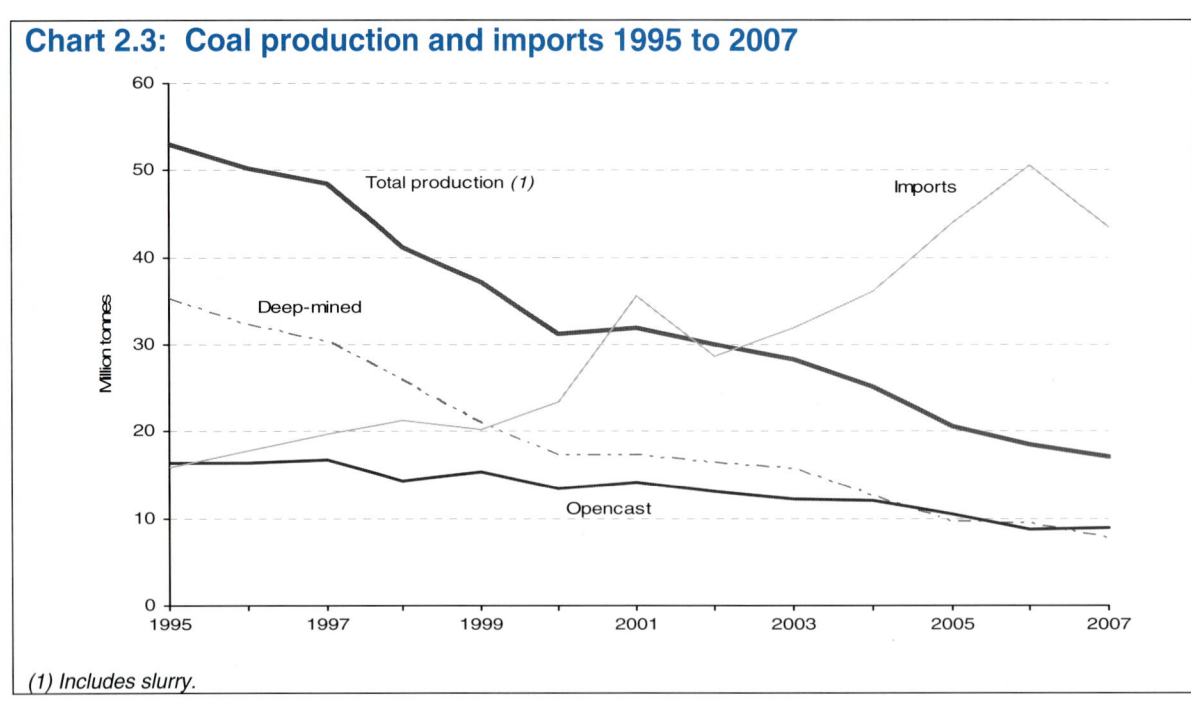

Chart 2.3: Coal production and imports 1995 to 2007

Total production (1)

Imports

Deep-mined

Opencast

Million tonnes

(1) Includes slurry.

Supply and consumption of coal (Table 2.7)

2.18 **Production** - Figures for 2007 show that coal production (including slurry) fell by 8.2 per cent compared to production in 2006. Deep-mined production fell by 19 per cent while opencast production rose by 2.7 per cent. Overall demand for coal fell by 6.8 per cent in 2007, driven largely by lower consumption by major power producers. Imports fell by 14 per cent in 2007 to 43 million tonnes. Longer-term trends in production are illustrated in Chart 2.3 above.

2.19 Table 2A shows how production of coal is divided between England, Wales and Scotland on a financial year basis. In 2007/08, 56 per cent of coal output was in England, 36 per cent in Scotland and 8 per cent in Wales.

Table 2A: Output from UK coal mines and employment in UK coal mines [1][2]

	Million tonnes				Employment		Number
	Output						
	April 2005 to March 2006	April 2006 to March 2007	April 2007 to March 2008		end March 2006	end March 2007	end March 2008
Deep-mined							
England	9.8	7.7	7.3		3,629	3,318	3,348
Wales	0.6	0.4	0.2		453	437	660
Total	10.3	8.2	7.5		4,082	3,755	4,008
Opencast							
England	1.2	1.0	1.8		206	334	425
Scotland	7.7	6.1	5.9		1,304	1,021	1,049
Wales	1.2	1.3	1.1		311	301	454
Total	10.2	8.4	8.8		1,821	1,656	1,928
Total							
England	11.0	8.7	9.1		3,835	3,652	3,773
Scotland	7.7	6.1	5.9		1,304	1,021	1,049
Wales	1.8	1.7	1.3		764	738	1,114
Total	20.5	16.6	16.3		5,903	5,411	5,936

Source: The Coal Authority

(1) Output is the tonnage declared by operators to the Coal Authority, including estimated tonnages. It excludes estimates of slurry recovered from dumps, ponds, rivers, etc.

(2) Employment includes contractors and is as declared by licensees to the Coal Authority at 31 March each year.

2.20 Table 2A also shows how numbers employed in the production of coal have changed over the last three years. During 2007/08, total employment, including contractors, was 9.7 per cent higher than in 2006/07. At 31 March 2008, 63 per cent of the 5,936 people employed in UK coal mining worked in England, while 18 per cent were employed in Scotland and 19 per cent in Wales. The closure of Longannet mine in 2002 brought employment in deep mining for coal in Scotland to an end.

2.21 **Foreign trade** - Imports of coal and other solid fuel in 2007 fell 14 per cent from 2006 levels to just over 44 million tonnes. Within the total, imports of steam coal fell by 18 per cent – largely due to sharp decreases in imports from the European Union (70 per cent) and South Africa (40 per cent). As Table 2B shows, in 2007, 74 per cent of the United Kingdom's imports of coal and other solid fuel came from just three countries: Australia, Russia and South Africa. A further 18 per cent of coal imports came from three additional countries: Colombia (mainly steam coal), Canada (coking coal) and the USA (mainly coking coal). Steam coal imports came mainly from Russia (57 per cent), South Africa (22 per cent) and Colombia (11 per cent). All but 5 per cent of UK coking coal imports came from just three countries: Australia (50 per cent), Canada (24 per cent) and the USA (19 per cent). For more details of imports and exports of solid fuels by country of origin, see Annex G on the BERR energy statistics web site at: www.berr.gov.uk/energy/statistics/publications/dukes/page45537.html

2.22 The proportion of coal consumed by major power producers from imports has continued to rise, from 20 per cent (8.1 million tonnes) in 1999 to a peak of 73.5 per cent (41 million tonnes) in 2006, with the exception of 2002, where the proportion fell to 42 per cent (19 million tonnes). In 2007, 66 per cent (34 million tonnes) of coal consumed by major power producers was from imports.

Table 2B: Imports of coal and other solid fuel in 2007[1]

Thousand tonnes

	Steam coal	Coking coal	Anthracite	Other solid fuel	Total
Russia	20,442	393	19	159	21,013
Republic of South Africa	7,706	-	36	-	7,742
Australia	527	3,733	-	-	4,260
Colombia	3,800	-	54	-	3,854
United States of America	1,121	1,402	-	-	2,523
Canada	-	1,771	-	-	1,771
Indonesia	1,455	-	-	-	1,455
People's Republic of China	175	68	22	550	815
European Union [2]	482	7	6	204	699
Other countries	38	108	-	163	309
Total all countries	**35,746**	**7,482**	**137**	**1,076**	**44,441**

Source: H M Revenue and Customs, ISSB
(1) Country of origin basis.
(2) Includes extra-EU coal routed through the Netherlands.

2.23   **Transformation** – The 7.2 per cent decline in total coal consumption between 2006 and 2007 reflected a decrease in electricity generation by major power producers of nearly 4.8 million tonnes. This fall was due to increased coal prices relative to gas, making coal-fired generation less competitive. In addition, in UK steel production coal use for coke making and for injection into blast furnaces rose by 1.8 per cent in 2007.

2.24   **Consumption** - Consumption by final consumers in 2007 rose by 5.7 per cent from 2006. Industry sector consumption rose by 3.1 per cent. Domestic demand increased by 19 per cent from the previous year.

2.25   Long term trends commentary and tables on the consumption of coal in the UK since 1970 onwards can be found on the BERR energy statistics web site:
www.berr.gov.uk/energy/statistics/publications/dukes/page45537.html

2.26   **Stocks** – Production and net imports together in 2007 were lower than the demand for coal. Consequently total stock levels were nearly 2.2 million tonnes lower at the end of 2007 compared to a year earlier. Total stocks at the end of 2007 were equivalent to around a quarter of the year's coal consumption. Stocks held at collieries and opencast sites at the end of 2007 were 0.1 thousand tonnes lower than a year earlier and stocks at major power stations and coke areas fell by 2.5 million tonnes. The recent changes in coal stocks are illustrated in Chart 2.4 below.

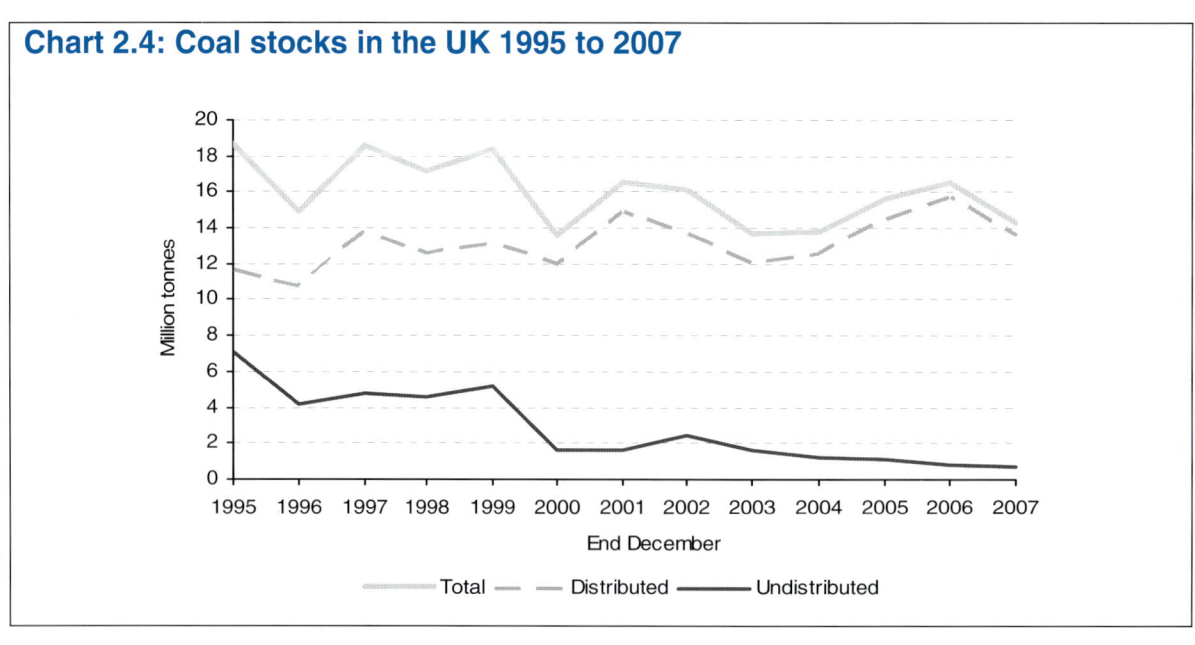

Chart 2.4: Coal stocks in the UK 1995 to 2007

Supply and consumption of coke oven coke, coke breeze and other manufactured fuels (Table 2.8)

2.27   This table presents figures for the most recent five years on the same basis as the balance tables. Figures for stocks are also included. Coal used to produce these manufactured fuels is shown in Table 2.7. For **coke oven coke,** demand rose by 4.0 per cent in 2007 and production rose by 1.5 per cent. In 2007 imports fell by 0.5 per cent from 2006 and exports rose by 12 per cent. Stock levels fell in 2007.

2.28   In 2007, the demand for **coke breeze** fell by 13 per cent compared to 2006 while re-screening rose by 17 per cent and imports rose by 32 per cent. Production of coke breeze fell by 92 per cent (due to the reclassification of the majority of coke breeze production to coke oven coke production). The net effect was an increase in stock levels. There was a 9.4 per cent decrease in the demand for **other manufactured solid fuels**, mainly for domestic use.   UK production was down 13 per cent on 2006 levels.

Supply and consumption of coke oven gas, blast furnace gas, benzole and tars (Table 2.9)

2.29   This table presents figures for the most recent five years on the same basis as the other balance tables. In 2007, production of **coke oven gas** fell by 1.8 per cent and demand fell by 1.2 per cent. Use in final consumption terms rose but demand fell for use in blast furnaces and electricity generation. Production of **blast furnace gas** rose 1.6 per cent and demand rose by 1.5 per cent in 2007 compared to 2006. Demand for benzole and tars fell by 2.7 per cent in 2007.

Structure of the coal industry (Tables 2.10 and 2.11)

2.30   At 31 March 2008, there were 6 major deep mines in production in England. These were Kellingley, Welbeck, Thoresby and Daw Mill, operated by UK Coal plc, plus Maltby, which UK Coal disposed of during 2006 to the Hargreaves Group, and Hatfield Colliery, operated by Powerfuel Mining Ltd, which restarted commercial production in January 2008. Tower Colliery in South Wales closed in early 2008 through exhaustion. Two medium-sized underground mines in South Wales – Aberpergwm, operated by Energybuild Mining Ltd and Unity, operated by Unity Mining Ltd – are in re-development and expect to restart commercial production in the next year.   Two small underground mines (Eckington and Hay Royds) were also in production. Surface mine output in England continued to decline owing to difficulty in obtaining necessary planning permissions, although two significant consents granted on appeal have been upheld in 2007 and the recovery in English surface mine output is expected to continue in 2008/09; H J Banks & Son and UK Coal plc remain the main operators in this sector. UK Coal plc is currently reviewing the potential of unworked reserves at Harworth Colliery, which has been in "care and maintenance" since its last working area was abandoned in 2006. The impact of new planning guidance on surface mine output in Scotland, (where Scottish Coal Co Ltd and ATH Resources are the largest operators) continues to be monitored, but the slight further decline in output in 2007/08 may be due to other factors. Wales is also considering adopting new planning guidance which could make it harder to obtain permission for new coal developments in future, but a major consent (Ffos-y-fran Reclamation scheme, operated by Miller Argent) upheld on appeal in 2007 is expected to lift Welsh output in 2008/09.

2.31   The deep mines that were in operation at the end of March 2008 are listed in Table 2.10. Opencast coal producers are similarly listed in Table 2.11, as well as those that are in development and not yet in operation. Further coal and slurry are supplied from recovery operations shown in the tables under "Other sources".

Technical notes and definitions

2.32   These notes and definitions are in addition to the technical notes and definitions covering all fuels and energy as a whole in Chapter 1, paragraphs 1.26 to 1.59. For notes on the commodity balances and definitions of the terms used in the row headings see Annex A, paragraphs A.7 to A.42. While the data in the printed and bound copy of this Digest cover only the most recent 5 years, these notes also cover data for earlier years that are available on the BERR web site.

Steam coal, coking coal and anthracite

2.33   **Steam coal** is coal classified as such by UK coal producers and by importers of coal. It tends to have calorific values at the lower end of the range.

2.34   **Coking coal** is coal sold by producers for use in coke ovens and similar carbonising processes. The definition is not therefore determined by the calorific value or caking qualities of each batch of coal sold, although calorific values tend to be higher than for steam coal.

2.35   **Anthracite** is coal classified as such by UK coal producers and importers of coal. Typically it has a high heat content making it particularly suitable for certain industrial processes and for use as a domestic fuel. Some UK anthracite producers have found a market for their lower calorific value output at power stations.

Coal production

2.36   **Deep-mined** - The statistics cover saleable output from deep mines including coal obtained from working on both revenue and capital accounts. All licensed collieries (and British Coal collieries prior to 1995) are included, even where coal is only a subsidiary product.

2.37   **Opencast** - The figures cover saleable output and include the output of sites worked by operators under agency agreements and licences, as well as the output of sites licensed for the production of coal as a subsidiary to the production of other minerals.

2.38   **Other** - Estimates of slurry etc recovered and disposed of from dumps, ponds, rivers, etc.

Imports and exports of coal and other solid fuels

2.39   Figures are derived from returns made to HM Revenue and Customs and are broken down in greater detail in Annex G on the BERR energy statistics web site at:
www.berr.gov.uk/energy/statistics/publications/dukes/page45537.html

2.40   However, in Tables 2.4, 2.5, 2.6 and 2.8, the export figures used for hard coke, coke breeze and other manufactured solid fuels for the years before 1998 (as reported on the BERR web site) are quantities of fuel exported as reported to BERR by the companies concerned, rather than quantities recorded by HM Revenue and Customs in their Trade Statistics. A long term trend commentary and tables on exports are on the BERR energy statistics web site at:
www.berr.gov.uk/energy/statistics/publications/dukes/page45537.html

Allocation of imported coal

2.41   Although data are available on consumption of home produced coal, and also on consumption of imported coal by secondary fuel producers, there is only very limited direct information on consumption of imported coal by final users. Surveys of the destination of steam coal imports (excluding those used by electricity generators) used to be carried out from time to time. The most recent was in 1998 and concluded that it was appropriate to allocate 60 per cent of such imports each year to industry, 15 per cent to the public administration sector and 25 per cent to the domestic sector. This was revised for 2002 and 2003 to 70, 25 and 5 per cent to industry, domestic and public administration respectively. These proportions were revised again in 2004 to 75 per cent to industry, 20 per cent to domestic and 5 per cent to public administration. In addition, 10 per cent of anthracite imports, excluding cleaned smalls, are allocated to industry, with 90 per cent to the domestic sector in all years shown in the tables. From 2000, imports have been allocated within the overall industry sector using the results of the Office for National Statistics Purchases Inquiry, and information derived from the EU Emissions Trading Scheme submissions. All imports of coking coal and cleaned anthracite smalls are allocated to coke and other solid fuel producers.

Stocks of coal

2.42   Undistributed stocks are those held at collieries and opencast sites. It is not possible to distinguish these two locations in the stock figures. Distributed stocks are those held at power stations and stocking grounds of the major power producing companies (as defined in Chapter 5, paragraph 5.50), coke ovens, low temperature carbonisation plants and patent fuel plants.

Transformation, energy industry use and consumption of solid fuels

2.43   Annex A of this Digest outlines the principles of energy and commodity balances and defines the activities that fall within these parts of the balances. However, the following additional notes relevant to solid fuels are given below:

Transformation: Blast furnaces - Coking coal injected into blast furnaces is shown separately within the balance tables.

Transformation: Low temperature carbonisation plants and patent fuel plants - Coal used at these plants for the manufacture of domestic coke such as Coalite and of briquetted fuels such as Phurnacite and Homefire.

Consumption: Industry - The statistics comprise sales of coal by the ten main coal producers to the iron and steel industry (excluding that used at coke ovens and blast furnaces) and to other industrial sectors and estimated proportions of anthracite and steam coal imports. The figures exclude coal used for industries' own generation of electricity, which appear separately under transformation.

Consumption: Domestic – Some coal is supplied free of charge to retired miners and other retired eligible employees through the National Concessionary Fuel Scheme (NCFS). The concessionary fuel provided in 2007 is estimated at 93 thousand tonnes (14 per cent). This estimate is included in the domestic steam coal and domestic anthracite figures.

Consumption of coke and other manufactured solid fuels - These are disposals from coke ovens to merchants. The figures also include estimated proportions of coke imports.

Coke oven coke (hard coke) and hard coke breeze

2.44   The statistics cover coke produced at coke ovens owned by Corus plc (formerly British Steel), Coal Products Ltd and other producers. Low temperature carbonisation plants are not included (see paragraph 2.47, below). Breeze (as defined in paragraph 2.45) is excluded from the figures for coke oven coke.

2.45   Breeze can generally be described as coke screened below 19 mm (¾ inch) with no fines removed, but the screen size may vary in different areas and to meet the requirements of particular markets. Coke that has been transported from one location to another is usually re-screened before use to remove smaller sizes, giving rise to further breeze.

2.46   In 1998, an assessment using industry data showed that on average over the last five years 91 per cent of imports have been coke and 9 per cent breeze and it is these proportions that have been used for 1998 and subsequent years in Tables 2.4, 2.5, 2.6 and 2.8.

2.47   Other manufactured solid fuels are mainly solid smokeless fuels for the domestic market for use in both open fires and in boilers. A smaller quantity is exported (although exports are largely offset by similar quantities of imports in most years). Manufacture takes place in patented fuel plants and low temperature carbonisation plants. The brand names used for these fuels include Homefire, Phurnacite, Ancit and Coalite.

Blast furnace gas, coke oven gas, benzole and tars

2.48   The following definitions are used in the tables that include these fuels:

Blast furnace gas - includes basic oxygen steel furnace (BOS) gas. Blast furnace gas is the gas produced during iron ore smelting when hot air passes over coke within the blast ovens. It contains carbon monoxide, carbon dioxide, hydrogen and nitrogen. In a basic oxygen steel furnace the aim is not to introduce nitrogen or hydrogen into the steel making process, so pure oxygen gas and suitable fluxes are used to remove the carbon and phosphorous from the molten pig iron and steel scrap. A similar fuel gas is thus produced.

Coke oven gas - is a gas produced during the carbonisation of coal to form coke at coke ovens.

Synthetic coke oven gas - is mainly natural gas that is mixed with smaller amounts of blast furnace and BOS gas to produce a gas with almost the same qualities as coke oven gas. The transfers row of Tables 2.4, 2.5, 2.6 and 2.8 show the quantities of blast furnace gas used for this purpose and the total input of gases to the synthetic coke oven gas process. There is a corresponding outward transfer from natural gas in Chapter 4, Table 4.1.

Benzole - a colourless, liquid, flammable, aromatic hydrocarbon by-product of the iron and steel making process. It is used as a solvent in the manufacture of styrenes and phenols but can also be used as a motor fuel.

Tars - viscous materials usually derived from the destructive distillation of coal, which are by-products of the coke and iron making processes.

Periods covered

2.49   Figures in this chapter (and figures for earlier years given in the tables on the BERR web site) generally relate to periods of 52 weeks or 53 weeks. For the most recent five years the periods are:

Year	52 weeks ended
2003	27 December 2003
2004	25 December 2004
	53 weeks ended
2005	31 December 2005
	52 weeks ended
2006	30 December 2006
2007	29 December 2007

The 53 week data for 2005 have been adjusted to 52 weeks by omitting data for an average week based on information provided by the largest companies for the first week in April 2005.

2.50   Data for coal used for electricity generation by major power producers follow the electricity industry calendar (see Chapter 5, paragraph 5.59) and coal use by other generators is for the 12 months ending 31 December each year. HM Revenue and Customs data on imports and exports are also for the 12 months ended 31 December each year. Data for coal and coke use in the iron and steel industry, and for gases, benzole and tars produced by the iron and steel industry follow the iron and steel industry calendar (see Chapter 5, paragraph 5.60).

Data collection

2.51   In 2007, aggregate data on coal production were obtained from the Coal Authority. In addition the largest producers (Celtic Energy, Energybuild, Goitre Tower Anthracite, H J Banks, Hall Construction Services Ltd (formerly known as Coal Contractors Limited), J D Flack & Sons Ltd, Scottish Coal Company Ltd, UK Coal plc, Maltby Colliery Ltd and Powerfuel Mining Ltd) have provided data in response to an annual BERR inquiry covering production (deep-mined and opencast), trade, stocks and disposals. The Iron and Steel Statistics Bureau (ISSB) provides BERR with an annual statement of coke and breeze production and use of coal, coke and breeze within that industry. The ISSB is also the source of data on gases produced by the iron and steel industry (coke oven gas, blast furnace gas and basic oxygen steel furnace gas). BERR directly surveys producers of manufactured fuels other than coke or breeze.

2.52   Trade in solid fuels is also covered by using data from HM Revenue and Customs (see Annex G on BERR energy statistics web site). Consumption of coal for electricity generation is covered by data collected by BERR from electricity generators as described in Chapter 5, paragraphs 5.62 to 5.64.

Monthly and quarterly data

2.53   Monthly data on coal production, foreign trade, consumption and stocks are available on BERR's Energy Statistics web site: www.berr.gov.uk/energy/statistics/source/coal/page18529.html in monthly Tables 2.4, 2.5, and 2.6. Quarterly commodity balances for coal, coke oven coke, coke breeze and other manufactured solid fuels; and coke oven gas, blast furnace gas, benzole and tars are published in BERR's quarterly statistical bulletin *Energy Trends*. These balances are also available on BERR's Energy Statistics web site. See Annex C for more information about *Energy Trends* and the BERR energy statistics web site.

Statistical differences

2.54   Tables 2.1 to 2.9 each contain a statistical difference term covering the difference between recorded supply and recorded demand. These statistical differences arise for a number of reasons. First, the data within each table are taken from varied sources, as described above, such as producers, intermediate consumers (such as electricity generators), final consumers and HM Revenue and Customs.   Second, some of these industries work to different statistical calendars (see paragraphs 2.49 and 2.50, above) and third, some of the figures are estimated either because data in the required detail are not readily available within the industry or because the methods of collecting the data do not cover the smallest members of the industry.

Contact: James Hemingway
     Energy Markets Unit
     james.hemingway@berr.gsi.gov.uk
     020 7215 2717

Mita Kerai
Energy Markets Unit
mita.kerai@berr.gsi.gov.uk
020 7215 3839

2.1 Commodity balances 2007
Coal

Thousand tonnes

	Steam coal	Coking coal	Anthracite	Total
Supply				
Production	..	266	..	16,540
Other sources	..	-	..	467
Imports	35,746	7,482	137	43,365
Exports	-428	-13	-80	-521
Marine bunkers	-	-	-	-
Stock change (1)	..	-565	..	+3,014
Transfers	-	-	-	-
Total supply	..	7,170	..	62,866
Statistical difference (2)	..	-5	..	-6
Total demand	54,375	7,175	1,323	62,872
Transformation	52,199	7,175	1,060	60,434
Electricity generation	51,742	-	815	52,558
Major power producers	50,202	-	815	51,017
Autogenerators	1,541	-	-	1,541
Heat generation	456	-	-	456
Petroleum refineries	-	-	-	-
Coke manufacture	-	5,933	-	5,933
Blast furnaces	-	1,242	-	1,242
Patent fuel manufacture and low temperature carbonisation	-	-	245	245
Energy industry use	5	-	1	5
Electricity generation	-	-	-	-
Oil and gas extraction	-	-	-	-
Petroleum refineries	-	-	-	-
Coal extraction	5	-	1	5
Coke manufacture	-	-	-	-
Blast furnaces	-	-	-	-
Patent fuel manufacture	-	-	-	-
Pumped storage	-	-	-	-
Other	-	-	-	-
Losses	-	-	-	-
Final consumption	2,171	-	262	2,433
Industry	1,683	-	76	1,759
Unclassified	-	-	-	-
Iron and steel	-	-	-	1
Non-ferrous metals	..	-	..	36
Mineral products	..	-	..	1,047
Chemicals	..	-	..	143
Mechanical engineering etc	..	-	..	10
Electrical engineering etc	..	-	..	6
Vehicles	..	-	..	49
Food, beverages etc	..	-	..	41
Textiles, leather, etc	..	-	..	74
Paper, printing etc	..	-	..	144
Other industries	..	-	..	209
Construction	-	-	-	-
Transport	-	-	-	-
Air	-	-	-	-
Rail	-	-	-	-
Road	-	-	-	-
National navigation	-	-	-	-
Pipelines	-	-	-	-
Other	..	-	..	674
Domestic	462	-	186	648
Public administration	..	-	..	14
Commercial	..	-	..	6
Agriculture	..	-	..	4
Miscellaneous	..	-	..	2
Non energy use	-	-	-	-

(1) Stock fall (+), stock rise (-).

(2) Total supply minus total demand.

2.2 Commodity balances 2006
Coal

	Steam coal	Coking coal	Anthracite	Thousand tonnes Total
Supply				
Production	..	266	..	18,079
Other sources	..	-	..	438r
Imports	43,609r	6,775r	145	50,529r
Exports	-349	-1	-94	-443
Marine bunkers	-	-	-	-
Stock change *(1)*	..	+4r	..	-1,262r
Transfers	-	-	-	-
Total supply	..	**7,044r**	..	**67,341r**
Statistical difference *(2)*	..	-6r	..	-109r
Total demand	**58,880r**	**7,050r**	**1,520r**	**67,450r**
Transformation	**56,832r**	**7,050r**	**1,264**	**65,146r**
Electricity generation	56,375r	-	988	57,363
Major power producers	54,817	-	988	55,805
Autogenerators	1,558r	-	-	1,558r
Heat generation	457r	-	-	457r
Petroleum refineries	-	-	-	-
Coke manufacture	-	5,929	-	5,929
Blast furnaces	-	1,121	-	1,121
Patent fuel manufacture and low temperature carbonisation	-	-	276	276
Energy industry use	**3**	**-**	**1**	**4**
Electricity generation	-	-	-	-
Oil and gas extraction	-	-	-	-
Petroleum refineries	-	-	-	-
Coal extraction	3	-	1	4
Coke manufacture	-	-	-	-
Blast furnaces	-	-	-	-
Patent fuel manufacture	-	-	-	-
Pumped storage	-	-	-	-
Other	-	-	-	-
Losses	-	-	-	-
Final consumption	**2,045r**	**-**	**255r**	**2,301r**
Industry	**1,660r**	**-**	**44**	**1,704r**
Unclassified	-	-	-	-
Iron and steel	-	-	-	-
Non-ferrous metals	..	-	..	60r
Mineral products	..	-	..	1,047
Chemicals	..	-	..	85r
Mechanical engineering etc	..	-	..	12
Electrical engineering etc	..	-	..	6
Vehicles	..	-	..	53
Food, beverages etc	..	-	..	25
Textiles, leather, etc	..	-	..	70
Paper, printing etc	..	-	..	141
Other industries	..	-	..	205r
Construction	-	-	-	-
Transport	-	-	-	-
Air	-	-	-	-
Rail	-	-	-	-
Road	-	-	-	-
National navigation	-	-	-	-
Pipelines	-	-	-	-
Other	..	-	..	**597r**
Domestic	349r	-	212	561r
Public administration	..	-	..	20r
Commercial	..	-	..	6
Agriculture	..	-	..	5
Miscellaneous	..	-	..	5
Non energy use	**-**	**-**	**-**	**-**

(1) Stock fall (+), stock rise (-).

(2) Total supply minus total demand.

2.3 Commodity balances 2005
Coal

Thousand tonnes

	Steam coal	Coking coal	Anthracite	Total
Supply				
Production	..	274	..	20,008
Other sources	..	-	..	490
Imports	37,230	6,551	187	43,968
Exports	-364	-3	-169	-536
Marine bunkers	-	-	-	-
Stock change (1)	..	-253r	..	-2,151r
Transfers	-	-	-	-
Total supply	..	**6,570r**	..	**61,780r**
Statistical difference (2)	..	-39r	..	-62r
Total demand	**53,333r**	**6,609r**	**1,900**	**61,842r**
Transformation	**51,225r**	**6,609r**	**1,558**	**59,392r**
Electricity generation	50,766r	-	1,292	52,058r
Major power producers	49,291	-	1,292	50,582
Autogenerators	1,476r	-	-	1,476r
Heat generation	459r	-	-	459r
Petroleum refineries	-	-	-	-
Coke manufacture	-	5,570r	-	5,570r
Blast furnaces	-	1,039	-	1,039
Patent fuel manufacture and low temperature carbonisation	-	-	266	266
Energy industry use	**5**	**-**	**1**	**6**
Electricity generation	-	-	-	-
Oil and gas extraction	-	-	-	-
Petroleum refineries	-	-	-	-
Coal extraction	5	-	1	6
Coke manufacture	-	-	-	-
Blast furnaces	-	-	-	-
Patent fuel manufacture	-	-	-	-
Pumped storage	-	-	-	-
Other	-	-	-	-
Losses	-	-	-	-
Final consumption	**2,104r**	**-**	**341**	**2,445r**
Industry	**1,756r**	**-**	**25**	**1,781r**
Unclassified	-	-	-	-
Iron and steel	-	-	-	-
Non-ferrous metals	..	-	..	41
Mineral products	..	-	..	1,120
Chemicals	..	-	..	132r
Mechanical engineering etc	..	-	..	12
Electrical engineering etc	..	-	..	5
Vehicles	..	-	..	55
Food, beverages etc	..	-	..	26r
Textiles, leather, etc	..	-	..	71
Paper, printing etc	..	-	..	142
Other industries	..	-	..	178r
Construction	-	-	-	-
Transport	-	-	-	-
Air	-	-	-	-
Rail	-	-	-	-
Road	-	-	-	-
National navigation	-	-	-	-
Pipelines	-	-	-	-
Other	..	-	..	**663r**
Domestic	298	-	316	614
Public administration	..	-	..	32r
Commercial	..	-	..	6
Agriculture	..	-	..	9
Miscellaneous	..	-	..	2
Non energy use	**-**	**-**	**-**	**-**

(1) Stock fall (+), stock rise (-).

(2) Total supply minus total demand.

2.4 Commodity balances 2007
Manufactured fuels

	Thousand tonnes				GWh		
	Coke oven coke	Coke breeze	Other manuf. solid fuel	Total manuf. solid fuel	Benzole and tars (5)	Coke oven gas	Blast furnace gas
Supply							
Production (1)	4,451	25	227	4,703	1,824	9,651	16,701
Other sources	-	-	-	-	-	-	-
Imports	745	324	8	1,076	-	-	-
Exports	-105	-152	-22	-279	-	-	-
Marine bunkers	-	-	-	-	-	-	-
Stock change (2)	+34	-80r	+2	-44	-	-	-
Transfers (3)	-1,115	+1,115	-	-	-	+81	-3
Total supply	**4,010**	**1,232**	**214**	**5,456**	**1,824**	**9,732**	**16,698**
Statistical difference (4)	-14	+3	-19	-29	-	+47	-113
Total demand	**4,024**	**1,228**	**233**	**5,485**	**1,824**	**9,685**	**16,811**
Transformation	**3,910**	**483**	**-**	**4,392**	**-**	**2,525**	**8,971**
Electricity generation	-	-	-	-	-	2,107	8,791
Major power producers	-	-	-	-	-	-	-
Autogenerators	-	-	-	-	-	2,107	8,791
Heat generation	-	-	-	-	-	418	179
Petroleum refineries	-	-	-	-	-	-	-
Coke manufacture	-	-	-	-	-	-	-
Blast furnaces	3,910	483	-	4,392	-	-	-
Patent fuel manufacture	-	-	-	-	-	-	-
Low temperature carbonisation	-	-	-	-	-	-	-
Energy industry use	**-**	**-**	**-**	**-**	**-**	**5,170**	**5,082**
Electricity generation	-	-	-	-	-	-	-
Oil and gas extraction	-	-	-	-	-	-	-
Petroleum refineries	-	-	-	-	-	-	-
Coal extraction	-	-	-	-	-	-	-
Coke manufacture	-	-	-	-	-	4,228	703
Blast furnaces	-	-	-	-	-	942	4,379
Patent fuel manufacture	-	-	-	-	-	-	-
Pumped storage	-	-	-	-	-	-	-
Other	-	-	-	-	-	-	-
Losses	**-**	**-**	**-**	**-**	**-**	**445**	**2,071**
Final consumption	**114**	**746**	**233**	**1,093**	**1,824**	**1,545**	**688**
Industry	**99**	**746**	**-**	**845**	**1,824**	**1,545**	**688**
Unclassified	76	12	-	88	1,824	221	-
Iron and steel	23	734	-	757	-	1,324	688
Non-ferrous metals	-	-	-	-	-	-	-
Mineral products	-	-	-	-	-	-	-
Chemicals	-	-	-	-	-	-	-
Mechanical engineering, etc	-	-	-	-	-	-	-
Electrical engineering, etc	-	-	-	-	-	-	-
Vehicles	-	-	-	-	-	-	-
Food, beverages, etc	-	-	-	-	-	-	-
Textiles, leather, etc	-	-	-	-	-	-	-
Paper, printing, etc	-	-	-	-	-	-	-
Other industries	-	-	-	-	-	-	-
Construction	-	-	-	-	-	-	-
Transport	**-**	**-**	**-**	**-**	**-**	**-**	**-**
Other	**15**	**-**	**233**	**248**	**-**	**-**	**-**
Domestic	15	-	233	248	-	-	-
Public administration	-	-	-	-	-	-	-
Commercial	-	-	-	-	-	-	-
Agriculture	-	-	-	-	-	-	-
Miscellaneous	-	-	-	-	-	-	-
Non energy use	**-**	**-**	**-**	**-**	**-**	**-**	**-**

(1) See paragraph 2.26
(2) Stock fall (+), stock rise (-).
(3) Coke oven gas and blast furnace gas transfers are for synthetic coke oven gas, see paragraph 2.48.
(4) Total supply minus total demand.
(5) Because of the small number of benzole suppliers, figures for benzole and tars cannot be given separately.

2.5 Commodity balances 2006
Manufactured fuels

	Thousand tonnes					GWh	
	Coke oven coke	Coke breeze	Other manuf. solid fuel	Total manuf. solid fuel	Benzole and tars (4)	Coke oven gas	Blast furnace gas
Supply							
Production	4,384	245r	260	4,889r	1,873	9,828	16,443
Other sources	-	-	-	-	-	-	-
Imports	748	261r	10	1,004	-	-	-
Exports	-94	-74	-12	-180	-	-	-
Marine bunkers	-	-	-	-	-	-	-
Stock change (1)	-237r	+25r	+2	-211	-	-	-
Transfers (2)	-955	+955	-	-	-	+55	-2
Total supply	3,846r	1,411r	260	5,502r	1,873	9,882	16,441
Statistical difference (3)	-21r	-4r	+3	-22r	-	+76	-119
Total demand	3,868	1,415r	257	5,524r	1,873	9,806	16,560
Transformation	3,745	688r	-	4,433r	-	2,593r	9,249
Electricity generation	-	-	-	-	-	2,175r	9,070
Major power producers	-	-	-	-	-	-	-
Autogenerators	-	-	-	-	-	2,175r	9,070
Heat generation	-	-	-	-	-	418	179
Petroleum refineries	-	-	-	-	-	-	-
Coke manufacture	-	-	-	-	-	-	-
Blast furnaces	3,745	688r	-	4,433r	-	-	-
Patent fuel manufacture	-	-	-	-	-	-	-
Low temperature carbonisation	-	-	-	-	-	-	-
Energy industry use	-	-	-	-	-	5,300	4,831
Electricity generation	-	-	-	-	-	-	-
Oil and gas extraction	-	-	-	-	-	-	-
Petroleum refineries	-	-	-	-	-	-	-
Coal extraction	-	-	-	-	-	-	-
Coke manufacture	-	-	-	-	-	4,282	536
Blast furnaces	-	-	-	-	-	1,019	4,294
Patent fuel manufacture	-	-	-	-	-	-	-
Pumped storage	-	-	-	-	-	-	-
Other	-	-	-	-	-	-	-
Losses	-	-	-	-	-	483	1,578
Final consumption	122	727r	257	1,091r	1,873	1,430r	902
Industry	111r	727r	-	822r	1,873	1,430	902
Unclassified	84r	26r	-	95r	1,873	194	-
Iron and steel	26	701	-	727	-	1,236r	902
Non-ferrous metals	-	-	-	-	-	-	-
Mineral products	-	-	-	-	-	-	-
Chemicals	-	-	-	-	-	-	-
Mechanical engineering, etc	-	-	-	-	-	-	-
Electrical engineering, etc	-	-	-	-	-	-	-
Vehicles	-	-	-	-	-	-	-
Food, beverages, etc	-	-	-	-	-	-	-
Textiles, leather, etc	-	-	-	-	-	-	-
Paper, printing, etc	-	-	-	-	-	-	-
Other industries	-	-	-	-	-	-	-
Construction	-	-	-	-	-	-	-
Transport	-	-	-	-	-	-	-
Other	12r	-	257	269r	-	-	-
Domestic	12r	-	257	269r	-	-	-
Public administration	-	-	-	-	-	-	-
Commercial	-	-	-	-	-	-	-
Agriculture	-	-	-	-	-	-	-
Miscellaneous	-	-	-	-	-	-	-
Non energy use	-	-	-	-	-	-	-

(1) Stock fall (+), stock rise (-).
(2) Coke oven gas and blast furnace gas transfers are for synthetic coke oven gas, see paragraph 2.48.
(3) Total supply minus total demand.
(4) Because of the small number of benzole suppliers, figures for benzole and tars cannot be given separately.

2.6 Commodity balances 2005
Manufactured fuels

| | Thousand tonnes | | | | | GWh | |
	Coke oven coke	Coke breeze	Other manuf. solid fuel	Total manuf. solid fuel	Benzole and tars (4)	Coke oven gas	Blast furnace gas
Supply							
Production	4,105	259	258	4,622	1,749	9,290	16,199
Other sources	-	-	-	-	-	-	-
Imports	674	235	6	915	-	-	-
Exports	-64	-55	-15	-134	-	-	-
Marine bunkers	-	-	-	-	-	-	-
Stock change (1)	-94	-59	+6	-147	-	-	-
Transfers (2)	-983	+983	-	-	-	+53	-2
Total supply	**3,638**	**1,363r**	**254**	**5,256**	**1,749**	**9,343**	**16,197**
Statistical difference (3)	-2r	-1	-2	-4	-	+64	-107r
Total demand	**3,639**	**1,364**	**256**	**5,259**	**1,749**	**9,279**	**16,304r**
Transformation	**3,516**	**568r**	**-**	**4,084r**	**-**	**2,625**	**9,490**
Electricity generation	-	-	-	-	-	2,207	9,310
Major power producers	-	-	-	-	-	-	-
Autogenerators	-	-	-	-	-	2,207	9,310
Heat generation	-	-	-	-	-	418	179
Petroleum refineries	-	-	-	-	-	-	-
Coke manufacture	-	-	-	-	-	-	-
Blast furnaces	3,516	568r	-	4,084r	-	-	-
Patent fuel manufacture	-	-	-	-	-	-	-
Low temperature carbonisation	-	-	-	-	-	-	-
Energy industry use	**-**	**-**	**-**	**-**	**-**	**5,064**	**4,474**
Electricity generation	-	-	-	-	-	-	-
Oil and gas extraction	-	-	-	-	-	-	-
Petroleum refineries	-	-	-	-	-	-	-
Coal extraction	-	-	-	-	-	-	-
Coke manufacture	-	-	-	-	-	4,321	285
Blast furnaces	-	-	-	-	-	743	4,189
Patent fuel manufacture	-	-	-	-	-	-	-
Pumped storage	-	-	-	-	-	-	-
Other	-	-	-	-	-	-	-
Losses	-	-	-	-	-	441	2,014
Final consumption	**123**	**796r**	**256**	**1,175r**	**1,749**	**1,149**	**326r**
Industry	**89**	**796r**	**-**	**885r**	**1,749**	**1,149**	**326r**
Unclassified	67	14	-	81	1,749	236	-
Iron and steel	22	782r	-	804r	-	913	326r
Non-ferrous metals	-	-	-	-	-	-	-
Mineral products	-	-	-	-	-	-	-
Chemicals	-	-	-	-	-	-	-
Mechanical engineering, etc	-	-	-	-	-	-	-
Electrical engineering, etc	-	-	-	-	-	-	-
Vehicles	-	-	-	-	-	-	-
Food, beverages, etc	-	-	-	-	-	-	-
Textiles, leather, etc	-	-	-	-	-	-	-
Paper, printing, etc	-	-	-	-	-	-	-
Other industries	-	-	-	-	-	-	-
Construction	-	-	-	-	-	-	-
Transport	-	-	-	-	-	-	-
Other	**34**	**-**	**256**	**290**	**-**	**-**	**-**
Domestic	34	-	256	290	-	-	-
Public administration	-	-	-	-	-	-	-
Commercial	-	-	-	-	-	-	-
Agriculture	-	-	-	-	-	-	-
Miscellaneous	-	-	-	-	-	-	-
Non energy use	**-**	**-**	**-**	**-**	**-**	**-**	**-**

(1) Stock fall (+), stock rise (-).
(2) Coke oven gas and blast furnace gas transfers are for synthetic coke oven gas, see paragraph 2.48.
(3) Total supply minus total demand.

(4) Because of the small number of benzole suppliers, figures for benzole and tars cannot be given separately.

2.7 Supply and consumption of coal

Thousand tonnes

	2003	2004	2005	2006	2007
Supply					
Production	27,759	24,535	20,008	18,079	16,540
Deep-mined	15,633	12,542	9,563	9,444	7,674
Opencast	12,126	11,993	10,445	8,635	8,866
Other sources (1)	520	561	490	438r	467
Imports	31,891	36,153	43,968	50,529r	43,365
Exports	-542	-622	-536	-443	-521
Stock change (2)	+3,237	-60	-2,151r	-1,262r	+3,014
Total supply	**62,865**	**60,567**	**61,780r**	**67,341r**	**62,866**
Statistical difference (3)	-159	+117	-62r	-109r	-6
Total demand	**63,024**	**60,451**	**61,842r**	**67,450r**	**62,872**
Transformation	**60,093**	**57,626**	**59,392r**	**65,146r**	**60,434**
Electricity generation	52,464	50,444	52,058r	57,363r	52,558
Major power producers	50,896	48,968	50,582	55,805	51,017
Autogenerators	1,568	1,476	1,476r	1,558r	1,541
Heat generation	622	473	459r	457r	456
Coke manufacture	5,729	5,487	5,570r	5,929	5,933
Blast furnaces	882	895	1,039	1,121	1,242
Patent fuel manufacture and low temperature carbonisation	396	327	266	276	245
Energy industry use	**6**	**8**	**6**	**4**	**5**
Coal extraction	6	8	6	4	5
Final consumption	**2,925**	**2,816**	**2,445r**	**2,301r**	**2,433**
Industry	**1,857**	**1,848**	**1,781r**	**1,704r**	**1,759**
Unclassified	-	-	-	-	-
Iron and steel	-	-	-	-	-
Non-ferrous metals	13	12	41	60r	36
Mineral products	1,199	1,127	1,120	1,047	1,047
Chemicals	70	148	132r	85r	143
Mechanical engineering etc	14	13	12	12	10
Electrical engineering etc	2	5	5	6	6
Vehicles	70	80	55	53	49
Food, beverages etc	50	38	26r	25	41
Textiles, clothing, leather, etc	86	82	71	70	74
Pulp, paper, printing etc	128	141	142	141	144
Other industries	225	203	178r	205	209
Construction	-	-	-	-	-
Transport	-	-	-	-	-
Other	**1,068**	**968**	**663r**	**597r**	**674**
Domestic	1,043	941	614	561r	648
Public administration	12	13	32r	20r	14
Commercial	5	5	6	6	6
Agriculture	6	8	9	5	4
Miscellaneous	2	2	2	5	2
Non energy use	-	-	-	-	-
Stocks at end of year (4)					
Distributed stocks	12,107	12,598	14,527r	15,738r	13,587
Of which:					
Major power producers	10,971	11,019	12,696	14,489	11,447
Coke ovens	1,086	1,291	1,317r	962r	1,489
Undistributed stocks	1,624	1,192	1,101	783r	691
Total stocks (5)	**13,731**	**13,791**	**15,628r**	**16,521r**	**14,278**

(1) *Estimates of slurry etc. recovered from ponds, dumps, rivers, etc.*

(2) *Stock fall (+), stock rise (-).*

(3) *Total supply minus total demand.*

(4) *Excludes distributed stocks held in merchants' yards, etc., mainly for the domestic market, and stocks held by the industrial sector.*

(5) *For some years, closing stocks may not be consistent with stock changes, due to additional stock adjustments*

2.8 Supply and consumption of coke oven coke, coke breeze and other manufactured solid fuels

Thousand tonnes

	2003	2004	2005	2006	2007
Coke oven coke					
Supply					
Production	4,286	4,038	4,105	4,384	4,451
Imports	929	847	674r	748	745
Exports	-74	-80	-64	-94	-105
Stock change (1)	-60	-88	-94	-237r	+34
Transfers	-1,095	-1,012	-983	-955	-1,115
Total supply	**3,986**	**3,704**	**3,638r**	**3,846**	**4,010**
Statistical difference (2)	-18r	-14	-2r	-21r	-14
Total demand	**4,004**	**3,718**	**3,639r**	**3,868**	**4,024**
Transformation	**3,716**	**3,569**	**3,516**	**3,745**	**3,910**
Blast furnaces	3,716	3,569	3,516	3,745	3,910
Energy industry use	-	-	-	-	-
Final consumption	**288**	**149**	**123r**	**122**	**114**
Industry	**159**	**98**	**89r**	**111r**	**99**
Unclassified	113	76	67r	84r	76
Iron and steel	23	22	22	26	23
Non-ferrous metals	24	-	-	-	-
Other	**129**	**51**	**34**	**22r**	**15**
Domestic	129	51	34	22r	15
Stocks at end of year (3)	**230**	**318**	**413**	**650**	**616**
Coke breeze					
Supply					
Production (4)	315	298	259	245r	25
Imports	49	199	235r	261r	324
Exports	-64	-62	-55	-74	-152
Stock change (1)	-83	-63	-59	25r	-80
Transfers	+1,095	+1,012	+983	+955	+1,115
Total supply	**1,311**	**1,363**	**1,337r**	**1,411r**	**1,232**
Statistical difference (2)	-21	-1	-	-4r	+3
Total demand	**1,332**	**1,364**	**1,337r**	**1,415r**	**1,228**
Transformation	**530**	**568**	**551**	**688r**	**483**
Coke manufacture	-	-	-	-	-
Blast furnaces	530	568	551	688r	483
Energy industry use	-	-	-	-	-
Final consumption	**802**	**827**	**796r**	**727r**	**746**
Industry	**802**	**827**	**796r**	**727r**	**746**
Unclassified	7	39	14r	26r	12
Iron and steel	795	788	782	701	734
Stocks at end of year (3)	**296**	**359**	**418**	**394**	**473**
Other manufactured solid fuels					
Supply					
Production	392	318	258	260	227
Imports	5	5	6	10	8
Exports	-55	-39	-15	-12	-22
Stock change (1)	-	+22	+6	+2	+2
Total supply	**342**	**305**	**254**	**260**	**214**
Statistical difference (2)	-20	-14	-2	+3	-19
Total demand	**363**	**320**	**256**	**257**	**233**
Transformation	-	-	-	-	-
Energy industry use	**4**	**4**	-	-	-
Patent fuel manufacture	4	4	-	-	-
Final consumption	**358**	**316**	**256**	**257**	**233**
Industry	**17**	**12**	-	-	-
Unclassified	17	12	-	-	-
Other	**341**	**303**	**256**	**257**	**233**
Domestic	341	303	256	257	233
Stocks at end of year (3)	**51**	**30**	**24**	**25**	**23**

(1) Stock fall (+), stock rise (-).

(2) Total supply minus total demand.

(3) Producers stocks and distributed stocks.

(4) See paragraph 2.26

2.9 Supply and consumption of coke oven gas, blast furnace gas, benzole and tars

GWh

	2003	2004	2005	2006	2007
Coke oven gas					
Supply					
Production	9,564	9,076	9,290	9,828	9,651
Imports	-	-	-	-	-
Exports	-	-	-	-	-
Transfers (1)	+86	+40	+53	+55	+81
Total supply	**9,650**	**9,116**	**9,343**	**9,882**	**9,732**
Statistical difference (2)	+36	+65	+64	+76	+47
Total demand	**9,614**	**9,051**	**9,279**	**9,806**	**9,685**
Transformation	**2,909**	**1,944**	**2,625**	**2,593r**	**2,525**
Electricity generation	1,854	1,526	2,207	2,175r	2,107
Heat generation	1,055	418	418	418	418
Other	-	-	-	-	-
Energy industry use	**5,630**	**5,273**	**5,064**	**5,300**	**5,170**
Coke manufacture	4,466	4,326	4,321	4,282	4,228
Blast furnaces	1,164	948	743	1,019	942
Other	-	-	-	-	-
Losses	**457**	**783**	**441**	**483**	**445**
Final consumption	**618**	**1,050**	**1,149**	**1,430r**	**1,545**
Industry	**618**	**1,050**	**1,149**	**1,430r**	**1,545**
Unclassified	53	265	236	194	221
Iron and steel	565	785	913	1,236r	1,324
Blast furnace gas					
Supply					
Production	15,790	15,770	16,199	16,443	16,701
Imports	-	-	-	-	-
Exports	-	-	-	-	-
Transfers (1)	-3	-2	-2	-2	-3
Total supply	**15,787**	**15,768**	**16,197**	**16,441**	**16,698**
Statistical difference (2)	-106	-103	-107r	-119	-113
Total demand	**15,893**	**15,872**	**16,304r**	**16,560**	**16,811**
Transformation	**9,301**	**9,370**	**9,490**	**9,249**	**8,971**
Electricity generation (3)	9,002	9,191	9,310	9,070	8,791
Heat generation	299	179	179	179	179
Other	-	-	-	-	-
Energy industry use	**4,771**	**4,570**	**4,474**	**4,831**	**5,082**
Coke manufacture	432	297	285	536	703
Blast furnaces	4,339	4,273	4,189	4,294	4,379
Other	-	-	-	-	-
Losses	**1,398**	**1,557**	**2,014**	**1,578**	**2,071**
Final consumption	**423**	**375**	**326r**	**902**	**688**
Industry	**423**	**375**	**326r**	**902**	**688**
Unclassified	-	-	-	-	-
Iron and steel (3)	423	375	326r	902	688
Benzole and tars (4)					
Supply					
Production	1,773	1,722	1,750r	1,873	1,824
Final consumption (5)	**1,773**	**1,722**	**1,750r**	**1,873**	**1,824**
Unclassified	1,773	1,722	1,750r	1,873	1,824
Iron and steel	-	-	-	-	-

(1) To and from synthetic coke oven gas, see paragraph 2.48.

(2) Total supply minus total demand.

(3) From 2003, a new method of calculating fuel use for CHP in the iron and steel industry has been used (see paragraph 6.30). This results in more blast furnace gas being allocated to electricity generation and less to final consumption than in previous years. It has not been possible to recalculate CHP use for previous years on this new basis.

(4) Because of the small number of benzole suppliers, figures for benzole and tars cannot be given separately.

(5) From 2000 Iron and steel under final consumption has been reclassified due to additional information being received.

2.10 Major deep mines in production at 31 March 2008 [1]

Licensee	Site	Location
Maltby Colliery Ltd	Maltby Colliery	Rotherham
Powerfuel Mining Ltd	Hatfield Colliery	Doncaster
UK Coal Mining Ltd	Daw Mill Colliery	Warwickshire
	Kellingley Colliery	North Yorkshire
	Thoresby Colliery	Nottinghamshire
	Welbeck Colliery	Nottinghamshire

(1) In addition, at 31 March 2008, there was:

Two medium sized mines producing or developing -
Aberpergwm Colliery operated by Energybuild Mining Ltd in Neath Port Talbot
Unity Mine, owned by Unity Mine Ltd, in Neath Port Talbot

Eight small mines producing or developing -
Blaentillery No.2 Colliery operated by Blaentillery Mining Partnership in Torfaen
Black Barn Colliery operated by Thomas Croft Mining Ltd in Torfaen
Cannop Drift Mine operated by S Harding & R Harding in Gloucestershire
Eckington Colliery operated by Eckington Colliery Partnerships in Derbyshire
Gleison Colliery, owned by Coal Direct Ltd, in Neath Port Talbot
Hay Royds Colliery operated by J Flack Ltd in Kirklees
Monument Colliery, owned by Ray Ashly, Richard Daniels and Neil Jones, in Gloucestershire
Nant Hir No. 2 Colliery operated by M and W A Fyfield in Neath Port Talbot

Source: The Coal Authority

2.11 Opencast sites in production at 31 March 2008[1]

Licensee	Site Name	Location
Aardvaark TMC Ltd	Glenmuckloch	Dumfries and Galloway
(trading as ATH Resources)	Leigh Glenmuir Site	East Ayrshire
	Skares Road	East Ayrshire
Celtic Energy Ltd	East Pit	Neath Port Talbot
	Margam Opencast	Neath Port Talbot
	Nant Helen	Powys
	Selar	Neath Port Talbot
Dynant Fach Colliery Company	Dynant Fawr	Carmarthenshire
Energybuild Ltd	Nant-y-Mynydd	Neath Port Talbot
H J Banks (Mining) Ltd	Delhi Site	Northumberland
Hall Construction Services Ltd	Earlseat	Fife
	Wilsontown	South Lanarkshire
Kier Minerals Ltd	Greenburn Project	East Ayrshire
Miller Argent (South Wales) Ltd	Ffos-y-Fran Land Reclamation Scheme	Merthyr Tydfil
The Scottish Coal Company Ltd	Chalmerston	East Ayrshire
	Chapelhill	South Lanarkshire
	Glentaggart	South Lanarkshire
	Greenbank (St Ninians)	Fife
	Thornton Wood (St Ninians)	Fife
	House of Water	East Ayrshire
	Powharnal	East Ayrshire
	Shewington	Midlothian
	Spireslack	East Ayrshire
UK Coal Mining Ltd	Cutacre	Bolton
	Long Moor	Leicestershire
	Maiden's Hall Extension	Northumberland
	Sharlston Colliery Reclamation	Wakefield
	Steadsburn	Northumberland
	Stobswood	Northumberland

(1) In addition, at 31 March 2008, there were:

Six mines under development -
Cwm Yr Onen Colliery Reclamation, owned by Bryn Bach Coal Ltd, in Carmarthenshire
Temple Quarry, owned by Holgate Aggregates Ltd, in Kirklees
Caughley Quarry, owned by Parkhill Estates Ltd, in Shropshire
Bwlch Ffos, owned by Tarmac Ltd, in Neath Port Talbot
Poniel, owned by The Scottish Coal Company Ltd, in South Lanarkshire
Lodge House, owned by UK Coal Mining Ltd, in Derbyshire

Source: The Coal Authority

Chapter 3
Petroleum

Introduction

3.1   This chapter contains commodity balances covering the supply and disposal of primary oils (crude oil and natural gas liquids), feedstocks (including partly processed oils) and petroleum products in the UK during the period 2005 to 2007. These balances are given in Tables 3.1 to 3.4. Additional data have been included in supplementary tables on areas not covered by the format of the balances. This extra information includes details on refinery capacities and aggregates for refinery operations, as well as additional detail on deliveries into consumption, including breakdowns by sector and industry.

3.2   Statistics of imports and exports of crude oil, other refinery feedstocks and petroleum products, refinery receipts, refinery throughput and output and deliveries of petroleum products are obtained from the joint United Kingdom oil industry and Department for Business, Enterprise and Regulatory Reform's (BERR) Petroleum Production Reporting System.

3.3   The annual figures relate to calendar years or the end of calendar years. Unless otherwise stated, the data in the tables cover the United Kingdom.

3.4   Information on long-term trends (Tables 3.1.1 and 3.1.2) and the annex on the oil and gas resources in the UK (Annex F) provide a more complete picture of the UK oil and gas production sector and are only available in the internet version of this publication which can be found on BERR's energy web site at www.berr.gov.uk/energy/statistics/publications/dukes/page45537.html

3.5   An energy flow chart for 2007 showing the movement of crude oil, other refinery feedstocks and petroleum products is included for the first time, overleaf. This is a way of simplifying the figures that can be found in the commodity balance for primary oils and petroleum products in Table 3.1 and 3.2. It illustrates the flow of crude oil, other refinery feedstocks and petroleum products from the point at which they become available from indigenous production or imports (on the left) to the eventual final uses (on the right).

Commodity balances for primary oil (Table 3.1)

3.6   This table shows details of the production, supply and disposals of primary oils (crude oil and natural gas liquids (NGLs)) and feedstocks in 2005, 2006 and 2007. The table examines the supply chain from the production of oil and NGLs, recorded by individual oil terminals and oil fields, to their disposal to export or to UK refineries (see Annex F, Table F.2 on BERR's energy statistics web site). It also shows the aggregate use of the primary oils as recorded by the refineries. The statistical difference in the tables represents the differences between data reported by these different sources and the sites of production and consumption.

3.7   Production from the United Kingdom Continental Shelf peaked in 1999 but has been in general decline since. Chart 3.1 illustrates recent trends in production, imports and exports of crude oil, NGLs and feedstocks. It shows that crude oil exports have fallen in line with the decline in production whilst imports have risen steadily.

3.8   Gross production of crude oil and NGLs in 2007 was 76.8 million tonnes, virtually unchanged when compared to 2006 but 44 per cent lower than the peak production level of 137 million tonnes in 1999. Production from older established fields continued to decline but this fall was offset by nine new fields which started production in 2007, including the very large Buzzard field. These new fields produced 10 million tonnes of crude oil during 2007 and without these new fields production would have been 13 per cent lower than a year ago.

3.9   In 2007 about two-thirds of the United Kingdom's primary oil production was exported with imported crude oil accounting for about two-thirds of refinery intake. Feedstocks (including partly processed oils) made up 13 per cent of total imports of oil in 2007. Total primary oil imports in 2007 decreased by 4 per cent in comparison to 2006, while exports rose by 2 per cent to 51 million tonnes.

Petroleum flow chart 2007 (million tonnes)

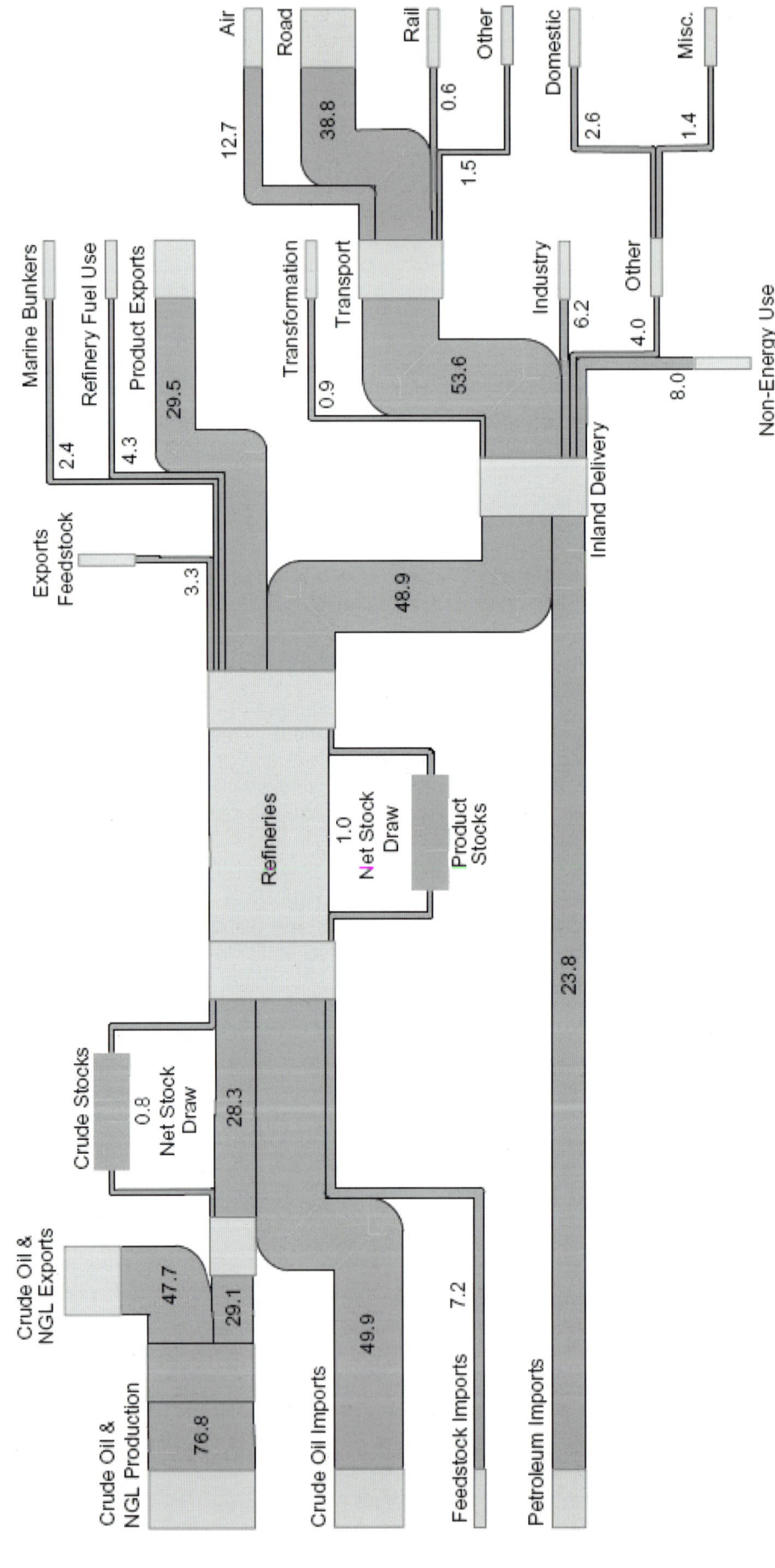

Note:
This flow chart is based on the data that appear in Tables 3.1 and 3.2.

In 2005 the UK returned to being a net-importer of primary oils for the first time since 1992, which has continued into 2007. Exports in 2007 were 11 per cent lower than imports, similar to 2006 when exports were 16 per cent lower. Further declines in exports and increases in imports will be seen in future years as indigenous production continues to decline. Even so, primary oil exports will continue to make a significant contribution to the UK economy (see Annex G on BERR's energy statistics web site).

Chart 3.1: Production, imports and exports of primary oils 2000 to 2007

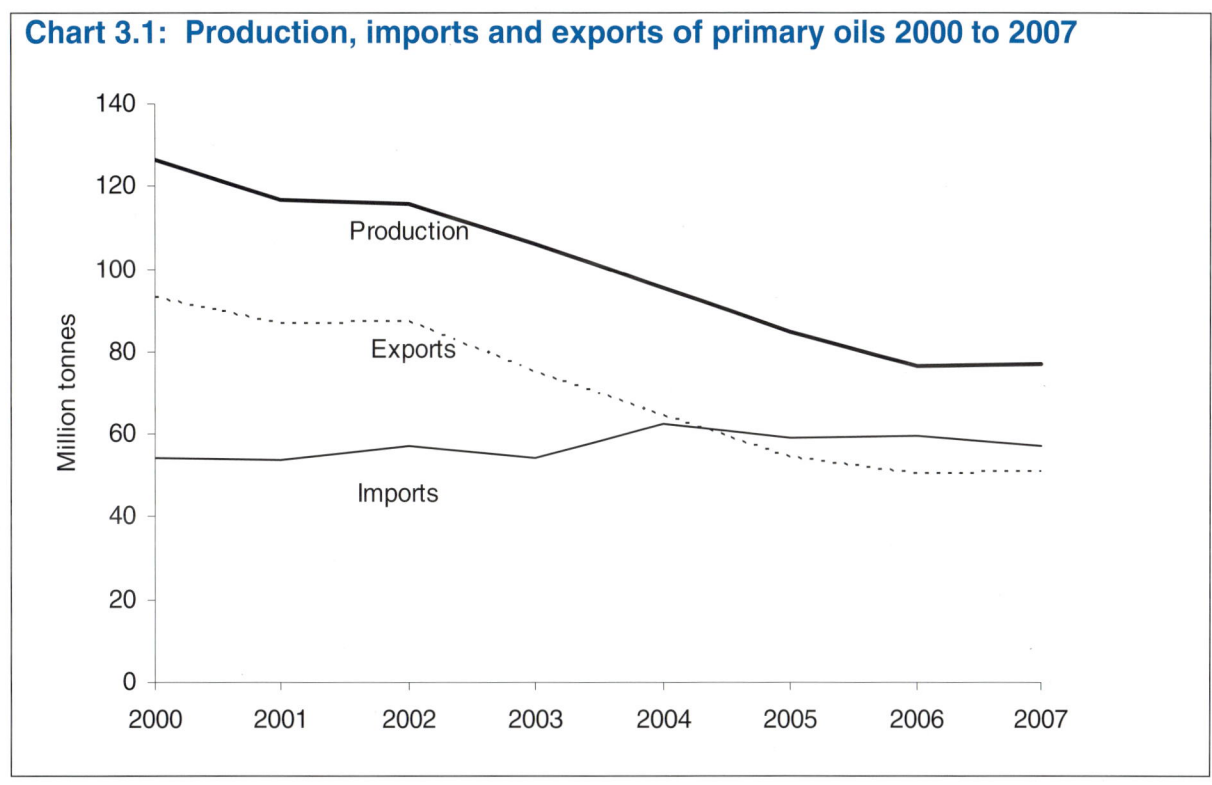

3.10 The UK imports crude oil for various commercial reasons. Primarily refineries consider the type of crude oil rather than its source origin. Most UK refineries use North Sea 'type' crude and do not differentiate between the UK and Norwegian sectors of the North Sea. Indeed, some UK refiners have production interests in both UK and Norwegian waters so the company may own the imported crude at the point of production. The close proximity of some UK and Norwegian oil fields mean that they may use the same pipeline infrastructure, for example the Norpipe oil terminal in Teesside receives both UK and Norwegian crude from the North Sea. Some crude oils are specifically imported for the heavier hydrocarbons which they contain as these are needed for the manufacture of various petroleum products such as bitumen and lubricating oils. This is in contrast to most North Sea type crude which contains a higher proportion of the lighter hydrocarbon fuels resulting in higher yields of products such as motor spirit and other transport fuels.

3.11 The 2007 energy balance in Table 3.1 shows that the overall statistical difference in the primary oil balance is minus 79 thousand tonnes. This means that the total quantities of crude oil and NGLs reported as being produced by the individual UK production fields are 79 thousand tonnes less than the totals reported by UK oil companies as being received by refineries or going for export. The reasons for this are discussed later in paragraphs 3.40 to 3.44.

Commodity balances - Petroleum products (Tables 3.2 to 3.4)

3.12 These tables show details of the production, supply and disposals of petroleum products into the UK market in 2007, 2006 and 2005. The upper half of the table represents the supply side and calculates overall availability of the various products in the UK by combining production at refineries with trade (imports and exports), stock changes, product transfers and deliveries to international marine bunkers. The lower half of the table reports the demand side and covers the uses made of the differing products, including the uses made within refineries as fuels in the refining process, and details of the amounts reported by oil companies within the UK as delivered for final consumption.

Supply of petroleum products

3.13   Total petroleum products output from UK refineries in 2007 was 81.2 million tonnes, which was 2 per cent (1.6 million tonnes) lower than the level in 2006 and 5 per cent lower than the 2005 level, although fluctuations in refinery output have tended to result from routine maintenance work.

3.14   In terms of output of individual products, production of burning oil fell by 0.4 million tonnes and fuel oil fell by 0.5 million tonnes. These falls were partially offset by increased production of gas/diesel oil, up by 0.3 million tonnes.

3.15   UK domestic production of individual petroleum products is increasingly no longer aligned with the domestic market demand. While the UK has surplus production of motor spirit and fuel oil, it produces insufficient aviation turbine fuel. Aviation turbine fuel and gas/diesel oil are extracted from the same fraction of crude oil (middle distillates), though to different quality criteria. Therefore, as the production of one increases there is less of this fraction of crude oil available for production of the other. More information on refinery capacity in the UK and refinery capacity utilisation is given in paragraphs 3.45 and 3.47.

Production, imports and exports of petroleum products

3.16   Chart 3.2 shows imports and exports of petroleum products along with production. Imports of products have followed a similar pattern to that of crude imports by generally increasing. Imports of oil products into the UK were 24 million tonnes in 2007, 11 per cent lower than in 2006 but 6 per cent higher than in 2005. The UK has been a net exporter of oil products every year since 1974, with the exception of 1984 which was due to the effects of the industrial action in the coal-mining sector. The UK is still a net exporter of petroleum products and refinery infrastructure suggests that exports will continue to be significant for sometime. Exports of petroleum products were 29 million tonnes in 2007, 2 per cent higher than in 2006 but 1 per cent lower than in 2005. Overall, the UK net exports were 6 million tonnes in 2007, up from 2 million tonnes reported in 2006 though down from 7 million tonnes reported in 2005. Additional analysis of the exports and imports of oil products is given in the long term trends internet section (paragraphs 3.1.2 to 3.1.9) and additional details about trends in UK oil production are given in Annex F on BERR's energy statistics web site.

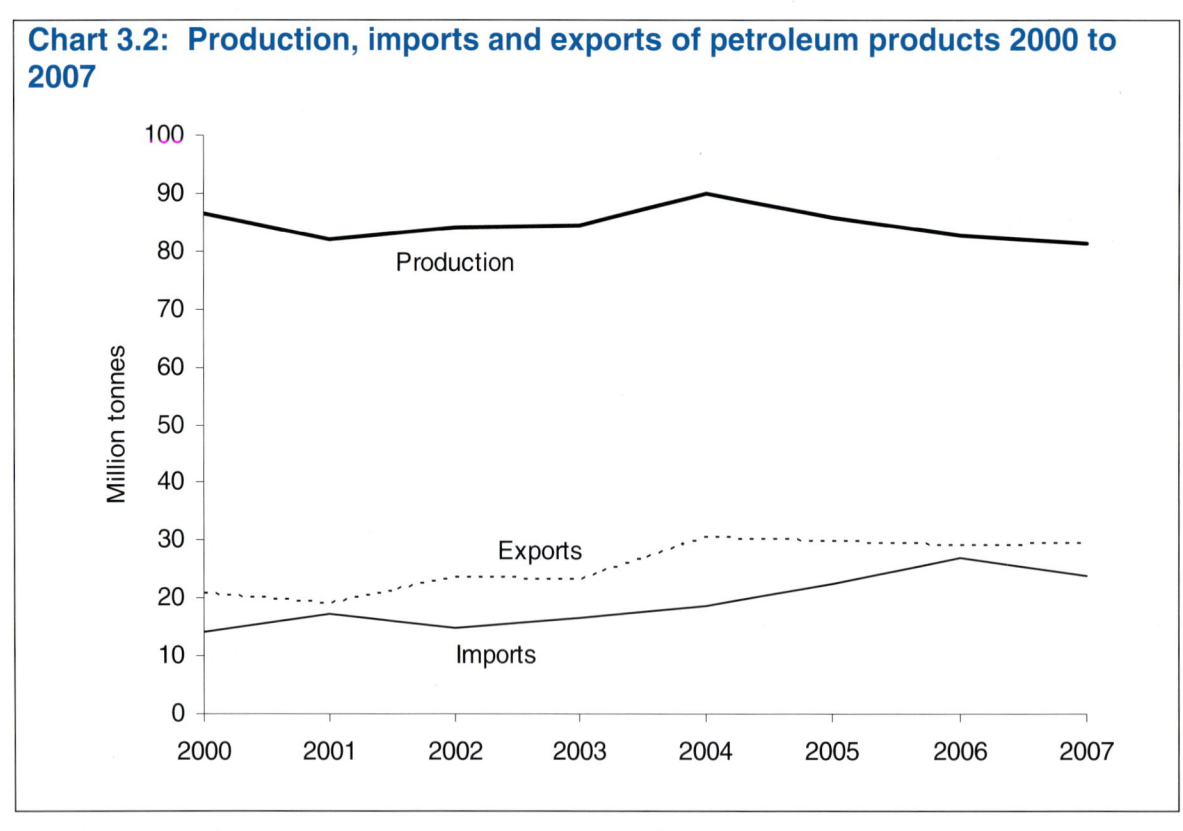

Chart 3.2: Production, imports and exports of petroleum products 2000 to 2007

3.17   The United States remains one of the key markets for UK exports of oil products, with 7 million tonnes being exported there in 2007. These exports made up 24 per cent of total UK exports of oil products in 2007, with the other main countries receiving UK exports of petroleum products being Ireland, the Netherlands, France, Spain, Germany and Belgium. The main sources of the UK's imports of petroleum products in 2007 were the Netherlands, Saudi Arabia, France, Kuwait, Belgium and Latvia (source: International Energy Agency).

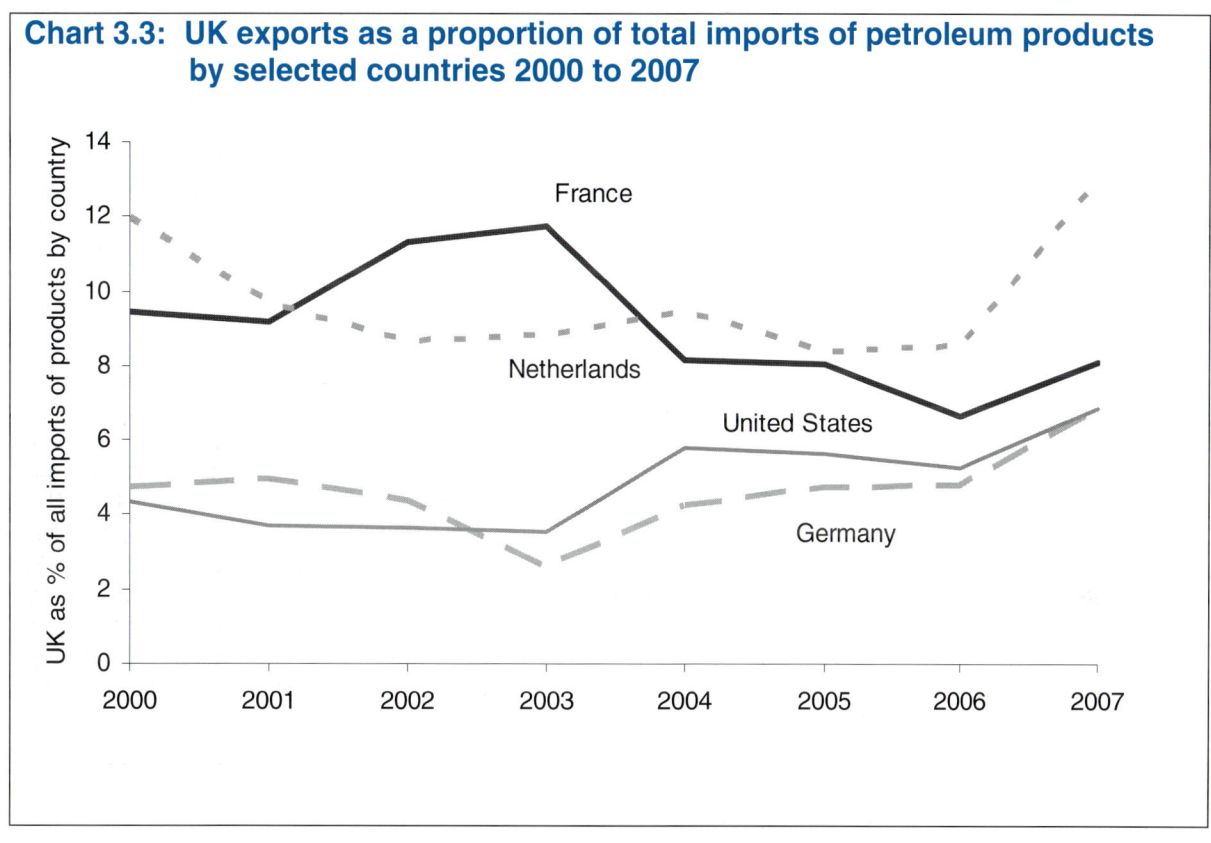

Chart 3.3: UK exports as a proportion of total imports of petroleum products by selected countries 2000 to 2007

3.18   Chart 3.3 shows how the UK has penetrated selected overseas markets for its petroleum products. The UK supplied 7 per cent by volume of all US imports of petroleum products in 2007 (mostly in the form of motor spirit) and 8, 13 and 7 per cent respectively of the total imports of petroleum products into France, the Netherlands and Germany (mostly as gas oil for heating, motor spirit and fuel oil). The UK regularly supplies the vast majority of total oil products imported into Ireland (mostly motor spirit and DERV fuel for transport, and gas oil and burning oil for heating).

3.19   Differences in product types partly explain why the UK imports petroleum products when there is an overall surplus available for export. Exports in 2007 were mainly made up of motor spirit (7 million tonnes), gas/diesel oil (7 million tonnes) and fuel oil (8 million tonnes). Imports, whereas, consisted of aviation turbine fuel (8 million tonnes), motor spirit (3 million tonnes) and gas/diesel oil (8 million tonnes). However, imports also take place to cover specific periods of heavy demand within the UK such as seasonally high demand or to cover production shortfalls in refinery shutdowns for maintenance.

3.20   For gas/diesel oil, exports from the UK tend to be of lower grades for use as heating fuels, while imports tend to be of higher-grade gas/diesel oil with a low sulphur content. With the introduction of low sulphur DERV fuel and motor spirit into the UK market and the required increased production capacity at UK refineries for these fuels, UK imports of these products increased to meet the shortfall. As noted above, aviation turbine fuel is imported simply because the UK cannot make enough of it to meet demand, since it is derived from the same sort of hydrocarbons as gas/diesel oil, and as such there is a physical limit to how much can be made from the amount of oil processed in the UK.

3.21   More information on the structure of refineries in the UK and trends in imports and exports of crude oil and oil products is given in the long-term trends section on BERR's energy statistics web site (Table 3.1.1).

3.22   In 2007, 12 per cent of UK production of fuel oil and 3 per cent of gas/diesel oil production went into international marine bunkers, totalling 2 million tonnes of products, which was 3 per cent of the total UK refinery production in the year.   These are fuel sales destined for consumption on ocean going vessels and therefore cannot be classified as being consumed within the UK.   Correspondingly these quantities are treated in a similar way to exports in the commodity balances.   It should be noted that these quantities do not include deliveries of fuels for use in UK coastal waters, which are counted as UK consumption and are given in the figures of the transport section of the commodity balances.

Overall oil net imports/exports

3.23   Chart 3.4 compares the level of imports and exports of crude oil with those for petroleum products over the period 2000 to 2007.   The chart shows that in 2005 the UK became a net-importer of crude oil, which has continued into 2007.   For petroleum products, it can be seen that in 2007 the UK was a net-exporter.   It can also be seen that overall the UK was a net-importer in 2007, influenced mainly by the scale of crude oil imports.

Chart 3.4:   Imports and exports of crude oil and petroleum products 2000 to 2007

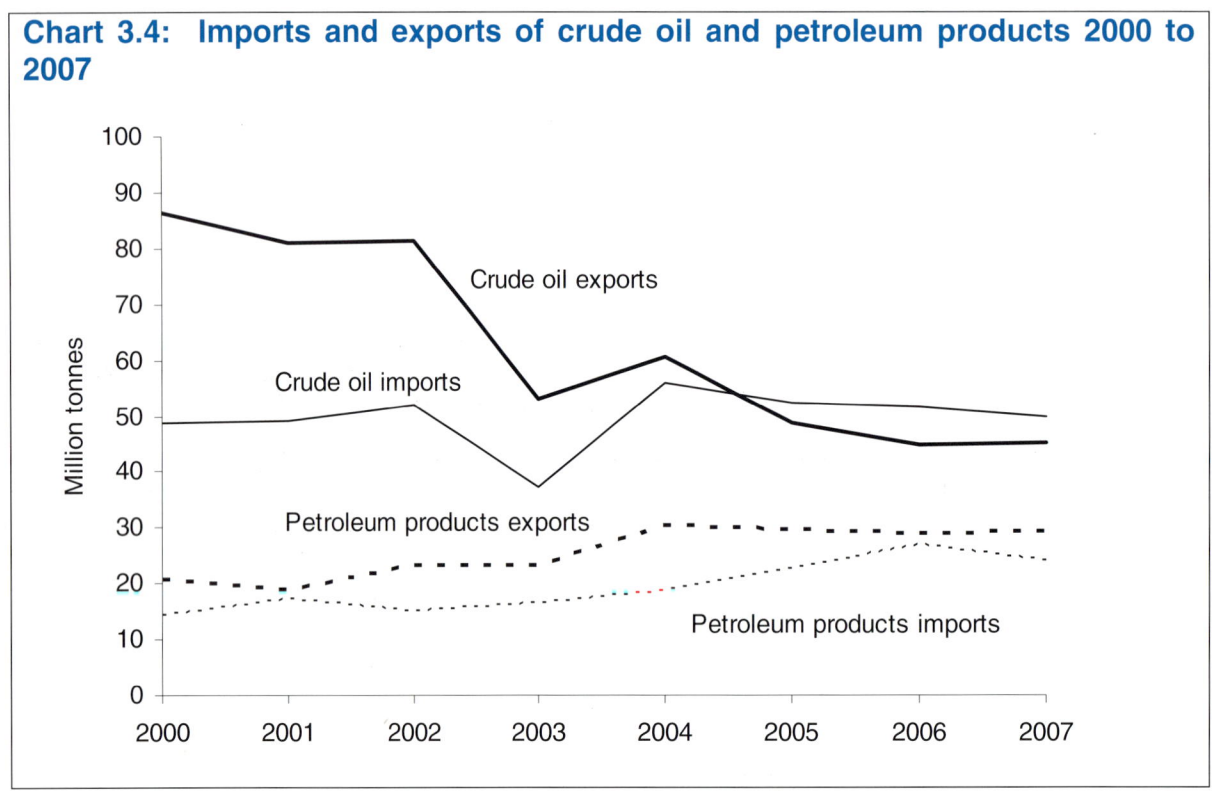

3.24   Two articles have been published, in the March and June 2007 editions of Energy Trends, which look at this analysis in greater detail, taking some of the data back as far as 1870.   The publications are available on the energy statistics website at
www.berr.gov.uk/energy/statistics/publications/trends/index.html

Stocks of petroleum products

3.25   Details are given in the balances of stocks of products held within the UK either at refineries or oil distribution centres such as coastal oil terminals (undistributed stocks).   In addition, some information is available on stocks of oil products held by major electricity generators (distributed stocks).   However, these figures exclude any details of stocks held by distributors of fuels or stocks held at retail sites, such as petrol stations.   The figures for stocks in the balances also solely relate to those stocks currently present in the UK and specifically exclude any stocks that might be held by UK oil companies in other countries under bilateral agreements.

3.26   In order for the UK to be prepared for any oil emergency, companies supplying oil products into final consumption in the UK are obligated by the UK Government to maintain a certain level of stocks of oil products used as fuels.   As part of this, companies are allowed to hold stocks in other EU countries subject to bilateral agreements between Governments and count these stocks towards their stocking obligations. The stocks figures in Table 3.7 take account of these bilateral stocks (see paragraphs 3.71 to 3.76) to give a true picture of the amount of stocks available to the UK.

Consumption of petroleum products

3.27 Tables 3.2 to 3.4 examine the data given on the consumption of oil products during the period 2005 to 2007. The main sectors of consumers will be looked at first (going down the tables) before the data for individual products (going across the tables) are considered.

3.28 Table 3.2 shows how overall deliveries of petroleum products for consumption in the UK in 2007, including those used by the UK refining industry as fuels within the refining process and all other uses, totalled 77 million tonnes. This was 4 per cent lower than 2006 and 5 per cent lower than in 2005.

3.29 As seen in the tables, one of the most significant changes in deliveries of products in recent years has been the decline in use for electricity generation. (See long term trends, Table 3.1.2 on BERR's energy statistics web site). This change is primarily a result of major electricity producers using natural gas as their fuel of choice for electricity generation rather than oil-based fuels. This trend is also reflected in the declining level of usage by auto-producers of electricity over the period, despite the growth in auto-generation of electricity by industry as a whole, and in the significant declining use in heat generation. Figures in tables 3.2 to 3.4 of fuel oil used for electricity generation differ to those in chapter 5. There are two causes for this difference. Firstly, figures in this chapter relate to deliveries while those in chapter 5 report consumption – fuel oil at electricity generating plants can be stored for varying time periods. Secondly, the consumption figures in chapter 5 include some recovered fuel oil that are recycled lubricants – see paragraph 3.65 for more information.

3.30 The data included under the blast furnaces heading of the Transformation sector represents fuel oil used in the manufacture of iron and steel which is directly injected into blast furnaces, as opposed to being used as a fuel to heat the blast furnaces. The fuel used for the latter (mostly gas oil) is included under the blast furnaces heading of the Energy Industry Use sector.

3.31 Other figures in the Energy Industry Use sector relate to uses within the UK refining industry in the manufacture of oil products. These are products either used as fuels during refining processes or products used by the refineries themselves as opposed to being sold to other consumers. It excludes any fuels used for the generation of electricity since these amounts are included in the Transformation sector totals. Given the interest in the total amounts of fuels used within refineries, Table 3.5 includes data on total refinery fuel usage (i.e. including that used in the generation of electricity) over the period 2003 to 2007. The data under the other headings of the Energy Industry Use sector represent fuels used by the gas supply industry.

3.32 Chart 3.5 shows the breakdown of consumption for energy uses by each sector in 2007, including energy industry use and transformation and excluding non energy use.

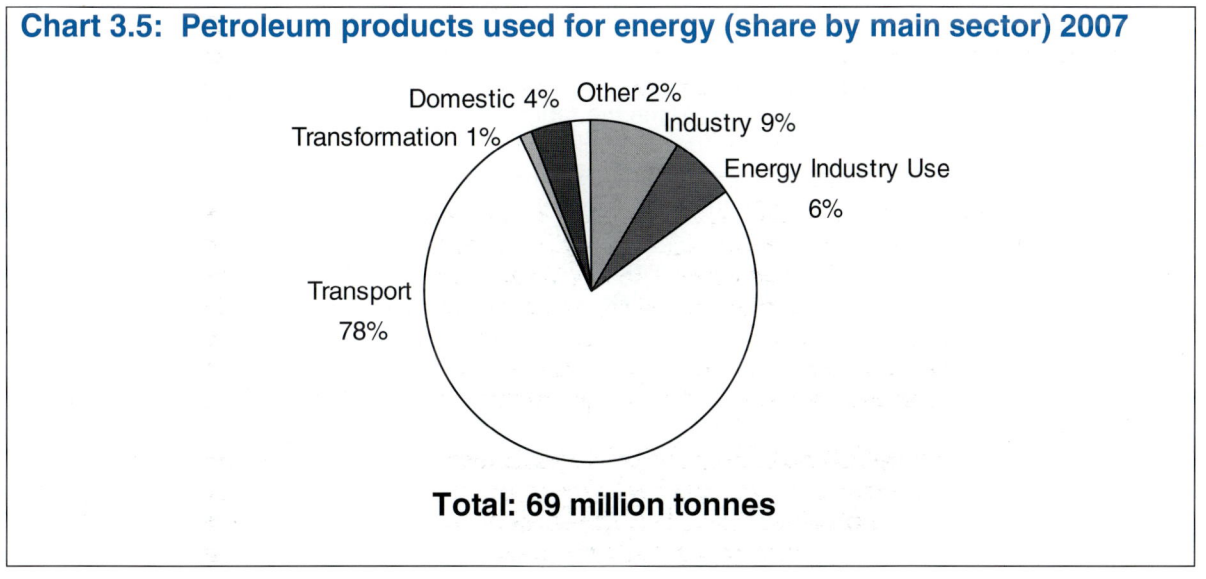

Chart 3.5: Petroleum products used for energy (share by main sector) 2007

Domestic 4% Other 2%

Transformation 1%

Industry 9%

Energy Industry Use 6%

Transport 78%

Total: 69 million tonnes

3.33   The total amount of oil products used by industry was in decline in the middle-1990s as industry moved away from using oil as an energy source. During the period 1998 to 2003 oil use by industry was fairly constant at around 5.5 million tonnes. Since then industrial usage has grown, to approximately 6 million tonnes in 2004 and remained at this level in 2005, 2006 and 2007.

3.34   Transport sector consumption in 2007 was virtually the same as in 2006 and 2 per cent higher than in 2005. The increase in transport sector consumption has largely been driven by increased use of DERV fuels that were up by 4 per cent in 2007 when compared to 2006. Total transport usage in 2007 was 53.6 million tonnes and accounted for 75 per cent of total final consumption of oil products. In contrast, final energy consumption by sectors other than transport and industry decreased by 8 per cent in 2007 compared with 2006.

3.35   Final consumption of oil products in 2007, i.e. excluding any uses by the energy industries themselves or for transformation purposes, amounted to 72 million tonnes, 2 million tonnes less than both 2005 and 2006.

3.36   Consumption of non-energy products fell to around 8 million tonnes in 2007, down by 20 per cent when compared to 2006, with non-energy products making up 11 per cent of final consumption of oil products. Additional detail on the non-energy uses of oil products, by product and by type of use where such information is available, is given in Table 3D and paragraphs 3.62 to 3.69 later in this text.

3.37   Final consumption is dominated by three individual products; aviation turbine fuel, motor spirit and gas/diesel oil, which between them made up 80 per cent of total final energy consumption in 2007. These three products are predominantly used as transport fuels although some gas oil is used for power generation. Changes in the consumption of aviation fuel are discussed in more detail in paragraphs 3.58 to 3.60 below while more detailed information on consumption of road fuels over the period 2003 to 2007 is given in Table 3.6 and discussed in paragraphs 3.48 to 3.55.

3.38   In 2007 total final consumption of fuel oil was 1.6 million tonnes, up by 10 per cent on 2006. This was mainly due to an increase in industrial usage of 10 per cent between 2006 and 2007. Further detail on the consumption of fuel oil broken down by grade is given in Table 3.6.

3.39   Tables 3.2 to 3.4 include estimates for the use of gas for road vehicles. These estimates are based on information on the amounts of duty received by HM Revenue and Customs from the tax on gas used as a road fuel. It is estimated that some 119 thousand tonnes of gas (mostly butane or propane) was used in road vehicles in the UK in 2007. Although this is a very small when compared to overall consumption of these fuels or the total consumption of fuels for road transport, the consumption of these gases for road transport grew rapidly until 2003, but has since remained broadly flat.

Supply and disposal of products (Table 3.5)

3.40   Table 3.5 brings together the commodity balances for primary oils and for petroleum products into a single overall balance table.

3.41   The statistical difference for primary oils in the table includes own use in onshore terminals and gas separation plants, losses, platform and other field stock changes. Another factor is the time lag that can exist between production and loading onto tankers being reported at an offshore field and the arrival of these tankers at onshore refineries and oil terminals. This gap is usually minimal and works such that any effect of this at the start of a month is balanced by a similar counterpart effect at the end of a month. However, there can be instances where the length of this interval is considerable and, if it happens at the end of a year, there can be significant effects on the statistical differences seen for the years involved.

3.42   With the downstream sector, the statistical differences can similarly be used to assess the validity and consistency of the data. From the tables, these differences are generally a very small proportion of the totals involved.

3.43   Paragraphs 3.81 to 3.92 provide details on the reasons why statistical differences occur for the upstream and downstream sectors. Following the identification of some discrepancies in the reporting of refinery production data in 2001, significant changes have been made to the downstream oil data reporting system that culminated in the launch of a revised system in 2005.

3.44   Table 3.5 shows a refining process gain of 92 thousand tonnes; however a loss is usually expected.   One possible reason for this is a time lag in the movement of primary oils from the oil terminals into the refineries over the end of the year.   BERR will be investigating this over the next year.

Refinery capacity

3.45   Data for refinery capacity as at the end of 2007 are presented in Table 3A, with the location of these refineries illustrated in Map 3A.   These figures are collected annually by BERR from individual oil companies.   Capacity per annum for each refinery is derived by applying the rated capacity of the plant per day when on-stream by the number of days the plant was on stream during the year. Fluctuations in the number of days the refinery is active are usually the main reasons for annual changes in the level of capacity. Reforming capacity covers catalytic reforming, and cracking/conversion capacity covers processes for upgrading residual oils to lighter products, e.g. catalytic, thermal or hydro-cracking, visbreaking and coking.

Table 3A: UK refinery processing capacity as at end 2007 [1]

	Million tonnes per annum		
(Symbols relate to Map 3A)	Distillation	Reforming	Cracking and Conversion
❶ Stanlow – Shell UK Ltd	11.5	1.5	3.8
❷ Fawley – ExxonMobil Co. Ltd	16.4	3.0	5.2
❸ Coryton – Petroplus International Ltd	8.8	1.8	3.4
❹ Grangemouth – Ineos Refining Ltd	9.9	2.1	3.5
❺ Lindsey Oil Refinery Ltd – Total (UK)	10.9	1.5	4.1
❻ Pembroke – Texaco Refining Co. Ltd	10.1	1.5	6.1
❼ Killingholme – Conoco Ltd	10.2	2.2	9.2
❽ Milford Haven - Murco Pet. Ltd	5.1	0.8	1.9
❾ North Tees – Petroplus International Ltd	5.7	-	-
① Harwich – Petrochem Carless Ltd	0.4	-	-
② Eastham – Eastham Refinery Ltd	1.1	-	-
③ Dundee (Camperdown) – Nynas UK AB	0.7	-	-
Total all refineries	**90.7**	**14.2**	**37.1**

(1) Rated design capacity per day on stream multiplied by the average number of days on stream.

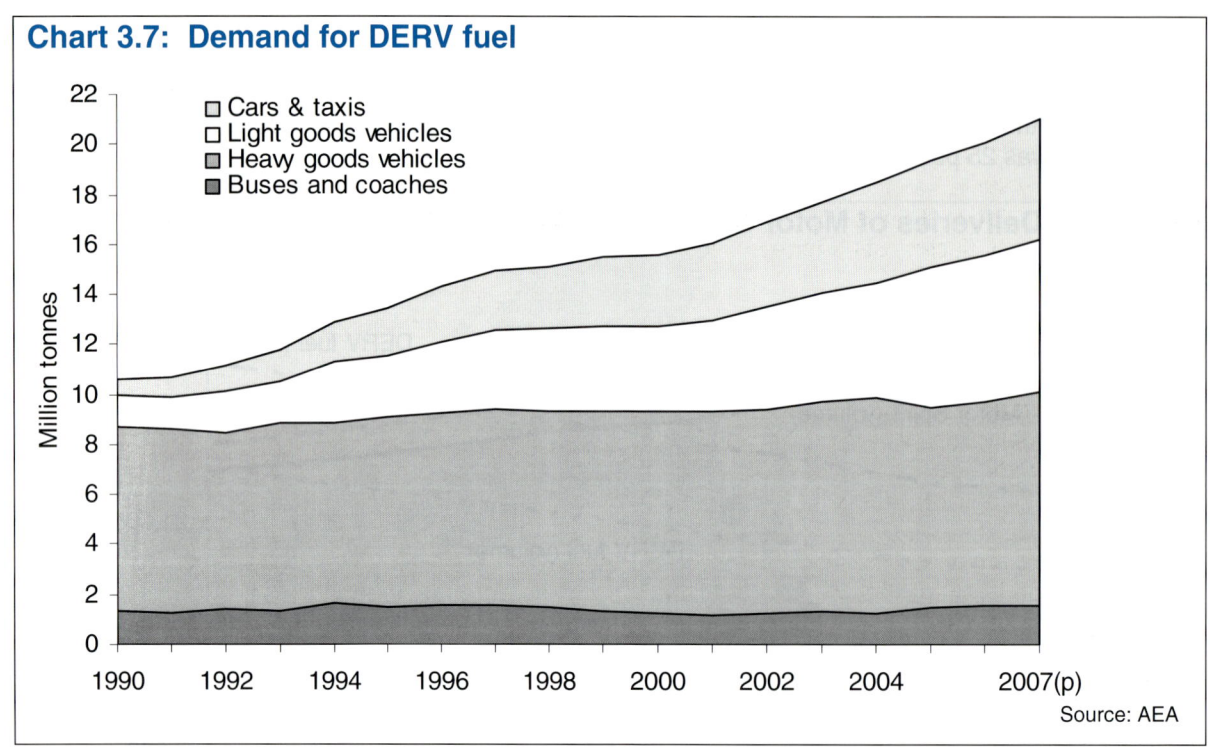

Chart 3.7: Demand for DERV fuel

Million tonnes

Legend:
□ Cars & taxis
□ Light goods vehicles
■ Heavy goods vehicles
■ Buses and coaches

Source: AEA

3.54   Chart 3.8 illustrates the switch from petrol to diesel engines in terms of vehicle licence registrations based on DVLA data. It contains details of vehicle licence registrations for private cars during each year for the period 1997 to 2007, broken down by type of engine. Whilst the number of petrol engined vehicles licensed only grew by 5 per cent, the number of diesel engined vehicles licensed has increased nearly 171 per cent in the same period. Petrol engined vehicle stock remains greater than diesel although its share has decreased from 89 per cent in 1997 to 76 per cent in 2007.

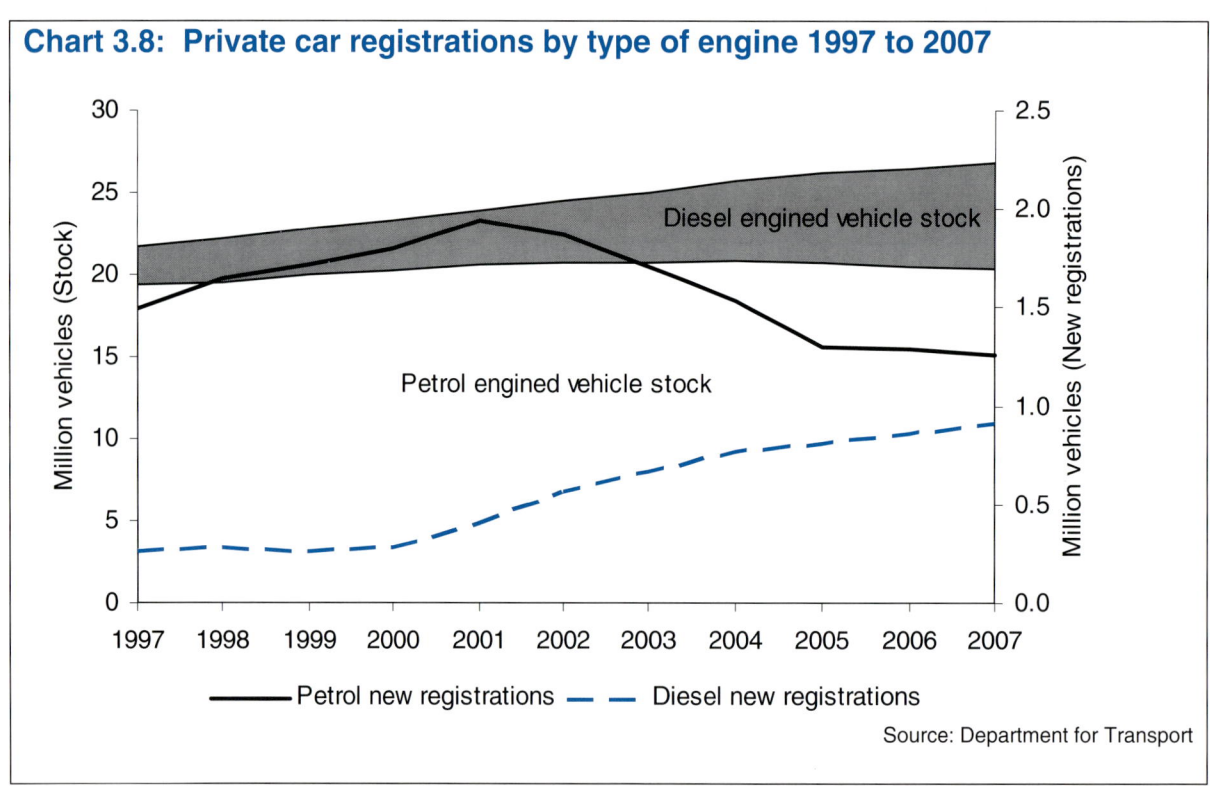

Chart 3.8: Private car registrations by type of engine 1997 to 2007

Million vehicles (Stock)

Million vehicles (New registrations)

Diesel engined vehicle stock

Petrol engined vehicle stock

—— Petrol new registrations — — Diesel new registrations

Source: Department for Transport

3.55   The chart also shows new registrations for petrol and diesel cars.   Although more new petrol cars are registered each year than diesel, registrations for new petrol cars peaked in 2001 and have declined since in contrast to registrations for new diesel cars that have increased since 2000.   As the actual number of petrol engined vehicles on the road is relatively flat, new registrations are simply replacing losses.   This turnover in the vehicle stock would suggest that the average fuel efficiency of the UK petrol vehicle stock would improve as newer vehicles tend to be more fuel efficient due to improved technology.   This is supported by a comparison of the static petrol car population in Chart 3.8 with the steady decline of motor spirit consumption illustrated in Chart 3.6.

Biofuels

3.56   Information regarding biofuels is currently limited and so has not been included in the commodity balances or the supplementary tables.   HM Revenue and Customs has the best available information covering duty clearances or volumes on which excise duty has been paid that effectively equate to the deliveries reported in this chapter. Biodiesel clearances began in 2002 with bioethanol clearances commencing in 2005. The amounts reported are very small in comparison with overall DERV and petrol deliveries but have been increasing rapidly.

3.57   Table 3B shows biofuel clearances since they began in 2002.   In 2007 biodiesel clearances were 347 million litres (up 106 per cent from 2006), while bioethanol clearances were 153 million litres (up by 61 per cent).   However, biodiesel represented just 1 per cent of the total DERV deliveries in 2007, bioethanol represented 1 per cent of the total motor spirit deliveries and overall, biofuels represented just 1 per cent of the total road fuel deliveries.   Further information about biofuels and the Renewable Transport Fuel Obligation (RTFO) can be found in Chapter 7 paragraphs 7.29 to 7.32 and 7.70 to 7.71.

Table 3B: Biofuel Clearances, 2002 to 2007

	Bioethanol		Biodiesel	
	Thousand tonnes	Million litres	Thousand tonnes	Million litres
2002	-	-	2.8	3.3
2003	-	-	17.3	19.9
2004	-	-	18.2	21.0
2005	62.7	85.4	28.5	32.9
2006	69.9	95.2	146.3	168.9
2007	112.3	152.8	300.5	347.1

Source: HM Customs and Revenue

Super/hypermarket share

3.58   Sales by super/hypermarkets have accounted for an increasing share of retail deliveries (i.e. deliveries to dealers) of motor spirit and DERV fuel in recent years as Table 3C shows. These figures have been derived from a survey of super/hypermarket companies to collect details of their sales of motor spirit and DERV fuel.   The share of total deliveries of both fuels, including deliveries direct to commercial consumers, is shown in brackets. The percentage shares are also affected by the decline in the overall deliveries of motor spirit in the UK seen in these years as mentioned earlier.   Super/hypermarket deliveries of motor spirit decreased by 3 per cent from 2006 whilst their DERV deliveries increased by 5 per cent.

Table 3C Super/hypermarkets share of retail deliveries, 2003 to 2007

per cent

	Motor spirit			DERV fuel	
	Share of retail	Share of total		Share of retail	Share of total
2003	30.7	(30)		23.6	(12)
2004	32.9	(32)		26.0	(13)
2005	37.5	(36)		28.9	(16)
2006	40.5	(39)		34.2	(19)
2007	40.9	(39)		35.5	(21)

Figures in brackets are shares of total deliveries (i.e. including deliveries direct to commercial consumers).

Aviation fuel

3.59   Data in Tables 3.2 to 3.4 show the changing amounts of aviation turbine fuel (ATF) being consumed in the UK between 2005 and 2007. The long-term trends section on the internet discusses the trend seen since 1970 in the use of ATF in the UK. Overall, deliveries in 2007 were virtually unchanged when compared with 2006, but were 1 per cent higher than in 2005.

3.60   Chart 3.9 shows annual deliveries of ATF in the UK over the last decade. ATF consumption increased steadily until 1997 and then rapidly until 2000. The September 11th 2001 terrorist attacks on the United States had a significant impact on the global aviation industry and reversed the trend for a period lasting more than twelve months. The increase in ATF deliveries since 2002 illustrates the subsequent recovery of the global aviation industry.

Chart 3.9: Aviation Turbine Fuel deliveries over the last decade

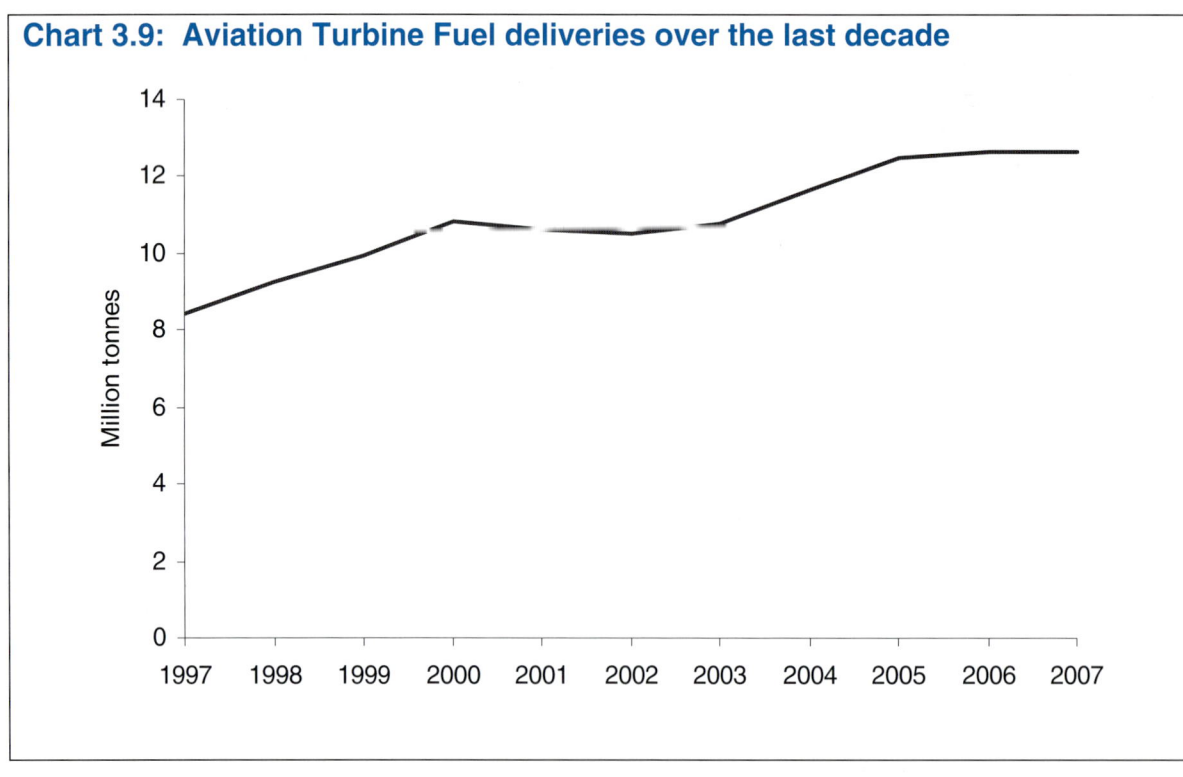

3.61   Chart 3.10 shows the aviation industry's activities over the last ten years split by the domestic and international passenger uplifts and cargo delivery categories. It is clear to see that the events of September 11th had varied effects on the different sections of the industry. International air passenger movements suffered a fall in the rate of growth, while domestic flights were largely unaffected. The most significant effect however is clearly visible within the cargo division of the industry, which saw a fall of 21 per cent in 2001. The differing ways in which the demand for the industry's separate services reacted to the terrorist attack implies that rather than causing a reluctance of passengers in the UK to fly, the negative effect on the economy, causing cargo demand to fall, had the greatest impact. Since 2001 the cargo division has steadily grown and in 2007 was similar to the level seen in 2000.

Chart 3.10: Aviation Turbine Fuel usage over the last decade by type

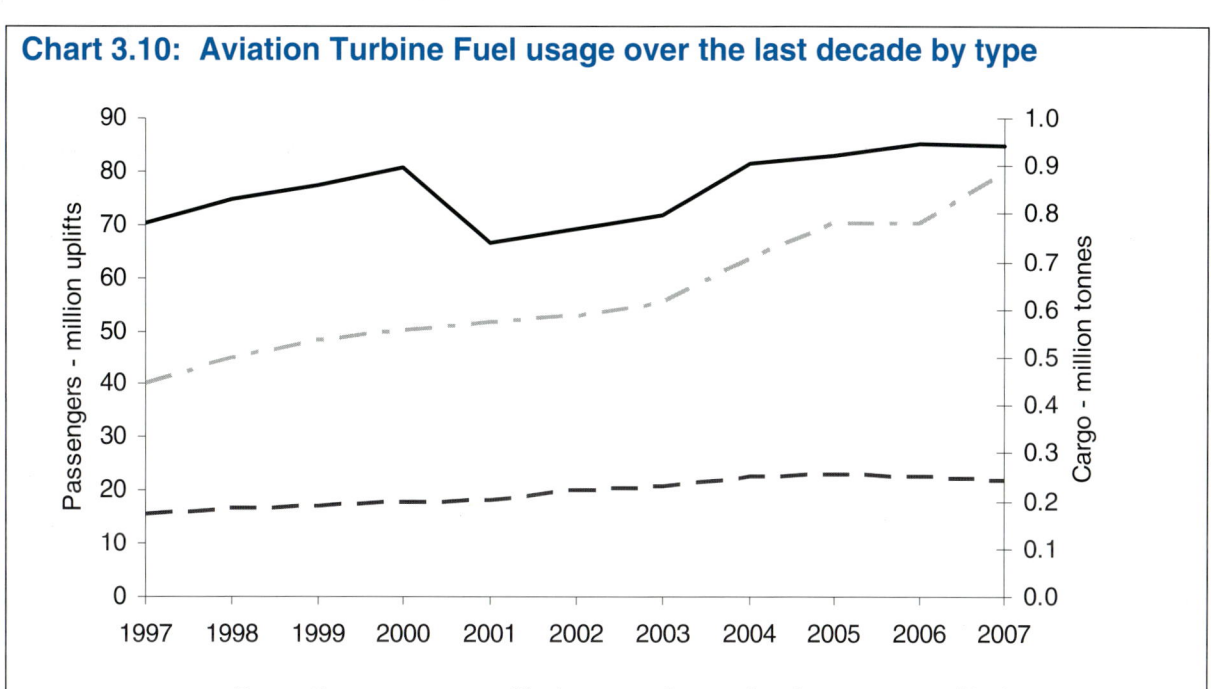

Additional information on inland deliveries for non-energy uses

3.62 Table 3D below summarises additional data on the non–energy uses made of the total deliveries of oil products included as the bottom line in the commodity balances in Tables 3.2 to 3.4. It provides extra information on the uses of lubricating oils and greases by use, and details of products used as petro-chemical feedstocks.

3.63 In 2007 there was a fall of 56 per cent for inland deliveries of naphtha. This fall has occurred predominately in the non-refining sector, which saw a decrease of 95 per cent in naphtha imports in 2007 when compared with 2006.

3.64 All inland deliveries of lubricating oils and petroleum coke have been classified as non-energy use only. However, some deliveries are used for energy purposes but it is difficult to estimate energy use figures for these products with any degree of accuracy, hence no such estimates appear in the commodity balance tables.

3.65 For lubricating oils, about 50 to 55 per cent of inland deliveries each year are consumed during use (e.g. burnt within engines), or lost as a film on manufactured goods, lost due to evaporation, or for products like process oils and white oils, the oil becomes part of the product. Work done by the Department for Environment, Food and Rural Affairs (Defra) suggested that each year about 45 to 50 per cent of inland deliveries become available as used lubricating oil. Big changes to waste legislation were introduced in 2005/6 and although some 85 per cent of the waste/used lubricating oil (between 270 and 300 thousand tonnes) is collected, it is now reused in a variety of different ways. During 2007 the Court of Appeal ruled that a non waste status fuel can be produced from waste oil subject to certain conditions being met. As a result, in 2007 post use oil was reprocessed for use as a chemical for steel making, a waste fuel for Waste Incineration Directive (WID) compliant plants, a non-waste fuel against an interim standard and a Re-refinery back to base oil. Some was also burnt locally in licensed and unlicensed small space heaters. However, the split for each of these is unclear. Previous studies have shown that the UK's 85 per cent collection rate of used lubricating oils was one of the highest in Europe, but the changes to legislation will make estimating this figure more difficult for some time.

Table 3D Additional information on inland deliveries for non-energy uses, 2005 to 2007

Thousand tonnes

	2005	2006	2007
Feedstock for petroleum chemical plants:			
Propane	1,088r	994	977
Butane	901	737	717
Other gases	1,767	1,648	1,514
Total gases	3,756r	3,379	3,208
Naphtha (LDF)	1,916	2,278	1,006
Middle Distillate Feedstock (MDF)	229	258	239
Other products	-	-	-
Total feedstock	5,901r	5,915	4,453
Lubricating oils and grease:			
Aviation	5	5	4
Industrial	503	478	450
Marine	27	26	24
Motors	210	200	189
Agricultural	5	5	5
Fuel oil sold as lubricant	-	-	-
Total lubricating oils and grease	750	713	672
Other non-energy products:			
Industrial spirit/white spirit	284	156	167
Bitumen	1,906	1,610	1,572
Petroleum wax	72	48	39
Petroleum coke	1,042	925	544
Miscellaneous products	484	628	551
Total non-energy use	10,439r	9,995	7,998

3.66 For petroleum coke, more information is available allowing more accurate estimates to be made. It has been possible to analyse the data available for the imports of petroleum coke to identify which type of company is importing the product. This work has shown that a significant proportion of petroleum coke imports each year are made by energy companies, such as power generators or fuel merchants, with another substantial proportion being imported by cement manufacturers. Whilst it cannot be certain that these imports are being used as a fuel, information on the use of petroleum coke in cement manufacture suggests that it is.

3.67 Using imports data, estimates have been constructed which show that around 500 thousand tonnes of petroleum coke were imported for inland deliveries in 2007. Around 410 thousand tonnes of petroleum coke were estimated to have been imported in 2006 for energy uses (for electricity generation, use as a fuel in the manufacture of cement, sold as a solid fuel or to be used in the manufacture of other solid fuels). Estimates of imports for energy uses fell in 2007 to 76 thousand tonnes or 16 per cent of total petroleum coke imports.

3.68 Analysis of the data on the quantity and value of imports of petroleum coke into the UK from HM Revenue and Customs provides some estimates for the cost of imports and gives some indication of the prices being paid. These are only indicative of the prices being paid in the port of importation, and do not include the extra transport costs from the port to the final destination that would be part of more rigorous price estimate. Details of these estimates are included in Annex G on trade in fuels, as part of Table G.3 on the BERR energy statistics website. A breakdown has been made by grade of petroleum coke and type of use for imports into the UK, which is given in Table 3E below. Calcined petroleum coke is virtually pure carbon, and as such is more valuable than non-calcined (otherwise known as "green") petroleum coke, as shown by the higher price per tonne it commands and the fact that it is not used simply as a fuel.

Table 3E Estimated £ per tonne for imports of petroleum coke into the UK

| | Non-calcined ("green") petroleum coke | | | Calcined petroleum coke |
	Energy	Non-energy	Total	Non-energy
2004	22.5	35.0	29.9	120.6
2005	30.8	35.3	32.9	131.1
2006	35.7	54.8	42.2	138.7
2007	43.2	49.5	47.5	137.0

3.69   Petroleum coke is a relatively low energy content fuel, having a calorific value of 35.8 GJ per tonne, compared with an average for petroleum products of 45.8 GJ per tonne, and 43.6 GJ per tonne for fuel oil. It is, however, higher than coal (26.9 GJ per tonne) and in certain areas is competing with coal as a fuel. It has the advantage of being a very cheap fuel, since it is often regarded as a waste product rather than a specific output from the refining process.

Inland deliveries by country

3.70   Over recent years BERR has been developing more detailed information on regional consumption data, which can be found at www.berr.gov.uk/energy/statistics/regional/index.html. Although this information is one year in arrears it is thought to be more comprehensive than the information provided in this chapter. Therefore, during 2007 BERR reviewed the regional information provided in this chapter of the Digest to assess its quality and usefulness with key stakeholders. The outcome of the review is that the regional data provided at the link above provides better information and so regional information will no longer be published in this chapter of the Digest; however, some information is available upon request.

Stocks of oil (Table 3.7)

3.71   This table shows stocks of crude oil, feedstocks (including partly processed oils) and products (in detail) at the end of each year. Stocks of crude oil and feedstocks decreased in 2007, with decreases in stocks held at terminals, refineries and offshore facilities.

3.72   The UK holds emergency stocks of oil to help reduce the adverse impact on the UK of any disruptions of supplies of oil arising from domestic or international incidents. EU legislation (EC Directive 2006/07) requires EU member states to hold oil stocks equivalent to 90 days worth of average daily consumption calculated from the previous calendar year. These stocks are held to deal with oil supply emergencies, not to manage or affect prices. The UK, as a producer, receives a derogation of 25 per cent on its obligation and is only required to hold stocks equivalent to 67½ days of consumption.

3.73   The International Energy Agency (IEA) also requires its members to hold stocks for use in the event of global disruption. Until 2007 the United Kingdom as a net exporter was exempt from this requirement. However, in 2006 the United Kingdom became a net importer and so since 2007 has had an IEA obligation to hold stocks as well as its EU obligation. The same stocks count towards meeting both sets of obligations, and as IEA obligations are based on net imports we do not expect a significant net increase in total UK obligations until about 2016 – 2018.

3.74   To meet these obligations the UK Government requires companies supplying oil products into final consumption in the UK to maintain a certain level of emergency stocks of oil products as fuels. As part of this, oil companies are allowed to hold stocks in other EU countries subject to bilateral agreements between Governments, and count these stocks towards their stocking obligations. The stock figures in Table 3.7 take account of these stocks to give a true picture of the amount of stocks available to the UK.

3.75   In the course of 2007 and 2008 BERR will be phasing out the system based on final consumption and phasing in a new system where company obligations will be based on their refinery production and imports. We believe that this system is better placed to ensure that we continue meeting our EU and IEA obligations in the long term.

3.76   Stocks of petroleum products at the end of 2007 were 7 per cent lower than a year earlier. The total stocks of crude oil and products held by UK companies at the end of 2007 were equivalent to approximately 73 days of UK consumption.

Technical notes and definitions

3.77   These notes and definitions are in addition to the technical notes and definitions covering all fuels and energy as a whole in Chapter 1, paragraphs 1.26 to 1.58. For notes on the commodity balances and definitions of the terms used in the row headings see the Annex A, paragraphs A.7 to A.42. While the data in the printed and bound copy of this Digest cover only the most recent 7 years, these notes also cover data for earlier years that are available on the BERR web site.

Indigenous production

3.78   The term indigenous is used throughout this chapter and includes oil from the UK Continental Shelf both offshore and onshore.

Deliveries

3.79   These are deliveries into consumption, as opposed to being estimates of actual consumption or use. They are split between inland deliveries and deliveries to marine bunkers. Inland deliveries will not necessarily be consumed in the United Kingdom (e.g. aviation fuels).

Sources of data

3.80   The majority of the data included in the text and tables of this chapter are derived from BERR's Downstream Oil Reporting System (DORS), which replaced the UK Petroleum Industry Association (UKPIA) reporting system in 2005. Data relating to the inland operations of the UK oil industry (i.e. information on the supply, refining and distribution of oil in the UK) are collected from companies. The data format and coverage have been designed to meet most of the needs of both the Government and the industry itself. Each member of UKPIA and a number of other contributing companies provides returns on its refining activities and deliveries of various products to the internal UK market. This information is supplemented whenever necessary to allow for complete coverage within the statistics, with separate exercises carried out on special topics (for example, the work on super/hypermarkets referred to in paragraph 3.58) or with the use of additional data (such as trade data from HM Customs and Revenue to cover import activity by non-reporting companies).

Statistical differences

3.81   In Tables 3.1 to 3.5, there are headings titled "statistical differences". These are differences between the separately observed figures for production and delivery of crude oil and products during the path of their movement from the point of production to the point of consumption.

3.82   The statistical differences headings listed in the primary oil commodity balances (Tables 3.1) are differences between the separately observed and reported figures for production from onshore or offshore fields and supply to the UK market that cannot be accounted for by any specific factors. Primarily they result from inaccuracies in the meters at various points along offshore pipelines. These meters vary slightly in their accuracy within accepted tolerances, giving rise to both losses and gains when the volumes of oil flowing are measured. Errors may also occur when non-standard conditions are used to meter the oil flow.

3.83   Another technical factor that can contribute to the statistical differences relates to the recording of quantities at the producing field (which is the input for the production data) and at oil terminals and refineries, since they are in effect measuring different types of oil. Terminals and refineries are able to measure a standardised, stabilised crude oil, that is, with its water content and content of NGLs at a standard level and with the amounts being measured at standard conditions. However, at the producing field they are dealing with a "live" crude oil that can have a varying level of water and NGLs within it. While offshore companies report live crude at field, the disposals from oil terminals and offshore loading fields are reported as stabilised crude oil. This effectively assumes that terminal disposals are stabilised crude production figures. These changes were introduced in the 2002 edition of this Digest.

3.84 Part of the overall statistical difference may also be due to problems with the correct reporting of individual NGLs at the production site and at terminals and refineries. It is known that there is some mixing of condensate and other NGLs in with what might otherwise be stabilised crude oil before it enters the pipeline. This mixing occurs as it removes the need for separate pipeline systems for transporting the NGLs and it also allows the viscosity of the oil passing down the pipeline to be varied as necessary. While the quantity figures recorded by terminals are in terms of stabilised crude oil, with the NGL component removed, there may be situations where what is being reported does not comply with this requirement.

3.85 Refinery data are collated from details of individual shipments received and made by each refinery and terminal operating company. Each year there are thousands of such shipments, which may be reported separately by two or three different companies involved in the movement. While intensive work is carried out to check these returns, it is possible that some double counting of receipts may occur.

3.86 Temperature, pressure and natural leakage also contribute to the statistical differences. In addition, small discrepancies can occur between the estimated calorific values used at the field and the more accurate values measured at the onshore terminal where data are shown on an energy basis. The statistical differences can also be affected by rounding, clerical errors or unrecorded losses, such as leakage. Other contributory factors are inaccuracies in the reporting of the amounts being disposed of to the various activities listed, including differences between the quantities reported as going to refineries and the actual amounts passing through refineries.

3.87 Similarly, the data under the statistical difference headings in Tables 3.2 to 3.4 are the differences between the deliveries of petroleum products to the inland UK market reported by the supplying companies and estimates for such deliveries. These estimates are calculated by taking the output of products reported by refineries and then adjusting it by the relevant factors (such as imports and exports of the products, changes in the levels of stocks etc.).

3.88 It may be thought that such differences should not exist as the data underlying both the observed deliveries into the UK market and the individual components of the estimates (i.e. production, imports, exports, stocks) come from the same source (the oil companies). While it is true that each oil company provides data on its own activities in each area, there are separate areas of operation within the companies that report their own part of the overall data. Table 3F below illustrates this.

Table 3F Sources of data within oil companies

Area covered	Source
Refinery production	Refinery
Imports and exports	Refinery, logistics departments, oil traders
Stocks	Refinery, crude and product terminals, major storage and distribution sites
Final deliveries	Sales, marketing and accounts departments

3.89 Each individual reporting source will have direct knowledge of its own data. For example, refineries will know what they produce and how much leaves the refinery gate as part of routine monitoring of the refinery operations. Similarly other data such as sales to final consumers or imports and exports will be closely monitored. Companies will ensure that each component set of data reported is as accurate as possible but their reporting systems may not be integrated, meaning that internal consistency checks across all reported data cannot be made. Each part of a company may also work to different timings as well, which may further add to the degree of differences seen.

3.90 The main area where there is known to be a problem is with the "Transfers" heading in the commodity balances. The data reported under this heading have two components. Firstly, there is an allowance for reclassification of products within the refining process. For example, butane can be added to motor spirit to improve the octane rating, aviation turbine fuel could be reclassified as domestic kerosene if its quality deteriorates, and much of the fuel oil imported into the UK is further refined into other petroleum products. Issues can arise with product flows between different reporting companies, for example when company A delivers fuel oil to company B who report a receipt of a feedstock. Secondly, and in addition to these inter-product transfers, the data also include an allowance to cover the receipt of backflows of products from petrochemical plants that are often very closely integrated with refineries (for example, Ineos' refinery at Grangemouth is right next to the petrochemical plant). A deduction for these backflows thus needs to be included under the "Transfers" heading so that calculated estimates reflect net output and are thus more comparable with the basis of the observed deliveries data.

3.91   However, there is scope for error in the recording of these two components. With inter-product transfers, the data are recorded within the refinery during the refining and blending processes where the usual units used to record the changes are volumes rather than masses. Different factors apply for each product when converting from a volume to mass basis, as shown by the conversion factors given in Annex A of this Digest. Thus, a balanced transfer in volume terms may not be equivalent when converted to a mass basis. This is thought to be the main source of error within the individual product balances.

3.92   With the backflows data, the scope for error results from the recording of observed deliveries data being derived from sales data on a "net" basis and will therefore exclude the element of backflows data as received at the refinery. For example, these could be seen simply as an input of fuel oils to be used as a feedstock, and thus recorded as an input without their precise nature being recorded – in effect a form of double-counting. This relationship between the petrochemical sector and refineries is thought to be one of the main sources of error in the overall oil commodity balances.

Imports and exports
3.93   The information given under the headings "imports" and "exports" in this chapter are the figures recorded by importers and exporters of oil. They can differ in some cases from the import and export figures provided by HM Revenue and Customs that are given in Annex G on the Internet. Such differences arise from timing differences between actual and declared movements but also result from the Customs figures including re-exports. These are products that may have originally entered the UK as imports from another country and been stored in the UK prior to being exported back out of the UK, as opposed to having been actually produced in the UK.

Marine bunkers
3.94   This covers deliveries to ocean going and coastal vessels under international bunker contracts. Other deliveries to fishing, coastal and inland vessels are excluded.

Crude and process oils
3.95   These are all feedstocks, other than distillation benzene, for refining at refinery plants. Gasoline feedstock is any process oil whether clean or dirty which is used as a refinery feedstock for the manufacture of gasoline or naphtha. Other refinery feedstock is any process oil used for the manufacture of any other petroleum products.

Refineries
3.96   Refineries distilling crude and process oils to obtain petroleum products. This excludes petrochemical plants, plants only engaged in re-distilling products to obtain better grades, crude oil stabilisation plants and gas separation plants.

Products used as fuel (energy use)
3.97   The following paragraphs define the product headings used in the text and tables of this chapter. The products are used for energy in some way, either directly as a fuel or as an input into electricity generation.

Refinery fuel - Petroleum products used as fuel at refineries.

Ethane – A naturally gaseous straight-chain hydrocarbon (C_2H_6) in natural gas and refinery gas streams. Primarily used, or intended to be used, as a chemical feedstock.

Propane - Hydrocarbon containing three carbon atoms (C_3H_8), gaseous at normal temperature but generally stored and transported under pressure as a liquid. Used mainly for industrial purposes, but also as transport LPG, and some domestic heating and cooking.

Butane - Hydrocarbon containing four carbon atoms (C_4H_{10}), otherwise as for propane. Additionally used as a constituent of motor spirit to increase vapour pressure and as a chemical feedstock.

Naphtha (Light distillate feedstock) - Petroleum distillate boiling predominantly below 200°C.

Aviation spirit - All light hydrocarbon oils intended for use in aviation piston-engine power units, including bench testing of aircraft engines.

Motor spirit - Blended light petroleum components used as fuel for spark-ignition internal-combustion engines other than aircraft engines:

(i)   Premium unleaded grade - all finished motor spirit, with an octane number (research method) not less than 95.

(ii) Lead Replacement petrol / Super premium unleaded grade - finished motor spirit, with an octane number (research method) not less than 97.

Aviation turbine fuel (ATF) - All other turbine fuel intended for use in aviation gas-turbine power units and including bench testing of aircraft engines.

Burning oil (kerosene or "paraffin") - Refined petroleum fuel, intermediate in volatility between motor spirit and gas oil, used primarily for heating. White spirit and kerosene used for lubricant blends are excluded.

Gas/diesel oil - Petroleum fuel having a distillation range immediately between kerosene and light-lubricating oil:

(i)   **DERV (Diesel Engined Road Vehicle) fuel** - automotive diesel fuel for use in high speed, compression ignition engines in vehicles subject to Vehicle Excise Duty.

(ii) **Gas oil** - used as a burner fuel in heating installations, for industrial gas turbines and as for DERV (but in vehicles not subject to Vehicle Excise Duty e.g. Agriculture vehicles, fishing vessels, construction equipment).

(iii) **Marine diesel oil** - heavier type of gas oil suitable for heavy industrial and marine compression-ignition engines.

Fuel oil - Heavy petroleum residue blends used in atomising burners and for heavy-duty marine engines (marine bunkers, etc.) with heavier grades requiring pre-heating before combustion. Excludes fuel oil for grease making or lubricating oil and fuel oil sold as such for road making.

Products not used as fuel (non-energy use)

3.98 The following paragraphs define the product headings used in the text and tables of this chapter, which are used for non-energy purposes.

Feedstock for petroleum chemical plants - All petroleum products intended for use in the manufacture of petroleum chemicals. This includes middle distillate feedstock of which there are several grades depending on viscosity. The boiling point ranges between 200°C and 400°C. (A deduction has been made from these figures equal to the quantity of feedstock used in making the conventional petroleum products that are produced during the processing of the feedstock. The output and deliveries of these conventional petroleum products are included elsewhere as appropriate.)

White spirit and specific boiling point (SBP) spirits – These are refined distillate intermediates with a distillation in the naphtha / kerosene range. **White spirit** has a boiling range of about 150°C to 200°C and is used as a paint or commercial solvent. **SBP spirit** is also known as **Industrial spirit** and has a wider boiling range that varies up to 200°C dependent upon its eventual use. It has a variety of uses that vary from use in seed extraction, rubber solvents and perfume.

Lubricating oils (and grease) - Refined heavy distillates obtained from the vacuum distillation of petroleum residues. Includes liquid and solid hydrocarbons sold by the lubricating oil trade, either alone or blended with fixed oils, metallic soaps and other organic and/or inorganic bodies. A certain percentage of inland deliveries are re-used as a fuel (see paragraphs 3.62 to 3.69).

Bitumen - The residue left after the production of lubricating oil distillates and vacuum gas oil for upgrading plant feedstock. Used mainly for road making and building construction purposes. Includes other petroleum products such as creosote and tar mixed with bitumen for these purposes and fuel oil sold specifically for road making.

Petroleum wax - Includes paraffin wax, which is a white crystalline hydrocarbon material of low oil content normally obtained during the refining of lubricating oil distillate, paraffin scale, slack wax, microcrystalline wax and wax emulsions. Used for candle manufacture, polishes, food containers, wrappings etc.

Petroleum cokes - Carbonaceous material derived from hydrocarbon oils, uses for which include metallurgical electrode manufacture. Quantities of imports of this product are used as a fuel, primarily in the manufacture of cement (see paragraphs 3.62 to 3.69).

Miscellaneous products - Includes aromatic extracts, defoament solvents and other minor miscellaneous products.

Main classes of consumer

3.99 The following are definitions of the main groupings of users of petroleum products used in the text and tables of this chapter.

Electricity generators - Petroleum products delivered for use by major power producers and other companies for electricity generation including those deliveries to the other industries listed below which are used for autogeneration of electricity (Tables 3.2 to 3.4). This includes petroleum products used to generate electricity at oil refineries and is recorded in the Transformation sector, as opposed to other uses of refinery fuels that are recorded in the Energy Industry Use sector. These numbers may not necessarily be the same as those reported in the **Electricity** chapter (Chapter 5), which gives **consumption** of petroleum products by electricity generators. Differences occur because delivered fuel may be put to stock and not used immediately.

Agriculture - Deliveries of fuel oil and gas oil/diesel for use in agricultural power units, dryers and heaters. Burning oil for farm use

Iron and steel - Deliveries of petroleum products to steel works and iron foundries. This is now based on information from the Iron and Steel Statistics Bureau.

Other industries - The industries covered correspond to the industrial groups shown in Table 1E excluding Iron and Steel of Chapter 1.

National navigation - Fuel oil and gas/diesel oil delivered, other than under international bunker contracts, for fishing vessels, UK oil and gas exploration and production, coastal and inland shipping and for use in ports and harbours.

Railways - Deliveries of fuel oil, gas/diesel oil and burning oil to railways now based on estimates produced by AEA as part of their work to compile the UK Greenhouse Gas Inventory.

Air transport - Total inland deliveries of aviation turbine fuel and aviation spirit. The figures cover deliveries of aviation fuels in the United Kingdom to international and other airlines, British and foreign Governments (including armed services) and for private flying. In order to compile the UK Greenhouse Gas Inventory, AEA need to estimate how aviation fuel usage splits between domestic and international consumption. Information from AEA suggests that virtually all aviation spirit is used domestically while just 6 per cent of civilian aviation turbine fuel use is for domestic consumption.

Road transport - Deliveries of motor spirit and DERV fuel for use in road vehicles of all kinds. Again as part of their work to compile the UK emissions inventory, AEA has constructed estimates for the consumption of road transport fuels by different vehicle classes, and these are shown in Table 3G The table shows the increasing share of DERV used by cars and light goods vehicles (vans).

Table 3G Estimated consumption of road transport fuels by vehicle class

	1990	1995	2000	2006
Motor spirit:				
Cars and taxis	90%	92%	95%	97%
Light goods vehicles	9%	7%	4%r	2%
Motor cycles etc	1%	1%	1%	1%
DERV:				
Cars and taxis	6%	14%	19%	23%
Light goods vehicles	12%r	18%r	22%r	29%
Heavy goods vehicles	69%r	56%r	52%r	41%
Buses and coaches	13%r	11%r	8%r	8%

Source: AEA

As part of the 2003 Energy White Paper remit to provide more regional data, BERR commissioned AEA to provide estimates for regional and local use of road transport fuels. This work was first published in the June 2005 edition of Energy Trends with figures for 2002 and 2003 and has been updated in the June 2007 edition of Energy Trends to include figures for 2005: www.berr.gov.uk/energy/statistics/index.html.

Domestic - Fuel oil and gas oil delivered for central heating of private houses and other dwellings and deliveries of kerosene (burning oil) and liquefied petroleum gases for domestic purposes (see Tables 3.2 to 3.4).

Public services - Deliveries to national and local Government premises (including educational, medical and welfare establishments and British and foreign armed forces) of fuel oil and gas oil for central heating and of kerosene (burning oil).

Miscellaneous - Deliveries of fuel oil and gas oil for central heating in premises other than those classified as domestic or public.

Monthly and quarterly data

3.100 Monthly or quarterly aggregate data for certain series presented in this chapter are available. This information can be obtained free of charge by following the links given in the Energy Statistics section of the BERR web site, at: www.berr.gov.uk/energy/statistics/index.html.

Contact: *Martin Young*
Energy Markets Unit
martin.young@berr.gsi.gov.uk
020 7215 5184

Clive Evans
Energy Markets Unit
clive.evans@berr.gsi.gov.uk
020 7215 5189

Lisa Vine
Energy Markets Unit
lisa.vine@berr.gsi.gov.uk
020 7215 6072

3.1 Commodity balances 2005 - 2007[1]

Primary oil

Thousand tonnes

	Crude oil	Ethane	Propane	Butane	Condensate	Total NGL	Feedstock	Total primary oil
2005								
Supply								
Production	77,179	1,414	2,181	1,648	2,300	7,543	-	84,721
Imports	52,210	-	-	-	-	-	6,675	58,885
Exports	-48,879	-14	-1,204	-760	-1,249	-3,227	-1,992	-54,099
Stock change *(2)*	-277	..	..	..	..	+73	-180	-385
Transfers	-	-1,398	-857	-500	-632	-3,387	+332	-3,054
Total supply	**80,233**	..	..	..	..	**1,001**	**4,835**	**86,069**
Statistical difference *(3)(4)*	+12	..	..	..	..	+8	-85	-65
Total demand *(4)*	**80,221**	..	..	..	..	**993**	**4,920**	**86,134**
Transformation (Petroleum refineries) *(4)*	80,221	..	..	..	..	993	4,920	86,134
Energy industry use	-	-	-	-	-	-	-	-
2006								
Supply								
Production	69,665	1,281	1,947	1,542	2,143	6,913	-	76,578
Imports	51,446	-	-	-	-	-	7,997	59,443
Exports	-44,923	-17	-891	-488	-1,232	-2,628	-2,643	-50,195
Stock change *(2)*	-354	..	..	..	..	-79	+78	-355
Transfers	-	-1,264	-848	-484	-427	-3,024	+683	-2,341
Total supply	**75,834**	..	..	..	..	**1,182**	**6,115**	**83,130**
Statistical difference *(3)(4)*	-10	..	..	..	..	+12	-85	-83
Total demand *(4)*	**75,844**	..	..	..	..	**1,169**	**6,200**	**83,213**
Transformation (Petroleum refineries) *(4)*	75,844	..	..	..	..	1,169	6,200	83,213
Energy industry use	-	-	-	-	-	-	-	-
2007								
Supply								
Production	70,357	1,215	1,879	1,462	1,919	6,475	-	76,832
Imports	49,893	-	-	-	-	-	7,206	57,100
Exports	-45,129	-13	-836	-548	-1,186	-2,584	-3,287	-50,999
Stock change *(2)*	+650	..	..	..	..	9	+125	+784
Transfers	-	-1,203	-861	-558	-603	-3,225	+547	-2,678
Total supply	**75,772**	..	..	..	..	**675**	**4,591**	**81,038**
Statistical difference *(3)(4)*	+66	..	..	..	..	-6	-139	-79
Total demand *(4)*	**75,707**	..	..	..	..	**681**	**4,730**	**81,117**
Transformation (Petroleum refineries) *(4)*	75,707	..	..	..	..	681	4,730	81,117
Energy industry use	-	-	-	-	-	-	-	-

(1) As there is no use made of primary oils and feedstocks by industries other than the oil and gas extraction and petroleum refining industries, other industry headings have not been included in this table. As such, this table is a summary of the activity of what is known as the Upstream oil industry.

(2) Stock fall (+), stock rise (-).

(3) Total supply minus total demand.

(4) Figures for total demand for the individual NGLs (and thus for the statistical differences as well) are not availble.

Note:

Differences between the upstream and downstream balance are currently being investigated.

3.2 Commodity balances 2007

Petroleum products

	Ethane	Propane	Butane	Other gases	Naphtha	Aviation spirit	Motor spirit	White Spirit & SBP	Aviation turbine fuel
Supply									
Production	-	1,697	601	2,526	2,561	0	21,313	70	6,176
Other sources	1,203	861	558	-	603	-	-	-	-
Imports	-	267	473	8	736	21	3,265	107	7,608
Exports	-0	-706	-578	-	-2,978	-4	-7,334	-7	-1,162
Marine bunkers	-	-	-	-	-	-	-	-	-
Stock change (2)	2	11	33	0	69	5	106	2	182
Transfers	-	-0	-40	8	14	8	59	-1	-338
Total supply	1,204	2,131	1,046	2,542	1,005	30	17,410	171	12,467
Statistical difference (3)	8	10	110	-86	-2	-3	-181	4	-167
Total demand	1,197	2,121	937	2,627	1,006	33	17,591	167	12,633
Transformation	-	-	-	251	-	-	-	-	-
Electricity generation	-	-	-	251	-	-	-	-	-
Major power producers	-	-	-	-	-	-	-	-	-
Autogenerators	-	-	-	251	-	-	-	-	-
Heat generation	-	-	-	-	-	-	-	-	-
Petroleum refineries	-	-	-	-	-	-	-	-	-
Coke manufacture	-	-	-	-	-	-	-	-	-
Blast furnaces	-	-	-	-	-	-	-	-	-
Patent fuel manufacture	-	-	-	-	-	-	-	-	-
Other	-	-	-	-	-	-	-	-	-
Energy industry use	-	39	-	2,009	-	-	-	-	-
Electricity generation	-	-	-	-	-	-	-	-	-
Oil & gas extraction	-	-	-	-	-	-	-	-	-
Petroleum refineries	-	39	-	2,009	-	-	-	-	-
Coal extraction	-	-	-	-	-	-	-	-	-
Coke manufacture	-	-	-	-	-	-	-	-	-
Blast furnaces	-	-	-	-	-	-	-	-	-
Patent fuel manufacture	-	-	-	-	-	-	-	-	-
Pumped storage	-	-	-	-	-	-	-	-	-
Other	-	-	-	-	-	-	-	-	-
Losses	-	-	-	-	-	-	-	-	-
Final consumption	1,197	2,082	937	367	1,006	33	17,591	167	12,633
Industry	49	664	194	-	-	-	-	-	-
Unclassified	49	664	194	-	-	-	-	-	-
Iron & steel	-	-	-	-	-	-	-	-	-
Non-ferrous metals	-	-	-	-	-	-	-	-	-
Mineral products	-	-	-	-	-	-	-	-	-
Chemicals	-	-	-	-	-	-	-	-	-
Mechanical engineering, etc	-	-	-	-	-	-	-	-	-
Electrical engineering, etc	-	-	-	-	-	-	-	-	-
Vehicles	-	-	-	-	-	-	-	-	-
Food, beverages, etc	-	-	-	-	-	-	-	-	-
Textiles, leather, etc	-	-	-	-	-	-	-	-	-
Paper, printing etc	-	-	-	-	-	-	-	-	-
Other industries	-	-	-	-	-	-	-	-	-
Construction	-	-	-	-	-	-	-	-	-
Transport	-	119	-	-	-	33	17,591	-	12,633
Air	-	-	-	-	-	33	-	-	12,633
Rail	-	-	-	-	-	-	-	-	-
Road	-	119	-	-	-	-	17,591	-	-
National navigation	-	-	-	-	-	-	-	-	-
Pipelines	-	-	-	-	-	-	-	-	-
Other	-	323	26	-	-	-	-	-	-
Domestic	-	225	26	-	-	-	-	-	-
Public administration	-	-	-	-	-	-	-	-	-
Commercial	-	-	-	-	-	-	-	-	-
Agriculture	-	98	-	-	-	-	-	-	-
Miscellaneous	-	-	-	-	-	-	-	-	-
Non energy use (4)	1,148	977	717	367	1,006	-	-	167	-

(1) Includes marine diesel oil.

(2) Stock fall (+), stock rise (-).

(3) Total supply minus total demand.

(4) For further details on non-energy usage see paragraphs 3.62 to 3.69.

3.2 Commodity balances 2007 (continued)
Petroleum products

Thousand tonnes

Burning oil	Gas/ Diesel Oil [1]	Fuel oils	Lubri -cants	Bitu -men	Petroleum wax	Petroleum coke	Misc. products	Total Products	
									Supply
2,968	26,397	11,809	547	1,628	12	1,810	1,095	81,209	Production
-	-	-	-	-	-	-	-	3,225	Other sources
551	8,250	1,131	359	386	37	485	162	23,846	Imports
-339	-6,533	-7,739	-194	-524	-21	-613	-759	-29,490	Exports
-	-901	-1,471	-	-	-	-	-	-2,371	Marine bunkers
33	457	137	-47	26	19	-4	-42	990	Stock change (2)
363	-241	-419	33	9	16	-0	-19	-547	Transfers
3,577	27,429	3,448	699	1,525	62	1,678	437	76,861	**Total supply**
-53	274	121	27	-47	23	-0	-114	-77	**Statistical difference (3)**
3,631	27,155	3,328	672	1,572	39	1,678	551	76,938	**Total demand**
-	30	593	-	-	-	-	-	874	**Transformation**
-	25	340	-	-	-	-	-	616	Electricity generation
-	10	192	-	-	-	-	-	202	Major power producers
-	15	148	-	-	-	-	-	414	Autogenerators
-	5	52	-	-	-	-	-	57	Heat generation
-	-	-	-	-	-	-	-	-	Petroleum refineries
-	-	-	-	-	-	-	-	-	Coke manufacture
-	-	201	-	-	-	-	-	201	Blast furnaces
-	-	-	-	-	-	-	-	-	Patent fuel manufacture
-	-	-	-	-	-	-	-	-	Other
1	5	1,119	-	-	-	1,134	-	4,307	**Energy industry use**
-	-	-	-	-	-	-	-	-	Electricity generation
-	-	-	-	-	-	-	-	-	Oil & gas extraction
1	5	1,119	-	-	-	1,134	-	4,307	Petroleum refineries
-	-	-	-	-	-	-	-	-	Coal extraction
-	-	-	-	-	-	-	-	-	Coke manufacture
-	-	-	-	-	-	-	-	-	Blast furnaces
-	-	-	-	-	-	-	-	-	Patent fuel manufacture
-	-	-	-	-	-	-	-	-	Pumped storage
-	-	-	-	-	-	-	-	-	Other
-	-	-	-	-	-	-	-	-	**Losses**
3,630	27,120	1,616	672	1,572	39	544	551	71,757	**Final Consumption**
1,424	2,957	931	-	-	-	-	-	6,218	**Industry**
1,424	-	-	-	-	-	-	-	2,330	Unclassified
-	-	19	-	-	-	-	-	19	Iron & steel
-	20	25	-	-	-	-	-	46	Non-ferrous metals
-	180	1	-	-	-	-	-	181	Mineral products
-	104	76	-	-	-	-	-	180	Chemicals
-	81	18	-	-	-	-	-	99	Mechanical engineering etc
-	25	8	-	-	-	-	-	33	Electrical engineering etc
-	91	23	-	-	-	-	-	114	Vehicles
-	208	40	-	-	-	-	-	248	Food, beverages etc
-	100	10	-	-	-	-	-	110	Textiles, leather, etc
-	29	32	-	-	-	-	-	61	Paper, printing etc
-	1,984	657	-	-	-	-	-	2,641	Other industries
-	135	21	-	-	-	-	-	156	Construction
12	22,613	569	-	-	-	-	-	53,571	**Transport**
-	-	-	-	-	-	-	-	12,666	Air
12	632	-	-	-	-	-	-	644	Rail
-	21,039	-	-	-	-	-	-	38,750	Road
-	942	569	-	-	-	-	-	1,511	National navigation
-	-	-	-	-	-	-	-	-	Pipelines
2,194	1,310	116	-	-	-	-	-	3,969	**Other**
2,170	173	-	-	-	-	-	-	2,594	Domestic
12	393	45	-	-	-	-	-	450	Public administration
-	323	55	-	-	-	-	-	378	Commercial
12	143	10	-	-	-	-	-	262	Agriculture
-	278	7	-	-	-	-	-	286	Miscellaneous
-	239	-	672	1,572	39	544	551	7,998	**Non energy use (4)**

3.3 Commodity balances 2006
Petroleum products

	Ethane	Propane	Butane	Other gases	Naphtha	Aviation spirit	Motor spirit	White Spirit & SBP	Aviation turbine fuel
Supply									
Production	0	1,737r	406	2,862	2,734	26	21,443	107	6,261
Other sources	1,264	848	484	-	427	-	-	-	-
Imports	12	275	545	0	2,003	16	3,790	82	7,983
Exports	-13	-683	-463	-	-2,925	-3	-6,997	-2	-995
Marine bunkers	-	-	-	-	-	-	-	-	-
Stock change (2)	-2	-1	-39	+0	-43	-6	-29	-27	-256
Transfers	+0	+0	-26	+0	+67	+15	+15	+0	-404
Total supply	1,262	2,176r	906	2,863	2,264	47	18,223	159	12,589
Statistical difference (3)	+5	-39	-44	-2	-14	+2	+79	+3	-52
Total demand	1,257	2,215r	950	2,865	2,278	46	18,144	156	12,641
Transformation	-	-	-	206r	-	-	-	-	-
Electricity generation	-	-	-	206r	-	-	-	-	-
Major power producers	-	-	-	-	-	-	-	-	-
Autogenerators	-	-	-	206r	-	-	-	-	-
Heat generation	-	-	-	-	-	-	-	-	-
Petroleum refineries	-	-	-	-	-	-	-	-	-
Coke manufacture	-	-	-	-	-	-	-	-	-
Blast furnaces	-	-	-	-	-	-	-	-	-
Patent fuel manufacture	-	-	-	-	-	-	-	-	-
Other	-	-	-	-	-	-	-	-	-
Energy industry use	-	38r	-	2,201r	-	-	-	-	-
Electricity generation	-	-	-	-	-	-	-	-	-
Oil & gas extraction	-	-	-	-	-	-	-	-	-
Petroleum refineries	-	38r	-	2,201r	-	-	-	-	-
Coal extraction	-	-	-	-	-	-	-	-	-
Coke manufacture	-	-	-	-	-	-	-	-	-
Blast furnaces	-	-	-	-	-	-	-	-	-
Patent fuel manufacture	-	-	-	-	-	-	-	-	-
Pumped storage	-	-	-	-	-	-	-	-	-
Other	-	-	-	-	-	-	-	-	-
Losses	-	-	-	-	-	-	-	-	-
Final consumption	1,257	2,177r	950	457	2,278	46	18,144	156	12,641
Industry	66	671r	179	-	-	-	-	-	-
Unclassified	66	671r	179	-	-	-	-	-	-
Iron & steel	-	-	-	-	-	-	-	-	-
Non-ferrous metals	-	-	-	-	-	-	-	-	-
Mineral products	-	-	-	-	-	-	-	-	-
Chemicals	-	-	-	-	-	-	-	-	-
Mechanical engineering, etc	-	-	-	-	-	-	-	-	-
Electrical engineering, etc	-	-	-	-	-	-	-	-	-
Vehicles	-	-	-	-	-	-	-	-	-
Food, beverages, etc	-	-	-	-	-	-	-	-	-
Textiles, leather, etc	-	-	-	-	-	-	-	-	-
Paper, printing etc	-	-	-	-	-	-	-	-	-
Other industries	-	-	-	-	-	-	-	-	-
Construction	-	-	-	-	-	-	-	-	-
Transport	-	126	-	-	-	46	18,144	-	12,641
Air	-	-	-	-	-	46	-	-	12,641
Rail	-	-	-	-	-	-	-	-	-
Road	-	126	-	-	-	-	18,144	-	-
National navigation	-	-	-	-	-	-	-	-	-
Pipelines	-	-	-	-	-	-	-	-	-
Other	-	386	34	-	-	-	-	-	-
Domestic	-	281	34	-	-	-	-	-	-
Public administration	-	-	-	-	-	-	-	-	-
Commercial	-	-	-	-	-	-	-	-	-
Agriculture	-	105	-	-	-	-	-	-	-
Miscellaneous	-	-	-	-	-	-	-	-	-
Non energy use (4)	1,191	994	737	457	2,278	-	-	156	-

(1) Includes marine diesel oil.
(2) Stock fall (+), stock rise (-).
(3) Total supply minus total demand.
(4) For further details on non-energy usage see paragraphs 3.62 to 3.69.

3.3 Commodity balances 2006 (continued)
Petroleum products

Thousand tonnes

Burning oil	Gas/ Diesel Oil [1]	Fuel oils	Lubri -cants	Bitu -men	Petroleum wax	Petroleum coke	Misc. products	Total Products	
									Supply
3,374	26,080	12,277	617	1,749	16	1,964	1,189	82,839r	Production
-	-	-	-	-	-	-	-	3,024	Other sources
670	8,063	1,332	505	404	77	869	200	26,828	Imports
-314	-5,819	-8,368	-401	-628	-39	-559	-801	-29,009	Exports
-	-1,035	-1,313	-	-	-	-	-	-2,348	Marine bunkers
-107	-283	-140	+18	+11	+4	-15	+73	-840	Stock change (2)
+403	-205	-573	+1	+22	+0	+0	+3	-682	Transfers
4,027	26,800	3,216	740	1,558	57	2,260	664	79,812r	**Total supply**
+10	+49	+68	+27	-52	+9	-23	+36	+63r	**Statistical difference (3)**
4,017	26,751	3,148	713	1,610	48	2,283	628	79,749r	**Total demand**
-	89	688	-	-	-	-	-	984r	**Transformation**
-	83	405	-	-	-	-	-	695r	Electricity generation
-	61	202	-	-	-	-	-	263	Major power producers
-	22	204	-	-	-	-	-	432r	Autogenerators
-	6	53	-	-	-	-	-	59	Heat generation
-	-	-	-	-	-	-	-	-	Petroleum refineries
-	-	-	-	-	-	-	-	-	Coke manufacture
-	-	230	-	-	-	-	-	230	Blast furnaces
-	-	-	-	-	-	-	-	-	Patent fuel manufacture
-	-	-	-	-	-	-	-	-	Other
1	44	997	-	1	-	1,358	-	4,639	**Energy industry use**
-	-	-	-	-	-	-	-	-	Electricity generation
-	-	-	-	-	-	-	-	-	Oil & gas extraction
1	44	997	-	1	-	1,358	-	4,639	Petroleum refineries
-	-	-	-	-	-	-	-	-	Coal extraction
-	-	-	-	-	-	-	-	-	Coke manufacture
-	-	-	-	-	-	-	-	-	Blast furnaces
-	-	-	-	-	-	-	-	-	Patent fuel manufacture
-	-	-	-	-	-	-	-	-	Pumped storage
-	-	-	-	-	-	-	-	-	Other
-	-	-	-	-	-	-	-	-	**Losses**
4,016	26,618	1,463	713	1,610	48	925	628	74,126r	**Final Consumption**
1,540	3,047	844	-	-	-	-	-	6,348r	**Industry**
1,540	-	-	-	-	-	-	-	2,456r	Unclassified
-	-	19	-	-	-	-	-	19	Iron & steel
-	22	28	-	-	-	-	-	50	Non-ferrous metals
-	183	1	-	-	-	-	-	184	Mineral products
-	105	72	-	-	-	-	-	176	Chemicals
-	81	18	-	-	-	-	-	98	Mechanical engineering etc
-	70	9	-	-	-	-	-	78	Electrical engineering etc
-	92	22	-	-	-	-	-	115	Vehicles
-	219	42	-	-	-	-	-	261	Food, beverages etc
-	110	11	-	-	-	-	-	121	Textiles, leather, etc
-	23	33	-	-	-	-	-	56	Paper, printing etc
-	2,002	571	-	-	-	-	-	2,573	Other industries
-	141	19	-	-	-	-	-	161	Construction
12	21,985	504	-	-	-	-	-	53,457	**Transport**
-	-	-	-	-	-	-	-	12,686	Air
12	654	-	-	-	-	-	-	666	Rail
-	20,146	-	-	-	-	-	-	38,416	Road
-	1,185	504	-	-	-	-	-	1,689	National navigation
-	-	-	-	-	-	-	-	-	Pipelines
2,464	1,327	114	-	-	-	-	-	4,326	**Other**
2,440	171	-	-	-	-	-	-	2,927	Domestic
12	393	46	-	-	-	-	-	451	Public administration
-	314	50	-	-	-	-	-	364	Commercial
12	145	10	-	-	-	-	-	272	Agriculture
-	304	9	-	-	-	-	-	312	Miscellaneous
-	258	-	713	1,610	48	925	628	9,995	**Non energy use (4)**

3.4 Commodity balances 2005
Petroleum products

Thousand tonnes

	Ethane	Propane	Butane	Other gases	Naphtha	Aviation spirit	Motor spirit	White Spirit & SBP	Aviation turbine fuel
Supply									
Production	5	1,704r	518	2,996	3,023	32	22,620	136	5,167
Other sources	1,397	857	500	-	632	-	-	-	-
Imports	-	281	502	137	1,380	13	2,376	224	9,083
Exports	-	-748	-550	-	-3,167	-3	-6,586	-63	-1,397
Marine bunkers	-	-	-	-	-	-	-	-	-
Stock change (2)	+0	+8	+13	+1	+63	-2	+366	-15	+96
Transfers	+0	-5	+2	-4	+32	+14	-4	+3	-343
Total supply	1,402	2,097r	986	3,130	1,964	53	18,772	285	12,606
Statistical difference (3)	-57	-184	-85	-6	+45	+1	+40	+0	+109
Total demand	1,459	2,281r	1,071	3,136	1,919	52	18,732	284	12,497
Transformation	-	-	-	182	-	-	-	-	-
Electricity generation	-	-	-	182	-	-	-	-	-
Major power producers	-	-	-	-	-	-	-	-	-
Autogenerators	-	-	-	182	-	-	-	-	-
Heat generation	-	-	-	-	-	-	-	-	-
Petroleum refineries	-	-	-	-	-	-	-	-	-
Coke manufacture	-	-	-	-	-	-	-	-	-
Blast furnaces	-	-	-	-	-	-	-	-	-
Patent fuel manufacture	-	-	-	-	-	-	-	-	-
Other	-	-	-	-	-	-	-	-	-
Energy industry use	5	38	-	2,569	3	-	-	-	-
Electricity generation	-	-	-	-	-	-	-	-	-
Oil & gas extraction	-	-	-	-	-	-	-	-	-
Petroleum refineries	5	38	-	2,569	3	-	-	-	-
Coal extraction	-	-	-	-	-	-	-	-	-
Coke manufacture	-	-	-	-	-	-	-	-	-
Blast furnaces	-	-	-	-	-	-	-	-	-
Patent fuel manufacture	-	-	-	-	-	-	-	-	-
Pumped storage	-	-	-	-	-	-	-	-	-
Other	-	-	-	-	-	-	-	-	-
Losses	-	-	-	-	-	-	-	-	-
Final consumption	1,454	2,243r	1,071	384	1,916	52	18,732	284	12,497
Industry	71	631	161	-	-	-	-	-	-
Unclassified	71	631	161	-	-	-	-	-	-
Iron & steel	-	-	-	-	-	-	-	-	-
Non-ferrous metals	-	-	-	-	-	-	-	-	-
Mineral products	-	-	-	-	-	-	-	-	-
Chemicals	-	-	-	-	-	-	-	-	-
Mechanical engineering, etc	-	-	-	-	-	-	-	-	-
Electrical engineering, etc	-	-	-	-	-	-	-	-	-
Vehicles	-	-	-	-	-	-	-	-	-
Food, beverages, etc	-	-	-	-	-	-	-	-	-
Textiles, leather, etc	-	-	-	-	-	-	-	-	-
Paper, printing etc	-	-	-	-	-	-	-	-	-
Other industries	-	-	-	-	-	-	-	-	-
Construction	-	-	-	-	-	-	-	-	-
Transport	-	120	-	-	-	52	18,732	-	12,497
Air	-	-	-	-	-	52	-	-	12,497
Rail	-	-	-	-	-	-	-	-	-
Road	-	120	-	-	-	-	18,732	-	-
National navigation	-	-	-	-	-	-	-	-	-
Pipelines	-	-	-	-	-	-	-	-	-
Other	-	404	9	-	-	-	-	-	-
Domestic	-	289	9	-	-	-	-	-	-
Public administration	-	-	-	-	-	-	-	-	-
Commercial	-	-	-	-	-	-	-	-	-
Agriculture	-	115	-	-	-	-	-	-	-
Miscellaneous	-	-	-	-	-	-	-	-	-
Non energy use (4)	1,383	1,088r	901	384	1,916	-	-	284	-

(1) Includes marine diesel oil.
(2) Stock fall (+), stock rise (-).
(3) Total supply minus total demand.
(4) For further details on non-energy usage see paragraphs 3.62 to 3.69.

3.4 Commodity balances 2005 (continued)

Petroleum products

Thousand tonnes

Burning oil	Gas/ Diesel Oil [1]	Fuel oils	Lubri -cants	Bitu -men	Petroleum wax	Petroleum coke	Misc. products	Total Products	
									Supply
3,325	28,691	11,728	936	1,912	98	1,867	1,005	85,763r	Production
-	-	-	-	-	-	-	-	3,386	Other sources
407	4,920	1,530	424	216	28	947	42	22,512	Imports
-282	-6,314	-8,452	-709	-242	-33	-570	-606	-29,722	Exports
-	-889	-1,166	-	-	-	-	-	-2,055	Marine bunkers
+24	+284	+136	+73	+1	-20	-26	+45	+1046	Stock change (2)
+333	-262	-92	+0	+24	-8	+0	-22	-333	Transfers
3,807	26,430	3,683	725	1,911	65	2,217	464	80,598r	**Total supply**
-63	-5	+145	-25	+6	-6	-32	-20	-137r	**Statistical difference (3)**
3,870	26,435	3,538	750	1,906	72	2,249	484	80,735r	**Total demand**
-	72	723	-	-	-	-	-	978	**Transformation**
-	66	402	-	-	-	-	-	650	Electricity generation
-	37	215	-	-	-	-	-	252	Major power producers
-	29	187	-	-	-	-	-	398	Autogenerators
-	6	52	-	-	-	-	-	59	Heat generation
-	-	-	-	-	-	-	-	-	Petroleum refineries
-	-	-	-	-	-	-	-	-	Coke manufacture
-	-	269	-	-	-	-	-	269	Blast furnaces
-	-	-	-	-	-	-	-	-	Patent fuel manufacture
-	-	-	-	-	-	-	-	-	Other
1	206	1,573	-	-	-	1,207	-	5,601	**Energy industry use**
-	-	-	-	-	-	-	-	-	Electricity generation
-	-	-	-	-	-	-	-	-	Oil & gas extraction
1	206	1,573	-	-	-	1,207	-	5,601	Petroleum refineries
-	-	-	-	-	-	-	-	-	Coal extraction
-	-	-	-	-	-	-	-	-	Coke manufacture
-	-	-	-	-	-	-	-	-	Blast furnaces
-	-	-	-	-	-	-	-	-	Patent fuel manufacture
-	-	-	-	-	-	-	-	-	Pumped storage
-	-	-	-	-	-	-	-	-	Other
-	-	-	-	-	-	-	-		**Losses**
3,869	26,158	1,242	750	1,906	72	1,042	484	74,156r	**Final Consumption**
1,490	3,433	786	-	-	-	-	-	6,572	**Industry**
1,490	-	-	-	-	-	-	-	2,353	Unclassified
-	-	14	-	-	-	-	-	14	Iron & steel
-	28	22	-	-	-	-	-	50	Non-ferrous metals
-	197	1	-	-	-	-	-	198	Mineral products
-	109	72	-	-	-	-	-	181	Chemicals
-	90	19	-	-	-	-	-	109	Mechanical engineering etc
-	26	7	-	-	-	-	-	33	Electrical engineering etc
-	109	19	-	-	-	-	-	128	Vehicles
-	259	39	-	-	-	-	-	298	Food, beverages etc
-	93	9	-	-	-	-	-	101	Textiles, leather, etc
-	51	29	-	-	-	-	-	81	Paper, printing etc
-	2,312	539	-	-	-	-	-	2,852	Other industries
-	159	16	-	-	-	-	-	175	Construction
12	20,992	355	-	-	-	-	-	52,760	**Transport**
-	-	-	-	-	-	-	-	12,549	Air
12	636	-	-	-	-	-	-	648	Rail
-	19,436	-	-	-	-	-	-	38,288	Road
-	920	355	-	-	-	-	-	1,274	National navigation
-	-	-	-	-	-	-	-	-	Pipelines
2,368	1,503	101	-	-	-	-	-	4,385	**Other**
2,344	141	-	-	-	-	-	-	2,782	Domestic
12	443	43	-	-	-	-	-	498	Public administration
-	315	43	-	-	-	-	-	358	Commercial
12	192	5	-	-	-	-	-	323	Agriculture
-	413	10	-	-	-	-	-	423	Miscellaneous
-	229	-	750	1,906	72	1,042	484	10,439r	**Non energy use (4)**

3.5 Supply and disposal of petroleum[1]

Thousand tonnes

	2003	2004	2005	2006	2007
Primary oils (Crude oil, NGLs and feedstocks)					
Indigenous production (2)	106,073	95,374	84,721	76,578	76,832
Imports	54,177	62,517	58,885	59,443	57,100
Exports (3)	-74,898	-64,504	-54,099	-50,195	-50,999
Transfers - Transfers to products (4)	-2,661	-3,724	-3,387	-3,024	-3,225
Product rebrands (5)	+1,653	+181	+332	+683	547
Stock change (6)	+469	-133	-385	-355	784
Use during production (7)	-	-2	-1	-	-
Calculated refinery throughput (8)	84,814	89,710	86,069	83,130	81,038
Overall statistical difference (9)	+229	-110	-65	-83	-79
Actual refinery throughput	**84,585**	**89,821**	**86,134**	**83,213**	**81,117**
Petroleum products					
Losses in refining process (10)	56	-7	371r	374r	-92
Refinery gross production (11)	84,529	89,828	85,763r	82,839r	81,209
Transfers - Transfers to products (4)	+2,661	+3,724	+3,386	+3,024	3,225
Product rebrands (5)	-1,652	-203	-333	-682	-547
Imports	16,472	18,545	22,512	26,828	23,846
Exports (12)	-23,323	-30,495	-29,722	-29,009	-29,490
Marine bunkers	-1,764	-2,085	-2,055	-2,348	-2,371
Stock changes (6) - Refineries	-233	-232	+1,043	-890	984
Power generators	-29	-57	+3	+51	5
Calculated total supply	76,661	79,025	80,598r	79,812r	76,861
Statistical difference (9)	-492	-41	-137r	+63r	-77
Total demand (4)	**77,154**	**79,066**	**80,735r**	**79,749r**	**76,938**
Of which:					
Energy use	66,743	68,482	70,296	69,753r	68,940
Of which, for electricity generation (13)	536	593	650	695r	616
total refinery fuels (13)	5,456	5,417	5,601	4,639r	4,307
Non-energy use	10,411	10,584	10,439r	9,995	7,998

(1) Aggregate monthly data on oil production, trade, refinery throughput and inland deliveries are available - see paragraph 3.100 and Annex C.

(2) Crude oil plus condensates and petroleum gases derived at onshore treatment plants.

(3) Includes NGLs, process oils and re-exports.

(4) Disposals of NGLs by direct sale (excluding exports) or for blending.

(5) Product rebrands (inter-product blends or transfers) represent petroleum products received at refineries/ plants as process for refinery or cracking unit operations.

(6) Impact of stock changes on supplies. A stock fall is shown as (+) as it increases supplies, and vice-versa for a stock rise (-).

(7) Own use in onshore terminals and gas separation plants. These figures ceased to be available from January 2001 with the advent of the new PPRS system.

(8) Equivalent to the total supplies reported against the upstream transformation sector in Table 3.1.

(9) Supply greater than (+) or less than (-) recorded throughput or disposals.

(10) Calculated as the difference between actual refinery throughput and gross refinery production.

(11) Includes refinery fuels.

(12) Excludes NGLs.

(13) Figures cover petroleum used to generate electricity by all major power producers and by all other generators, including petroleum used to generate electricity at refineries. These quantities are also included in the totals reported as used as refinery fuel, so there is thus some overlap in these figures.

3.6 Additional information on inland deliveries of selected products[(1)(2)(3)]

Thousand tonnes

	2003	2004	2005	2006	2007
Motor spirit					
Retail deliveries (4)					
Hypermarkets (5)					
Lead Replacement Petrol/Super premium unleaded (6)	131	119	130	229	254
Premium unleaded	5,803	6,019	6,580	6,838	6,625
Total hypermarkets	5,935	6,138	6,710	7,067	6,879
Refiners/other traders					
Lead Replacement Petrol/Super premium unleaded (6)	912	765	818	509	560
Premium unleaded	12,488	11,776	10,374	9,866	9,364
Total Refiners/other traders	13,400	12,541	11,193	10,375	9,924
Total retail deliveries					
Lead Replacement Petrol/Super premium unleaded (6)	1,044	884	949	738	814
Premium unleaded	18,291	17,795	16,954	16,704	15,989
Total retail deliveries	19,335	18,679	17,903	17,442	16,803
Commercial consumers (7)					
Lead Replacement Petrol/Super premium unleaded (6)	41	40	16	65	22
Premium unleaded	542	765	812	637	766
Total commercial consumers	583	805	828	702	788
Total motor spirit	**19,918**	**19,484**	**18,731**	**18,144**	**17,591**
Unleaded as % of Total motor spirit	99.0	99.5	99.9	99.9	100
Gas oil/diesel oil					
DERV fuel:					
Retail deliveries (4):					
Hypermarkets (5)	2,135	2,474	3,091	3,917	4,360
Refiners/other traders	6,922	7,043	7,587	7,536	7,925
Total retail deliveries	9,057	9,517	10,678	11,453	12,285
Commercial consumers (7)	8,655	8,997	8,757	8,693	8,754
Total DERV fuel	17,712	18,514	19,435	20,146	21,039
Other gas oil (8)	6,325	6,030	6,794	6,567	6,110
Total gas oil/diesel oil	**24,037**	**24,544**	**26,229**	**26,713**	**27,150**
Fuel oils (9)					
Light	169	214	209r	287	348
Medium	582	961	918r	875r	863
Heavy	788	888	837r	989r	997
Total fuel oils	**1,539**	**2,062**	**1,965**	**2,151**	**2,209**

(1) Aggregate monthly data for inland deliveries of oil products are available - see paragraph 3.100 and Annex C.

(2) The end use section analyses are based partly on recorded figures and on estimates. They are intended for general guidance only. See also the notes in the main text of this chapter.

(3) For a full breakdown of the end-uses of all oil products, see Commodity Balances in Tables 3.2 to 3.4.

(4) Retail deliveries - deliveries to garages, etc. mainly for resale to final consumers.

(5) Data for sales by super and hypermarket companies are collected via a separate reporting system, but are consistent with the main data collected from UKPIA member companies - see paragraph 3.80.

(6) Sales of Leaded Petrol ceased on 31 December 1999. Separate breakdowns for lead replacement and super premium unleaded petrol are no longer provided, see Digest of UK Energy Statistics 2007 chapter 3 paragraph 3.49 for details.

(7) Commercial consumers - direct deliveries for use in consumer's business.

(8) Includes marine diesel oil.

(9) Inland deliveries excluding that used as a fuel in refineries, but including that used for electricity generation by major electricity producers and other industries.

Gas flow chart 2007 (TWh)

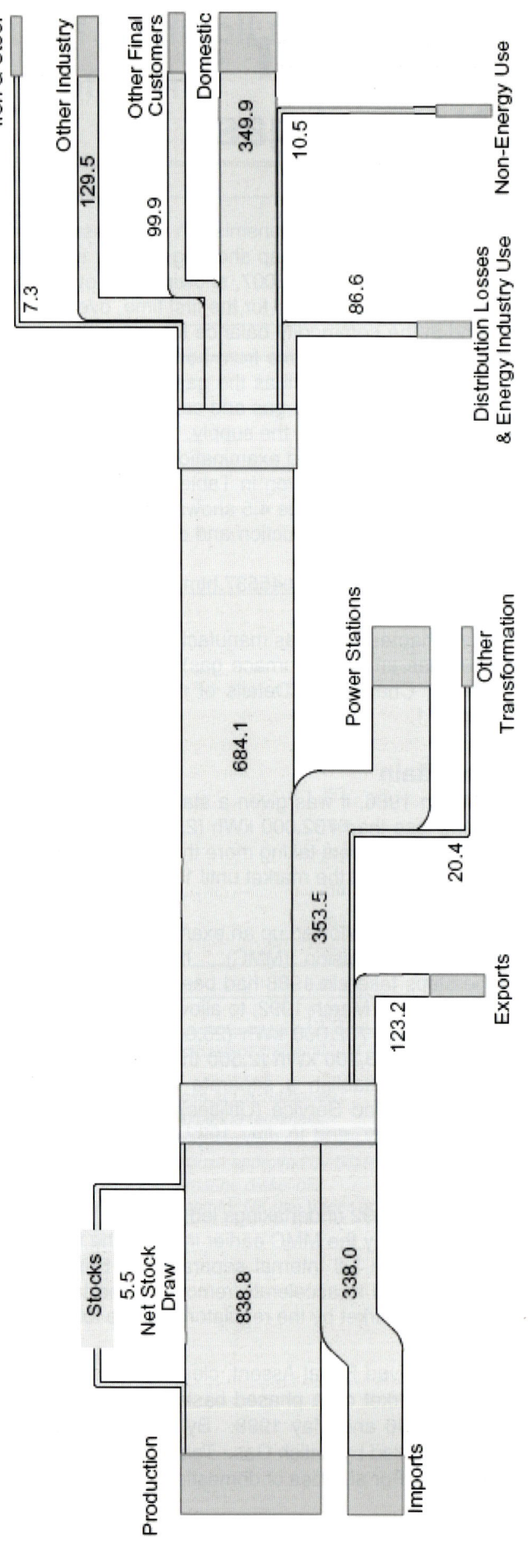

Iron & Steel 7.3

Other Industry 129.5

Other Final Customers 99.9

Domestic 349.9

Non-Energy Use 10.5

Distribution Losses & Energy Industry Use 86.6

Power Stations 353.5

Other Transformation 20.4

684.1

Stocks
5.5
Net Stock Draw

Production 838.8

Imports 338.0

Exports 123.2

Notes:
This flow chart is based on the data that appear in Table 4.1.

in Northern England and Wales that new suppliers have had most success. Since the market has opened up, British Gas had lost around 44 per cent of the credit and 63 per cent of the direct debit market, compared to 42 per cent of the pre-payment market. Historically, British Gas's pre-payment prices have tended to be below the average of new suppliers; however, since 2004, this trend has reversed. At the end of May 2008, 38 suppliers were licensed to supply gas to domestic customers.

Table 4A: Domestic gas market penetration (in terms of percentage of customers supplied) by local distribution zone and payment type, fourth quarter of 2007

Region	British Gas Trading			Non-British Gas		
	Credit	Direct Debit	Prepayment	Credit	Direct Debit	Prepayment
Wales	45	33	37	55	67	63
Northern	44	28	39	56	72	61
Scotland	55	34	70	45	66	30
North West	56	38	71	44	62	29
East Midlands	53	35	60	47	65	40
Eastern	53	38	59	47	62	41
South East	57	38	54	43	62	46
Southern	52	36	54	48	64	46
North East	57	38	58	43	62	42
North Thames	63	47	62	37	53	38
South Western	59	40	51	41	60	49
West Midlands	61	40	65	39	60	35
Great Britain	56	37	58	44	63	42

4.7   For the non-domestic market, about three-quarters (by volume) in the United Kingdom was opened to competition at the end of 1982 and the remainder in August 1992 (with the reduction in the tariff threshold). However, no other suppliers entered the market until 1990. After 1990, there was a rapid increase in the number of independent companies supplying gas, although from 1999 there were signs of some consolidation and in recent years sales of gas have become more concentrated in the hands of the largest companies in the domestic, industrial and commercial sectors. This came about through larger companies absorbing smaller suppliers and through mergers between already significant suppliers. After an increase in competition in 2006, during a period of rising gas prices, 2007 saw competition decrease slightly. In 2007, the three largest suppliers in the domestic market accounted for 71 per cent of sales, no change on 2006. However, in the industrial sector, the share of the largest three suppliers increased from 54 per cent in 2006 to 62 per cent in 2007. For commercial sector sales, the three largest suppliers accounted for 57 per cent of sales in 2006, rising to 60 per cent in 2007.

4.8   Following the 1995 Act, the business of British Gas was fully separated into two corporate entities. The supply and shipping businesses were devolved to a subsidiary, British Gas Trading Limited, while the transportation business (Transco) remained within British Gas plc. In February 1997, Centrica plc was demerged from British Gas plc (which was itself renamed as BG plc) completing the division of the business into two independent entities. Centrica became the holding company for British Gas Trading, British Gas Services, the Retail Energy Centres and the company producing gas from the North and South Morecambe fields. BG plc comprised the gas transportation and storage business of Transco, along with British Gas's other exploration and production, international downstream, research and technology and property activities. In October 2000, BG plc demerged into two separately listed companies, of which Lattice Group plc was the holding company for Transco, while BG Group plc included the international and gas storage businesses. On 21 October 2002, Transco and the National Grid Company merged to form National Grid.

4.9   From 1 October 2001, under the Utilities Act, gas pipeline companies have been able to apply for their own national Gas Transporter Licences so that they can compete with Transco. In some areas, low pressure spur networks had already been developed by new transporters competing with Transco to bring gas supplies to new customers (mainly domestic). In addition, some very large loads (above 60 GWh) are serviced by pipelines operated independently, some by North Sea producers.

4.10   By the end of 1994, competitors had exceeded the target 45 per cent of the market above 73,200 kWh (2,500 therms), but virtually all of this was in the firm gas market. From 1995, British Gas's competitors made inroads into the interruptible market, and in 2005 Centrica's share of the industrial and commercial market fell to 13 per cent. However, by 2007, Centrica had increased its market share to 20 per cent. At the end of 2007, there were about 28 suppliers active in the UK gas market, although some companies own or part own more than one supplier. The structure of the gas industry in Great Britain, as it stood at the end of 2007, is shown in Chart 4.1.

Chart 4.1: Structure of the gas industry in Great Britain in 2007

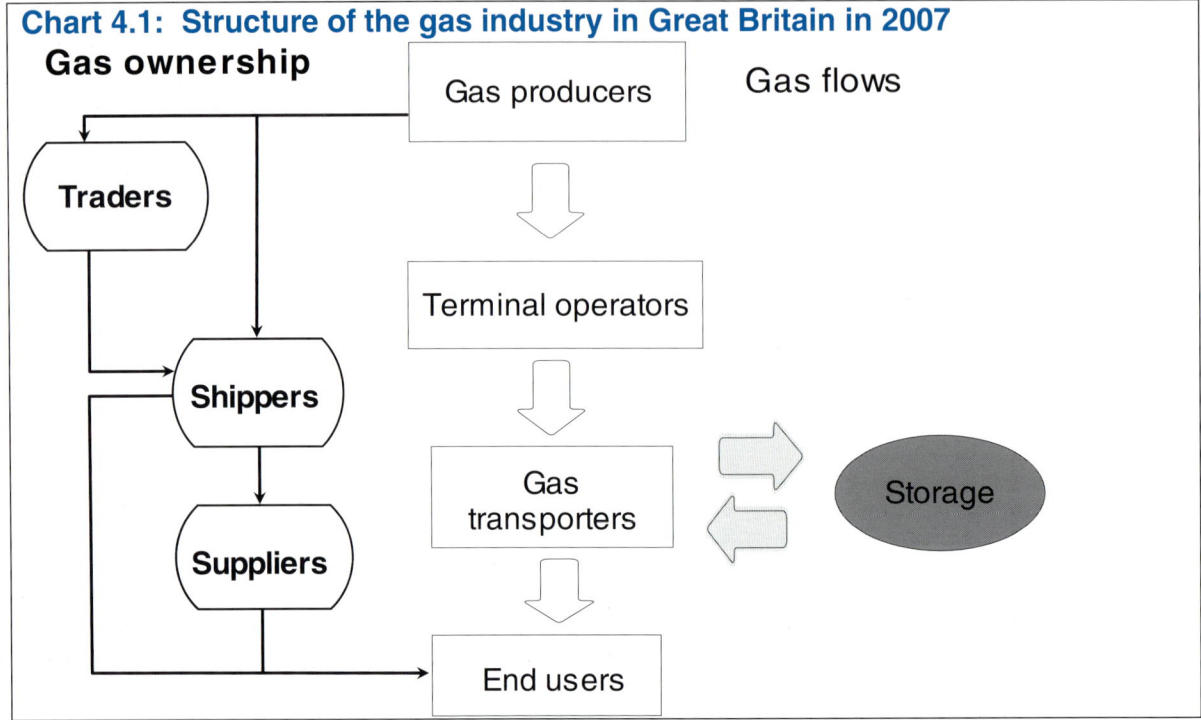

Regional analysis

4.11   Table 4B gives the number of consumers with a gas demand below 73,200 kWh per year in 2006 and the total number of gas consumers. It covers customers receiving gas from the national transmission system. The below 73,200 kWh category covers both domestic and small business customers, and it was this section of the market that was progressively opened up to competition between April 1996 and May 1998. Note that the data are for 2006, one year in arrears of the other data presented in this chapter.

Table 4B: Consumption by gas customers by region in 2006

Government Office Region	Consumption by customers below 73,200 kWh (2,500 therms) annual demand		Consumption by all customers	
	Number of consumers (thousands)	Gas sales 2006 (GWh)	Number of consumers (thousands)	Gas sales 2006 (GWh)
Wales	1,060	19,599	1,075	32,401
Scotland	1,785	34,342	1,817	58,811
North East	1,051	20,018	1,067	32,335
North West	2,764	51,567	2,811	82,209
Yorkshire and the Humber	2,021	37,954	2,056	64,852
East Midlands	1,649	30,484	1,676	47,976
West Midlands	2,013	36,676	2,047	57,848
East of England	1,923	34,679	1,955	54,445
Greater London	2,923	50,943	2,982	76,950
South East	3,008	55,121	3,065	78,790
South West	1,686	27,785	1,712	42,090
Great Britain	21,884	399,179	22,263	628,733

Source: xoserve and the independent gas transporters

4.12   In December 2007, BERR published in Energy Trends and on its regional energy web site (www.berr.gov.uk/energy/statistics/regional/index.html) gas consumption data at both regional and local level. The local level data are at "NUTS4" level (see article in December 2007 Energy Trends for definition) and the regional data at "NUTS1" level. Data for earlier years are presented on the web site but only 2006 data appear in the article. Domestic sector sales are shown separately from commercial and industrial sales, along with the numbers of consumers. BERR has now produced electricity and gas consumption estimates for 2006 at Middle Layer Super Output Area (MLSOA) level and, for Scotland, intermediate geography zones. MLSOAs are a statistical geography developed by the Office for National Statistics (ONS) as part of the 2001 census. There are 7,193 MLSOAs (plus the Isles of Scilly) which are areas containing a minimum population of 5,000 or around 2,000 households. In addition to this in Scotland there are 1,235 intermediate geography zones which are designed to contain between 2,500 and 6,000 people. Eleven Excel workbooks have been produced, relating to the nine English Government Office Regions, Scotland and Wales. Each worksheet on the eleven individual workbook files contains MLSOA (or intermediate geography for Scotland) electricity and gas consumption for each local authority within the relevant region. The information published includes total consumption, the total number of meters and average consumption for domestic and commercial/industrial consumers. Combined data for electricity and gas consumption are also displayed. Due to data disclosure issues, consumption relating to larger commercial/industrial consumers could not be disaggregated below local authority level. To assist data users interpret the estimates, a guidance note has also been released that provides further background information and details of potential issues with the data. Users can access this guide and all eleven workbooks along with all other regional datasets produced on BERR's regional energy web site.

Northern Ireland

4.13   Before 1997, Northern Ireland did not have a public natural gas supply. The construction of a natural gas pipeline from Portpatrick in Scotland to Northern Ireland was completed in 1996 and provided the means of establishing such a system. The initial market was Ballylumford power station, which was purchased by British Gas in 1992 and converted from oil to gas firing (with a heavy fuel oil back up). A second gas-fired power station was built at Coolkeeragh in 2005. The onshore line has been extended to serve wider industrial, commercial and domestic markets and this extension is continuing. In late 2007, the South-North gas pipeline was completed, to allow gas to be imported to Northern Ireland from the Republic of Ireland. In 2007, 83 per cent of all gas supplies in Northern Ireland were used to generate electricity.

Commodity balances for gas (Table 4.1)

4.14   In 2005, the UK was a net importer of gas, with imports of natural gas 77 TWh higher than exports. In 2006, these net imports increased to 123 TWh and accounted for 12 per cent of total natural gas supply, increasing further in 2007 to 215 TWh (20 per cent of total gas supply). Imports and exports of natural gas are described in greater detail below in paragraph 4.19.

4.15   Demand for natural gas is traditionally slightly less than supply because of the various measurement differences described in paragraphs 4.47 to 4.50. In 2007, demand was 1.5 TWh (just over 0.1 per cent) less than supply.

4.16   In 2007, 33 per cent of natural gas demand was for electricity generation (transformation sector), 3.8 percentage points more than in 2006. A further 7 per cent was consumed for heating purposes within the energy industries, while 1 per cent was accounted for by distribution losses within the gas network. (For an explanation of the items included under losses, see paragraphs 4.47 to 4.50). Of the remaining 59 per cent, 2 per cent was transformed into heat for sale to a third party, and 13 per cent was accounted for by the industrial sector, with the chemicals industry (excluding natural gas for petrochemical feedstocks), food, mineral products and paper making industries being the largest consumers. The chemicals sector accounted for over a quarter of the industrial consumption of natural gas.

4.17   Sales of gas to households (domestic sector) accounted for 33 per cent of gas demand, while public administration (including schools and hospitals) consumed 4 per cent of total demand, which was more than was sold to the chemicals sector. The commercial, agriculture and miscellaneous sectors together took up 5 per cent. Non-energy use of gas accounted for the remaining 1 per cent. (See the technical notes section, paragraph 4.40, for more details on non-energy use of gas).

4.18   Care should be exercised in interpreting the figures for individual industries in these commodity balance tables. As companies switch contracts between gas suppliers, it has not been possible to ensure consistent classification between and within industry sectors and across years. The breakdown of final consumption includes a substantial amount of estimated data. For 2007, the allocation of about 7 per cent of consumption is estimated.

4.19   Imports of natural gas from the Norwegian sector of the North Sea began to decline in the late 1980s as output from the Frigg field tailed off. Frigg finally ceased production in October 2004. The southern part of the Langeled pipeline from Sleipner to the UK became operational in October 2006 and has a potential capacity of 27 billion cubic metres (bcm/y). The interconnector linking the UK's transmission network with Belgium via a Bacton to Zeebrugge pipeline began to operate in October 1998. Since 1998 there has been an increase in imports brought about by inflows through the Bacton to Zeebrugge interconnector. In November 2005, its import capacity was almost doubled from 8.5 bcm/y to 16.5 bcm/y. In July 2005 imports of liquefied natural gas (LNG) commenced at the Isle of Grain LNG import facility, the first time LNG had been imported to the UK since the early 1980s. A second interconnector linking the Netherlands to the UK began transporting gas to the UK in December 2006. The Balgzand-Bacton (BBL) pipeline comes ashore at Bacton in Norfolk and has a potential capacity of 16 bcm/y. Exports to mainland Europe from the United Kingdom's share of the Markham field began in 1992 with Windermere's output being added in 1997. Exports to the Republic of Ireland started in 1995. Exports of natural gas exceeded imports for the first time in 1997 and grew rapidly to peak in 2003 before falling by 36 per cent in 2004 to a lower level than imports. In 2007, net imports of gas accounted for 22 per cent of gas put into the National Transmission System (See Map 4.2).

Supply and consumption of natural gas and colliery methane (Table 4.2)

4.20   This table summarises the production and consumption of gas from these sources in the United Kingdom over the last five years.

4.21   Chart 4.2 shows that indigenous production has been in decline since reaching a peak in 2000 and that the UK returned to being a net importer of gas in 2004, and has been in each year since.

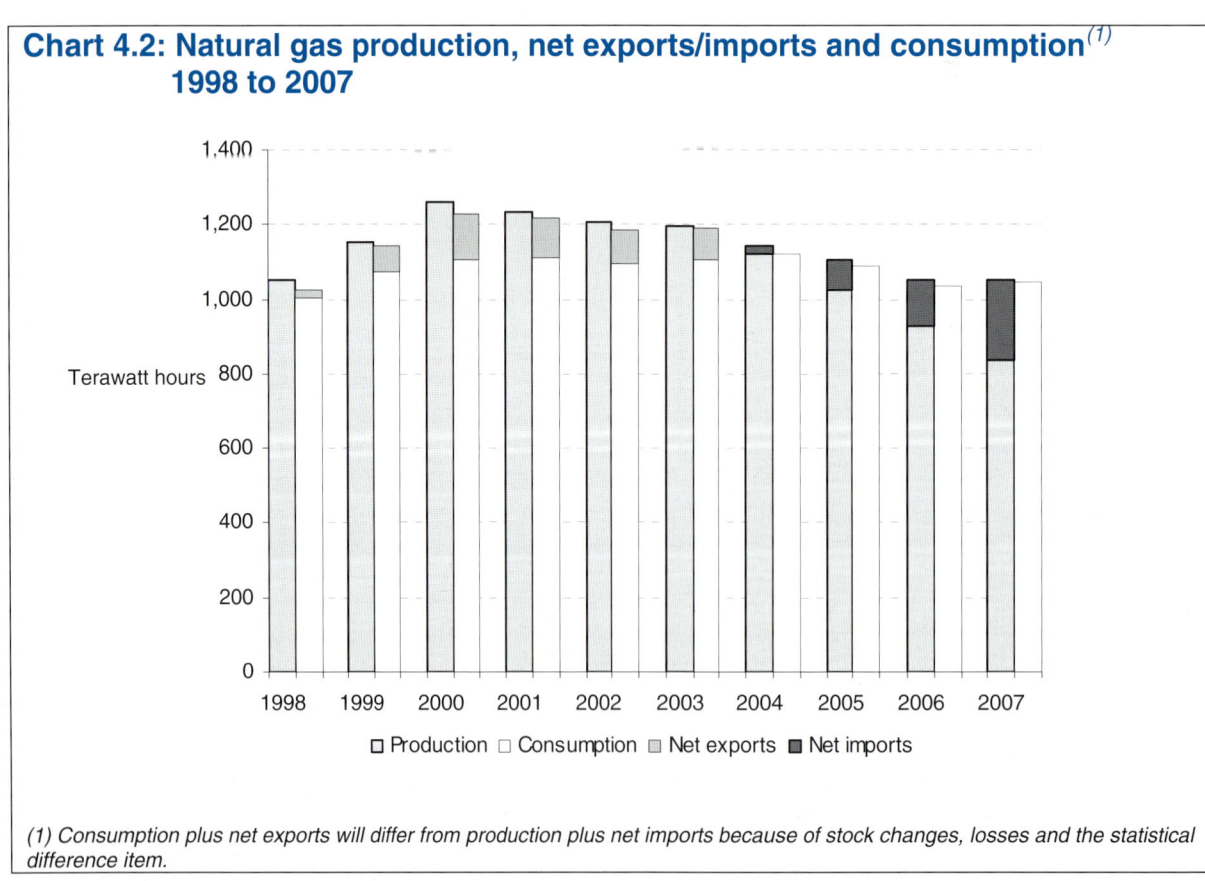

Chart 4.2: Natural gas production, net exports/imports and consumption[1] 1998 to 2007

(1) Consumption plus net exports will differ from production plus net imports because of stock changes, losses and the statistical difference item.

4.22   As Chart 4.3 shows, the growth in consumption for electricity generation has dominated the growth in natural gas consumption over the last ten years. Most of this gas was used in Combined Cycle Gas Turbine (CCGT) stations, although the use of gas in dual fired conventional steam stations was a growth area in 1997 and 1998. However, gas use for electricity generation fell by 3.5 per cent in 2001 as higher gas prices made it more difficult for gas fired stations to compete with large coal fired stations. This was reversed by a 5.5 per cent growth in 2002 when gas prices eased but fell again in 2003 by 1.5 per cent. Higher gas prices again meant that at times some generators found it more profitable to sell gas than use it for generation, particularly given plentiful supplies of inexpensive coal. In 2004, gas use for generation rose by 5 per cent as newly built power stations came on stream. Gas use then fell by 3.5 per cent in 2005, as prices paid by generators rose substantially in the second half of the year. Prices continued to rise in 2006, leading to a further 5.6 per cent fall in gas use. In 2007, however, gas use by generators rose by 13.9 per cent, while the transformation sector as a whole accounted for 35 per cent of gas demand, 3.6 per cent more than in 2006.

Chart 4.3: Consumption of town gas and natural gas 1970 to 2007

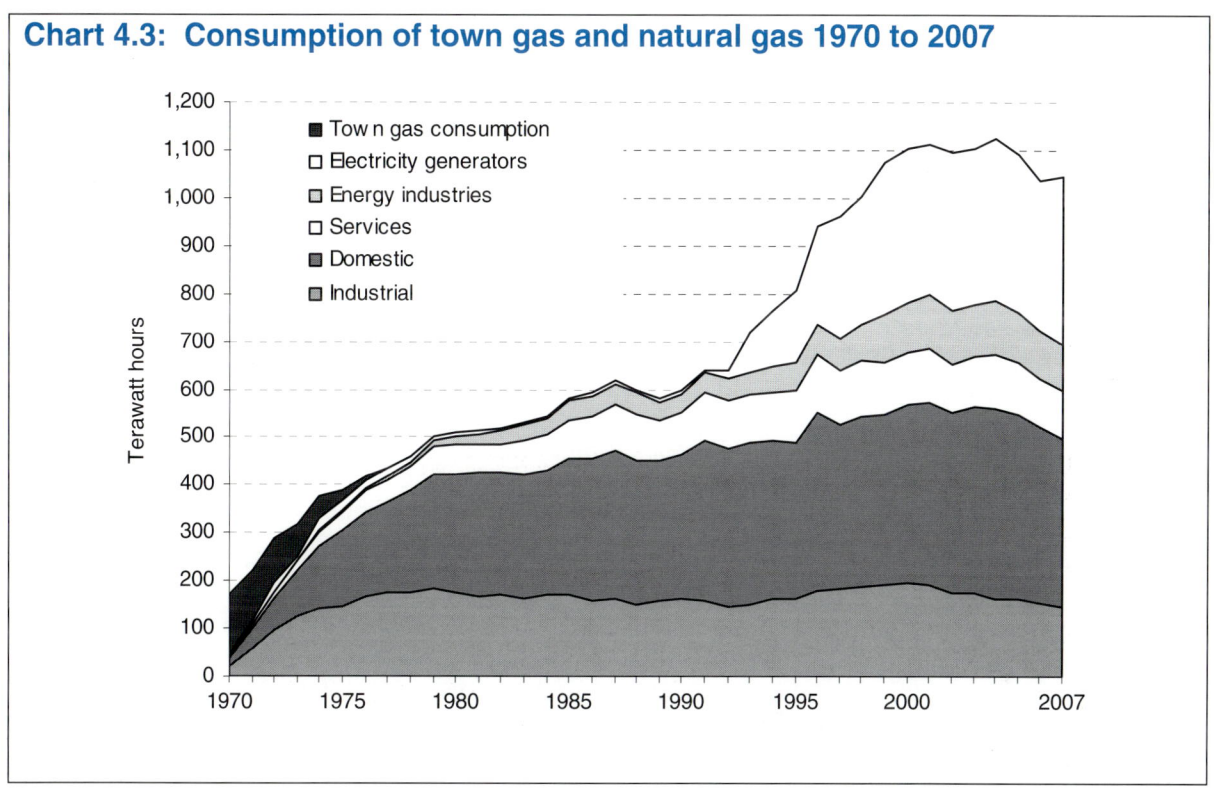

4.23   Since 2000, industrial use of gas has been on a downward trend apart from a small recovery in 2003. In 2007, the decline continued across all major industrial sectors. Overall industrial demand fell by 4.9 per cent in 2007. At the same time, there was a 7.2 per cent decrease in gas used for heat that was then sold to other companies. If heat use and total industrial use are combined then the decrease in gas use in 2007 is 5.2 per cent. Use by the public administration and commercial sectors was 8.7 per cent lower and 4.9 per cent higher, respectively, in 2007 than in 2006. Consumption in the energy industries other than electricity (and heat) generation fell by 7.1 per cent.

4.24   Gas use in the domestic sector is particularly dependent on winter temperatures. Whilst the second half of 2007 experienced lower than average temperatures than in 2006, the first half of the year, including the winter months of January and February, was warmer than in the previous year. This, along with domestic gas prices peaking, in real terms, in the first quarter of 2007, led to a 4.1 per cent fall in domestic sector consumption in 2007.

4.25   Maximum daily demand for natural gas through the National Transmission System in winter 2007/08 was 4,588 GWh on 17th December 2007. This total maximum daily demand was 4.0 per cent lower than the 2006/07 level, and 8.2 per cent lower than January 2003's record level.

4.26   It is estimated that sales of gas supplied on an interruptible basis accounted for around 24 per cent of total gas sales in 2007, roughly 3.0 percentage points higher than in 2006.

UK continental shelf and onshore natural gas (Table 4.3)

4.27   Table 4.3 shows the flows for natural gas from production through transmission to consumption. The footnotes to the table give more information about each table row. This table departs from the standard balance methodology and definitions in order to maintain the link with past data and with monthly data given on BERR's energy statistics web site (see paragraph 4.46). The relationship between total UK gas consumption shown in this table and total demand for gas given in the balance tables (4.1 and 4.2) is illustrated for 2007 as follows:

		GWh
Total UK consumption (Table 4.3)		974,773
plus Producers' own use		65,305
plus Operators' own use		4,698
equals		
"Consumption of natural gas" (see paragraph 4.34)		1,044,776
plus Other losses and metering differences (upstream)		-
plus Downstream losses - leakage assessment	5,117)	7,596
- own use gas	413)	
- theft	2,066)	
plus Metering differences (transmission)		4,472
equals		
Total demand for natural gas (Tables 4.1 and 4.2)		1,056,844

4.28   Gross production of natural gas rose steadily from the year indigenous production began in 1967 until it peaked in 2000 at 1.3 TWh. It has since fallen steadily as reserves on the UKCS deplete. In 2007, natural gas production was 33 per cent lower than the 2000 peak. Gas available at UK terminals has remained fairly constant over this period mainly due to the changes in exports and imports described in paragraph 4.19.   Producers' and operators' own use of gas have tended to change in proportion to the volumes of gas produced and transmitted. Gas input into the transmission system increased by 0.5 per cent between 2006 and 2007, while output of natural gas rose 1.9 per cent. Output from the transmission system increased by more than the input due to stock falls and metering differences.

4.29   For a discussion of the various losses and statistical difference terms in this table, see paragraphs 4.47 to 4.50 in the technical notes and definitions section below. The statistical difference between output from the National Transmission System and total UK consumption has been disaggregated using information obtained from Transco on leakage from local distribution zone pipes, theft and use regarded as own use by pipeline operators. The convention used is set out in paragraph 4.48.

4.30   Losses and metering differences attributable to the information provided on the upstream gas industry are zero from 2001 onwards because these data are no longer reported in the revised Petroleum Production Reporting System. This simplified system for reporting the production of crude oil, NGLs and natural gas in the UK was implemented from 1st January 2001; it reduced the burden on the respondents and improved the quality of data reported on gas production.

4.31   Table 4.3 also includes two rows showing gas stocks and gas storage capacity at the end of the year. Storage data are not currently available before 2004. Stocks data for 2005 onwards has been sourced from the National Grid's weekly brief, and storage data from its 2007 Ten Year Statement.

Gas storage sites and interconnector pipelines (Table 4.4)

4.32   This table details current and planned gas storage facilities in the UK as at 31 May 2008, and also the two operational interconnector pipelines that bring gas to the UK from continental Europe.

Natural gas imports and exports (Table 4.5)

4.33   This table is included for the first time, and shows how much gas is imported to, and exported from, the UK, via the interconnector pipelines and LNG. In 2007, 70 per cent of the UK's gross gas imports were from Norway, up from 67 per cent in 2006. In 2007, 53 per cent of our gas exported was to continental Europe, and 47 per cent to the Republic of Ireland. The flows of gas across Europe are illustrated in Map 4.1.

Map 4.1: Gas European Transit System

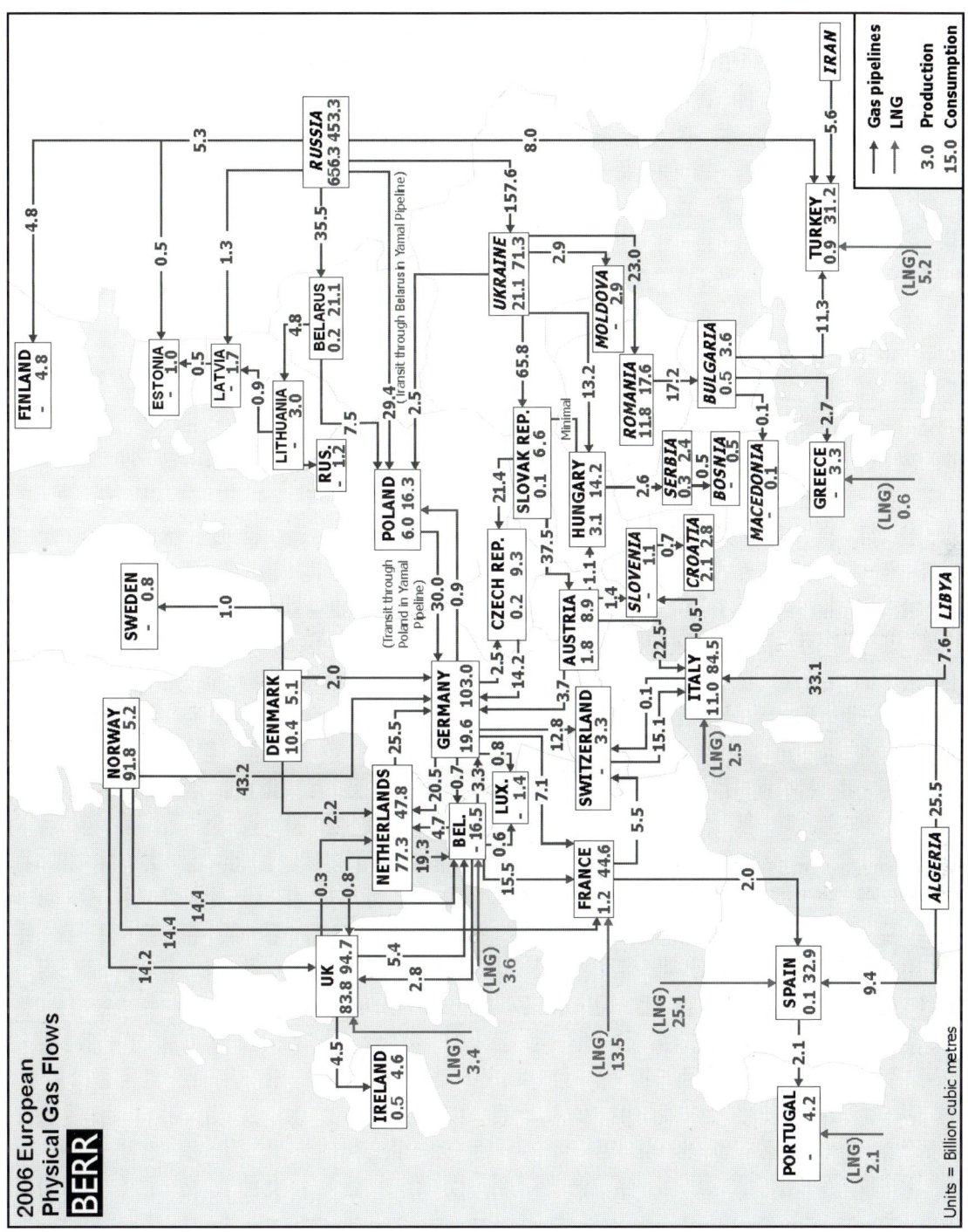

4.34   Gas data are less transparent at the wider European level given missing information on transit flows and incomplete trade information. The above map was produced using published International Energy Agency data to reconstruct the missing physical gas flow data and was prepared as part of BERR's contribution to a Eurostat project to improve gas data transparency and quality.

Technical notes and definitions

4.35   These notes and definitions are in addition to the technical notes and definitions covering all fuels and energy as a whole in Chapter 1, paragraphs 1.26 to 1.59. For notes on the commodity balances and definitions of the terms used in the row headings see Annex A, paragraphs A.7 to A.42. While the data in the printed and bound copy of this Digest cover only the most recent 5 years, these notes also cover data for earlier years that are available on the BERR web site.

Definitions used for production and consumption

4.36   **Natural gas** production in Tables 4.1 and 4.2 relates to the output of indigenous methane at land terminals and gas separation plants (includes producers' and processors' own use). For further explanation, see Annex F, paragraph F.10 on BERR's Energy Statistics web site under 'Production of gas' - www.berr.gov.uk/energy/statistics/publications/dukes/page45537.html. Output of the Norwegian share of the Frigg and Murchison fields is included under imports. A small quantity of onshore produced methane (other than colliery methane) is also included.

4.37   Table 4.3 shows production, transmission and consumption figures for UK continental shelf and onshore natural gas. Production includes waste and own use for drilling, production and pumping operations, but excludes gas flared. Gas available in the United Kingdom excludes waste, own use for drilling etc, stock change, and includes imports net of exports. Gas transmitted (input into inland transmission systems) is after stock change, own use, and losses at inland terminals. The amount consumed in the United Kingdom differs from the total gas transmitted by the gas supply industry because of losses in transmission, differences in temperature and pressure between the points at which the gas is measured, delays in reading meters and consumption in the works, offices, shops, etc of the undertakings. The figures include an adjustment to the quantities billed to consumers to allow for the estimated consumption remaining unread at the end of the year.

4.38   **Colliery methane** production is colliery methane piped to the surface and consumed at collieries or transmitted by pipeline to consumers. As the output of deep-mined coal declines so does the production of colliery methane, unless a use can be found for gas that was previously vented. The supply of methane from coal measures that are no longer being worked or from drilling into coal measures is licensed under the same legislation as used for offshore gas production.

4.39   **Transfers** of natural gas include natural gas use within the iron and steel industry for mixing with blast furnace gas to form a synthetic coke oven gas. For further details see paragraph 2.48 in Chapter 2.

4.40   **Non-energy gas**: Non-energy use is gas used as feedstock for petrochemical plants in the chemical industry as raw material for the production of ammonia (an essential intermediate chemical in the production of nitrogen fertilisers) and methanol. The contribution of liquefied petroleum gases (propane and butane) and other petroleum gases is shown in Tables 3.4 to 3.6 of Chapter 3. Firm data for natural gas are not available, but estimates for 2003 to 2007 are shown in Table 4.2 and estimates for 2005 to 2007 in Table 4.1. The estimates for the years up to 2006 have been obtained from AEA's work for the National Atmospheric Emissions Inventory; 2007 data are BERR extrapolations.

Sectors used for sales/consumption

4.41   For definitions of the various sectors used for sales and consumption analyses see Chapter 1 paragraphs 1.54 to 1.58 and Annex A, paragraphs A.31 to A.42. However, **miscellaneous** has a wider coverage than in the commodity balances of other fuels. This is because some gas supply companies are unable to provide a full breakdown of the services sector and the gas they supply to consumers is allocated to miscellaneous when there is no reliable basis for allocating it elsewhere. See also paragraph 4.44, below, for information on the source of the sectoral data for consumption of gas.

Data collection

4.42   Production figures are generally obtained from returns made under BERR's Petroleum Production Reporting System (PPRS) and from other sources. BERR obtain data on the transmission of natural gas from National Grid (who operate the National Transmission System) and from other pipeline operators. Data on consumption are based on returns from gas suppliers and UKCS producers who supply gas directly to customers.

4.43   The production data are for the United Kingdom (including natural gas from the UKCS - offshore and onshore). The restoration of a public gas supply to parts of Northern Ireland in 1997 (see paragraph 4.13 means that all tables in this chapter, except Tables 4A and 4B, cover the UK).

4.44   BERR carry out an annual survey of gas suppliers to obtain details of gas sales to the various categories of consumer. Estimates are included for the suppliers with the smallest market share since the BERR inquiry covers only the largest suppliers (ie those with more than about a 0.5 per cent share of the UK market up to 1997 and those known to supply more than 1,750 GWh per year for 1998 onwards). For 2000 and subsequent years, gas consumption for the iron and steel sector is based on data provided by the Iron and Steel Statistics Bureau (ISSB) rather than gas suppliers since gas suppliers were over estimating their sales to this sector. The difference between the ISSB and gas suppliers figures has been re-allocated to other sectors using the results of the Office for National Statistics' Purchases Inquiry and information derived from the EU Emissions Trading Scheme submissions.

Period covered
4.45   Figures generally relate to years ended 31 December. However, before 2004, data for natural gas for electricity generation relate to periods of 52 weeks as set out in Chapter 5, paragraphs 5.59 and 5.60.

Monthly and quarterly data
4.46   Monthly data on natural gas production and supply are available from BERR's Energy Statistics web site www.berr.gov.uk/energy/statistics/source/gas/page18525.html in monthly Table 4.2. A quarterly commodity balance for natural gas (which includes consumption data) is published in BERR's quarterly statistical bulletin *Energy Trends* and is also available from quarterly Table 4.1 on BERR's Energy Statistics web site. See Annex C for more information about *Energy Trends* and BERR's Energy Statistics web site.

Statistical and metering differences
4.47   In Table 4.3 there are several headings that refer to statistical or metering differences. These arise because measurement of gas flows, in volume and energy terms, takes place at several points along the supply chain. The main sub-headings in the table represent the instances in the supply chain where accurate reports are made of the gas flows at that particular key point in the supply process. It is possible to derive alternative estimates of the flow of gas at any particular point by taking the estimate for the previous point in the supply chain and then applying the known losses and gains in the subsequent part of the supply chain. The differences seen when the actual reported flow of gas at any point and the derived estimate are compared are separately identified in the table wherever possible, under the headings statistical or metering differences.

4.48   The differences arise from several factors:-

• Limitations in the accuracy of meters used at various points of the supply chain. While standards are in place on the accuracy of meters, there is a degree of error allowed which, when large flows of gas are being recorded, can become significant.

• Differences in the methods used to calculate the flow of gas in energy terms. For example, at the production end, rougher estimates of the calorific value of the gas produced are used which may be revised only periodically, rather than the more accurate and more frequent analyses carried out further down the supply chain. At the supply end, although the calorific value of gas shows day-to-day variations, for the purposes of recording the gas supplied to customers a single calorific value is used. Until 1997 this was the lowest of the range of calorific values for the actual gas being supplied within each LDZ, resulting in a "loss" of gas in energy terms. In 1997 there was a change to a "capped flow-weighted average" algorithm for calculating calorific values resulting in a reduction in the losses shown in the penultimate row of Table 4.3. This change in algorithm, along with improved meter validation and auditing procedures, also reduced the level of the "metering differences" row within the downstream part of Table 4.3.

• Differences in temperature and pressure between the various points at which gas is measured. Until February 1997 British Gas used "uncorrected therms" on their billing system for tariff customers when converting from a volume measure of the gas used to an energy measure. This made their supply figure too small by a factor of 2.2 per cent, equivalent to about 1 per cent of the wholesale market.

- Differences in the timing of reading meters. While National Transmission System meters are read daily, customers' meters are read less frequently (perhaps only annually for some domestic customers) and profiling is used to estimate consumption. Profiling will tend to underestimate consumption in a strongly rising market.
- Other losses from the system, for example theft through meter tampering by consumers.

4.49   The headings in Table 4.3 show where, in the various stages of the supply process, it has been possible to identify these metering differences as having an effect. Usually they are aggregated with other net losses as the two factors cannot be separated. Whilst the factors listed above can give rise to either losses or gains, losses are more common. However, the negative downstream gas metering difference within the transmission system in 2003 was an anomaly that was investigated by National Grid during 2004. They concluded that this unaccounted for element of National Transmission System shrinkage was due to an exceptional run of monthly negative figures between February and June 2003 within what is usually a variable but mainly positive series. However, after a comprehensive investigation of this exceptional period no causal factors were identified. It is probable that the meter error or errors that caused this issue were corrected during the validation of metering.

4.50   The box below shows how, in 2007, the wastage, losses and metering differences figures in Table 4.3 are related to the losses row in the balance Tables 4.1 and 4.2. It should be noted that losses from 2001 onwards are lower than in earlier years because figures for losses and metering differences in the upstream gas industry are no longer available (see above):

Table 4.3	GWh
Upstream gas industry:	
Other losses and metering differences	-
Downstream gas industry:	
Transmission system metering differences	4,472
Leakage assessment	5,117
Own use gas	413
Theft	2,066
Tables 4.1 and 4.2	
Losses	12,068

Similarly, the statistical difference row in Tables 4.1 and 4.2 is made up of the following components in 2007:

Table 4.3	GWh
Statistical difference between gas available from upstream and gas input to downstream	-1,029
plus Downstream gas industry:	
Distribution losses and metering differences	2,509
Tables 4.1 and 4.2	
Statistical difference	1,480

Contact: James Hemingway
Energy Markets Unit
james.hemingway@berr.gsi.gov.uk
020 7215 2717

Clive Evans
Energy Markets Unit
clive.evans@berr.gsi.gov.uk
020 7215 5189

Map 4.2: The National Gas Transmission System 2007

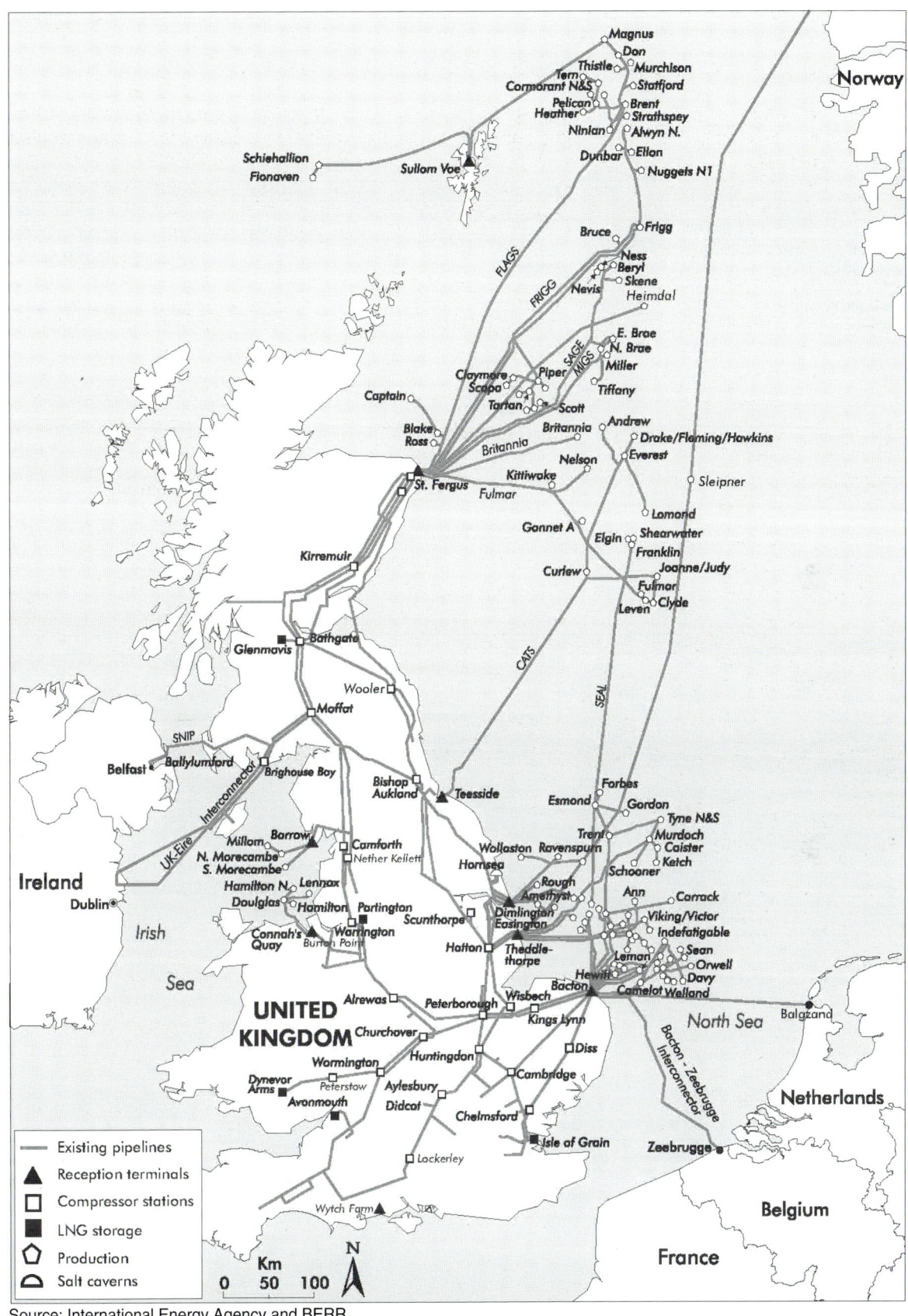

Source: International Energy Agency and BERR

4.1 Commodity balances

Natural gas

GWh

	2005			2006			2007		
	Natural gas	Colliery methane	Total Natural gas	Natural gas	Colliery methane	Total Natural gas	Natural gas	Colliery methane	Total Natural gas
Supply									
Production	1,025,232	757	1,025,989	929,784	754r	930,538r	838,092	717	838,809
Other sources	-	-	-	-	-	-	-	-	-
Imports	173,328	-	173,328	244,029	-	244,029	338,027	-	338,027
Exports	-96,181	-	-96,181	-120,591	-	-120,591	-123,158	-	-123,158
Marine bunkers	-	-	-	-	-	-	-	-	-
Stock change (1)	+1,321	-	+1,321	-6,435	-	-6,435	+5,480	-	+5,480
Transfers (2)	-51	-	-51	-52	-	-52	-78	-	-78
Total supply	**1,103,649**	**757**	**1,104,406**	**1,046,735**	**754r**	**1,047,489r**	**1,058,364**	**717**	**1,059,081**
Statistical difference (3)	+3,152r	-	+3,152r	-676r	-r	-676r	+1,480	-	+1,480
Total demand	**1,100,498**	**757**	**1,101,255**	**1,047,411r**	**754r**	**1,048,165r**	**1,056,884**	**717**	**1,057,601**
Transformation	**350,860r**	**588**	**351,448r**	**331,817r**	**595**	**332,412r**	**373,375**	**586**	**373,961**
Electricity generation	328,372	588	328,960	309,794	595	310,389	352,929	586	353,515
Major power producers	295,643	-	295,643	278,127	-	278,127	319,836	-	319,836
Autogenerators	32,729	588	33,317	31,667	595	32,262	33,093	586	33,679
Heat generation	22,488r	-	22,488r	22,023r	-	22,023r	20,446	-	20,446
Petroleum refineries	-	-	-	-	-	-	-	-	-
Coke manufacture	-	-	-	-	-	-	-	-	-
Blast furnaces	-	-	-	-	-	-	-	-	-
Patent fuel manufacture	-	-	-	-	-	-	-	-	-
Other	-	-	-	-	-	-	-	-	-
Energy industry use	**86,159**	**114**	**86,273**	**80,072**	**112r**	**80,184r**	**74,413**	**91**	**74,504**
Electricity generation	-	-	-	-	-	-	-	-	-
Oil and gas extraction	74,187	-	74,187	70,130	-	70,130	65,305	-	65,305
Petroleum refineries	4,274	-	4,274	3,225	-	3,225	3,381	-	3,381
Coal extraction	-	114	114	-	112r	112r	-	91	91
Coke manufacture	-	-	-	-	-	-	-	-	-
Blast furnaces	941	-	941	611	-	611	719	-	719
Patent fuel manufacture	-	-	-	-	-	-	-	-	-
Pumped storage	-	-	-	-	-	-	-	-	-
Other	6,757	-	6,757	6,106	-	6,106	5,008	-	5,008
Losses (4)	**10,964**	**-**	**10,964**	**12,012**	**-**	**12,012**	**12,068**	**-**	**12,068**
Final consumption	**652,515r**	**55**	**652,570r**	**623,510r**	**47r**	**623,557r**	**597,028**	**40**	**597,068**
Industry	**151,336r**	**55**	**151,391r**	**143,770r**	**47r**	**143,817r**	**136,771**	**40**	**136,811**
Unclassified	-	55	55	-	47r	47r	-	40	40
Iron and steel	8,469r	-	8,469r	8,406r	-	8,406r	7,337	-	7,337
Non-ferrous metals	3,218r	-	3,218r	3,206r	-	3,206r	3,392	-	3,392
Mineral products	13,299r	-	13,299r	12,302r	-	12,302r	11,212	-	11,212
Chemicals	41,073r	-	41,073r	38,246r	-	38,246r	37,231	-	37,231
Mechanical Engineering, etc	8,676r	-	8,676r	8,679r	-	8,679r	7,828	-	7,020
Electrical engineering, etc	4,333r	-	4,333r	4,159r	-	4,159r	3,981	-	3,981
Vehicles	10,258r	-	10,258r	9,270r	-	9,270r	8,814	-	8,814
Food, beverages, etc	27,918r	-	27,918r	27,279r	-	27,279r	26,621	-	26,621
Textiles, leather, etc	7,130r	-	7,130r	6,636r	-	6,636r	6,380	-	6,380
Paper, printing, etc	13,687r	-	13,687r	12,008r	-	12,008r	11,044	-	11,044
Other industries	10,599r	-	10,599r	11,022r	-	11,022r	10,249	-	10,249
Construction	2,675r	-	2,675r	2,555r	-	2,555r	2,683	-	2,683
Transport	-	-	-	-	-	-	-	-	-
Air	-	-	-	-	-	-	-	-	-
Rail	-	-	-	-	-	-	-	-	-
Road (5)	-	-	-	-	-	-	-	-	-
National navigation	-	-	-	-	-	-	-	-	-
Pipelines	-	-	-	-	-	-	-	-	-
Other	**491,500r**	**-**	**491,500r**	**467,983r**	**-**	**467,983r**	**449,798**	**-**	**449,798**
Domestic	384,009	-	384,009	364,850r	-	364,850r	349,943	-	349,943
Public administration	50,319r	-	50,319r	48,816r	-	48,816r	44,589	-	44,589
Commercial	35,097r	-	35,097r	34,277r	-	34,277r	35,943	-	35,943
Agriculture	2,261	-	2,261	2,013	-	2,013	1,999	-	1,999
Miscellaneous	19,814	-	19,814	18,027r	-	18,027r	17,323	-	17,323
Non energy use	**9,678r**	**-**	**9,678r**	**11,757r**	**-**	**11,757r**	**10,459**	**-**	**10,459**

(1) Stock fall (+), stock rise (-).
(2) Natural gas used in the manufacture of synthetic coke oven gas.
(3) Total supply minus total demand.

(4) See paragraphs 4.47 to 4.50.
(5) See footnote 5 to Table 4.2.

4.2 Supply and consumption of natural gas and colliery methane[1]

GWh

	2003	2004	2005	2006	2007
Supply					
Production	1,197,030	1,121,257	1,025,989	930,538r	838,809
Imports	86,298	133,033	173,328	244,029	338,027
Exports	-177,039	-114,112	-96,181	-120,591	-123,158
Stock change (2)	+3,532	-6,235	+1,321	-6,435	+5,480
Transfers	-82	-39	-51	-52	-78
Total supply	**1,109,739**	**1,133,904**	**1,104,406**	**1,047,489r**	**1,059,081**
Statistical difference (3)	+748	+702	+3,152r	-676r	+1,480
Total demand	**1,108,991**	**1,133,202**	**1,101,255r**	**1,048,165r**	**1,057,601**
Transformation	**344,410**	**362,668**	**351,448r**	**332,412r**	**373,961**
Electricity generation	324,580	340,824	328,960	310,389	353,515
Major power producers	284,662	304,497	295,643	278,127	319,836
Autogenerators	39,918	36,328	33,317	32,262	33,679
Heat generation	19,830	21,844	22,488r	22,023r	20,446
Other	-	-	-	-	-
Energy industry use	**88,907**	**88,468**	**86,273**	**80,184r**	**74,504**
Electricity generation	-	-	-	-	-
Oil and gas extraction	76,837	77,753	74,187	70,130	65,305
Petroleum refineries	2,773	3,076	4,274	3,225	3,381
Coal extraction	187	150	114	112r	91
Coke manufacture	1	-	-	-	-
Blast furnaces	539	728	941	611	719
Other	8,570	6,761	6,757	6,106	5,008
Losses (4)	**6,217**	**8,207**	**10,964**	**12,012**	**12,068**
Final consumption	**669,457**	**673,860**	**652,570r**	**623,557r**	**597,068**
Industry	**166,217**	**153,953**	**151,391r**	**143,817r**	**136,811**
Unclassified	75	65	55	47r	40
Iron and steel	10,327	9,715	8,469r	8,406r	7,337
Non-ferrous metals	4,781	3,199	3,218r	3,206r	3,392
Mineral products	14,105	13,401	13,299r	12,302r	11,212
Chemicals	45,048	42,002	41,073r	38,246r	37,231
Mechanical engineering, etc	9,126	8,611	8,676r	8,679r	7,828
Electrical engineering, etc	4,395	4,158	4,333r	4,159r	3,981
Vehicles	11,621	10,228	10,258r	9,270r	8,814
Food, beverages, etc	28,799	28,232	27,918r	27,279r	26,621
Textiles, leather, etc	7,901	7,120	7,130r	6,636r	6,380
Paper, printing, etc	15,898	13,879	13,687r	12,008r	11,044
Other industries	11,126	10,413	10,599r	11,022r	10,249
Construction	3,015	2,931	2,675r	2,555	2,683
Transport	-	-	-	-	-
Road (5)	-	-	-	-	-
Other	**493,219**	**509,886**	**491,500r**	**467,983r**	**449,798**
Domestic	386,486	396,411	384,009	364,850r	349,943
Public administration	44,362	51,934	50,319r	48,816r	44,589
Commercial	39,537	37,595	35,097r	34,277r	35,943
Agriculture	2,324	2,355	2,261	2,013	1,999
Miscellaneous	20,510	21,591	19,814	18,027r	17,323
Non energy use	**10,021**	**10,021**	**9,678r**	**11,757r**	**10,459**

(1) Colliery methane figures included within these totals are as follows:

	2003	2004	2005	2006	2007
Total production	**915**	**810**	**757**	**754r**	**717**
Electricity generation	653	595	588	595	586
Coal extraction	187	150	114	112r	91
Other industries	75	65	55	47r	40
Total consumption	**915**	**810**	**757**	**754r**	**717**

(2) Stock fall (+), stock rise (-).
(3) Total supply minus total demand.
(4) For an explanation of what is included under losses, see paragraphs 4.47 to 4.50.

(5) A small amount of natural gas is consumed by road transport, but gas use in this sector is predominantly of petroleum gas, hence road use of gas is reported in the petroleum products balances in Chapter 3.

4.3 UK continental shelf and onshore natural gas production and supply[1]

GWh

	2003	2004	2005	2006	2007
Upstream gas industry:					
Gross production [2]	1,196,115	1,120,447	1,025,232	929,784	838,092
Minus Producers' own use [3]	76,837	77,753	74,187	70,130	65,305
Exports	177,039	114,112	96,181	120,591	123,158
Plus Imports of gas	86,298	133,033	173,328	244,029	338,027
Gas available at terminals [4]	1,028,537	1,061,615	1,028,192	983,093	987,656
Minus Statistical difference [5]	-1,391	-2,310	-1,329	-731	-1,029
Downstream gas industry:					
Gas input into the national transmission system [6]	1,029,928	1,063,926	1,029,521	983,824	988,686
Minus Operators' own use [7]	7,475	6,560	6,555	5,831	4,698
Stock change (storage sites) [8]	-3,532	+6,235	-1,321	+6,435	-5,480
Metering differences [5]	-874	137	1,230r	4,544	4,472
Gas output from the national transmission system [9]	1,026,859	1,050,994	1,023,057r	967,014	984,996
Minus Leakage assessment [10]	4,452	5,433	5,291	5,030	5,117
Own use gas [11]	439	439	427r	406	413
Theft [12]	2,197	2,194	2,137r	2,031	2,066
Transfers [13]	82	39	51	52	51
Statistical difference and metering differences [5]	2,138r	3,016	6,359r	55r	2,576
Total UK consumption [14]	**1,017,547**	**1,039,873**	**1,008,792r**	**959,439r**	**974,773**
Stocks of gas (at end year)	30,565	36,800	35,479	41,914	36,434
Storage capacity [15]		45,309	48,092r	48,247r	48,247

(1) For details of where to find monthly updates of natural gas production and supply see paragraph 4.46.

(2) Includes waste and producers' own use, but excludes gas flared.

(3) Gas used for drilling, production and pumping operations.

(4) The volume of gas available at terminals for consumption in the UK as recorded by the terminal operators. The percentage of gas available for consumption in the UK from indigenous sources in 2007 was 78 per cent, compared with 79 per cent in 2006.

(5) Measurement of gas flows, in volume and energy terms, occurs at several points along the supply chain. As such, differences are seen between the actual recorded flow through any one point and estimates calculated for the flow of gas at that point. More detail on the reasons for these differences is given in the technical notes and definitions section of this chapter, paragraphs 4.47 to 4.50.

(6) Gas received as reported by the pipeline operators. The pipeline operators include National Grid, who run the national pipeline network, and other pipelines that take North Sea gas supplies direct to consumers.

(7) Gas consumed by pipeline operators in pumping operations and on their own sites.

(8) Stocks of gas held in specific storage sites, either as liquefied natural gas, pumped into salt cavities or stored by pumping the gas back into an offshore field. Stock rise (+), stock fall (-).

(9) Including public gas supply, direct supplies by North Sea producers, third party supplies and stock changes.

(10) This is a National Grid assessment of leakage through the local distribution system based on the National Leakage Reduction Monitoring Model.

(11) Equivalent to about 0.06 per cent of LDZ throughput, this is an assessment of the energy used to counter the effects of gas cooling on pressure reduction.

(12) Calculated by National Grid as 0.3 per cent of LDZ throughput, this is theft before the gas reaches customer meters.

(13) Transfers are the use within the iron and steel industry for the manufacture of synthetic coke oven gas.

(14) See paragraph 4.27 for an explanation of the relationship between these "Total UK consumption" figures and "Total demand" shown within the balance tables.

(15) Data compiled by BERR from individual storage site information. Converted from billion cubic metres to GWh assuming 11.055 kWh per cubic metre. See paragraph 4.31.

4.4 Gas storage sites and interconnector pipelines in the United Kingdom at 31 May 2008

Owner	Site	Location	Capacity (Mm3)	Flow rate (Mm3/day)	Type	Status (1)
Operational storage						
Centrica Storage Ltd	Rough	Southern North Sea	3,340	42	Depleted field	Long
National Grid	Avonmouth	Bristol	81	14	LNG	Short
	Dynevor Arms	Mid Glamorgan	28	5	LNG	Short
	Glenmavis	North Lanarkshire	47	9	LNG	Short
	Partington	Manchester	104	20	LNG	Short
Scottish and Southern Energy	Hornsea	East Yorkshire	326	20	Salt cavern	Medium
Energy Merchants Gas Storage	Hole House Farm	Cheshire	42-55(2)	8	Salt cavern	Medium
Scottish Power	Hatfield Moor	South Yorkshire	116	2	Depleted field	Medium
Star Energy Ltd	Humbly Grove	Hampshire	280-340(2)	7	Depleted field	Medium
Planned storage (3)						
Statoil and Scottish and Southern Energy	Aldbrough	East Yorkshire	140-840(4)	13-70(5)	Salt cavern	Medium
E.On	Holford	Cheshire	20-165(4)	12-16(5)	Salt cavern	Medium
Star Energy Ltd	Welton	Lincolnshire	225-340(4)	6-9(5)	Depleted field	Medium
	Albury	Surrey	170-850(4)	4-13(5)	Depleted field	Medium
	Bletchingley	Surrey	850	11	Depleted field	Medium
	Gainsborough	Lincolnshire	156	6	Depleted field	Medium
Portland Gas Ltd	Portland	Dorset	143-1,000(4)	3-20(5)	Salt cavern	Medium
GdF	Stublach	Cheshire	67-424(4)	11-34(5)	Salt cavern	Medium
Canatxx	Preesall	Lancashire	170-1,700(4)	28-51(5)	Salt cavern	Medium
Wingas	Saltfleetby	Lincolnshire	715	8	Depleted field	Medium
Warwick Energy	Caythorpe	East Yorkshire	210	9	Depleted field	Medium
E.On	Whitehill	East Yorkshire	210-420(4)	21-42(5)	Salt cavern	Medium
NPL Estates	King Street	Cheshire	16-160(4)	12	Salt cavern	Medium
Gateway Storage Co Ltd	Gateway	Irish Sea	500-1,140(4)	20	Salt cavern	Medium
Centrica Storage	Bains	Irish Sea	..(6)	..	Depleted Field	..

Owner	Between	Capacity (Bcm/year)
Operational pipelines		
Interconnector UK - Import	Zeebrugge-Bacton	25
Interconnector UK - Export	Bacton-Zeebrugge	21
Balgzand Bacton Line (BBL)	Balgzand-Bacton	15

(1) Long range, medium range or short range storage. Status is determined both by capacity size and injection, deliverability and storage re-cycling rates.

(2) Additional storage capacity expected to be commissioned summer 2008.

(3) Planned storage is defined as under construction, consented or planned (including those projects that are in the planning process; are in the pre-planning stage; have had their applications withdrawn or rejected and pending a decision about next steps).

(4) Range indicates the initial capacity followed by an (incremental) rise.

(5) Range indicates the initial withdrawal rate followed by an (incremental) rise.

(6) Capacity and flow rate figures are not available for the Bains facility at the time of publication.

4.5 Natural gas imports and exports [1]

GWh

	2003	2004	2005	2006	2007
Imports from:					
Belgium *(2)*	4,387	25,592	24,108	30,505	6,471
The Netherlands *(3)*	-	-	-	9,135	76,602
Norway *(4)*	71,753	95,359	127,895	157,035	225,764
Liquefied Natural Gas *(5)*	-	-	5,453	37,576	14,903
Total Imports	**76,140**	**120,951**	**157,456**	**234,251**	**323,741**
Exports to:					
Belgium *(2)*	122,648	60,060	36,641	60,195	51,390
The Netherlands *(6)*	3,424	2,887	4,261	3,371	6,358
Norway *(7)*	-	-	-	-	153
Republic of Ireland *(8)*	40,806	39,084	39,407	47,247	50,972
Total Exports	**166,878**	**102,031**	**80,309**	**110,813**	**108,872**
Net Imports *(9)*	**-90,738**	**18,920**	**77,147**	**123,438**	**214,869**

(1) In 2007 this table was Table G.6 of the Internet Annex G to the Digest.

(2) Physical flows of gas through the Bacton-Zeebrugge Interconnector. In tables 4.1 to 4.3 the nominated flows of gas through the pipeline are used. Nominated flows are the amounts of gas that companies requested be supplied through the pipeline. Net imports are the same whichever measurement is used.

(3) Via the Bacton-Balgzand (BBL) pipeline. Commissioned in November 2006.

(4) Currently via the Langeled and Vesterled pipelines, and the Tampen Link (from Statfjord to FLAGS). Prior to 2005 includes the Norwegian share of the Frigg field.

(5) From various sources to the Isle of Grain and Gasport Teesside.

(6) Direct exports from the Grove, Chiswick, Markham, Minke and Windermere offshore gas fields using the Dutch offshore gas pipeline infrastructure.

(7) With effect from September 2007, UK gas from the Blane field to the Norwegian Ula field for injection into the Ula reservoir.

(8) Includes gas to the Isle of Man for which separate figures are not available.

(9) A negative figure means the UK was a net exporter of gas.

Chapter 5
Electricity

Introduction

5.1   This Chapter presents statistics on electricity from generation through to sales. In addition, statistics on generating capacity, on fuel used for generation and on load factors and efficiencies are included along with a map showing the transmission system in Great Britain and the location of the main power stations (page 125). An energy flow chart for 2007, showing the flows of electricity from fuel inputs through to consumption, is included for the first time, overleaf. This is a way of simplifying the figures that can be found in the commodity balance for electricity in Table 5.1. It illustrates the flow of primary fuels from the point at which they become available for the production of electricity (on the left) to the eventual final use of the electricity produced or imported (on the right) as well as the energy lost in conversion, transmission and distribution.

5.2   Commodity balances for electricity, for each of the last three years, form the introductory table (Table 5.1). The supply and consumption elements of the electricity balance are presented as 5-year time series in Table 5.2. Table 5.3 separates out the public distribution system for electricity from electricity generated and consumed by autogenerators and uses a commodity balance format. Fuels used to generate electricity in the United Kingdom in each of the last five years are covered in Table 5.4. Table 5.5 shows the relationship between the commodity balance definitions and traditional Digest definitions for electricity, so that the most recent data can be linked to the long term trends data, which can be found on the BERR energy statistics web site. Table 5.6 shows the relationship between fuels used, generation and supply in each of the latest five years. Tables on plant capacity (Tables 5.7, 5.8 and 5.9) and on plant loads and efficiency (Table 5.10) have been included. Two of these contain data at a sub-national level. Table 5.11 lists individual power stations in operation and it is supplemented by a table showing large scale Combined Heat and Power (CHP) schemes in the United Kingdom (Table 5.12). The long term trends commentary and tables on fuel use, generation, supply and consumption back to 1970 are to be found on BERR's energy statistics web site and accessible from the Digest of UK Energy Statistics home page:
www.berr.gov.uk/energy/statistics/publications/dukes/page45537.html .

Structure of the industry

5.3   Up to March 2005 the electricity industries of Scotland, Northern Ireland and England and Wales operated independently although interconnectors joined all three grid systems together. From April 2005 under the British Electricity Trading and Transmission Arrangements (BETTA), introduced in the Energy Act 2004, the electricity systems of England and Wales and Scotland have been integrated. The paragraphs below describe the position up to March 2005 but indicate the further changes that have been made under BETTA.

5.4   From the period immediately after privatisation of the industry in 1990, when there were seven generating companies in England and Wales and 12 Regional Electricity companies distributing and supplying electricity to customers in their designated area, there were many structural and business changes and residual floatations. At the end of 2007 there were 30 major power producers operating in Great Britain[1]. Competition developed in mainland Britain as follows:

(a)   From 1 April 1990, customers with peak loads of more than 1 MW (about 45 per cent of the non-domestic market) were able to choose their supplier.

(b)   From 1 April 1994, customers with peak loads of more than 100 kW were able to choose their supplier.

---

[1] Some of these producers are joint ventures and so the number of generating companies involved is less than 30.

Electricity flow chart 2007 (TWh)

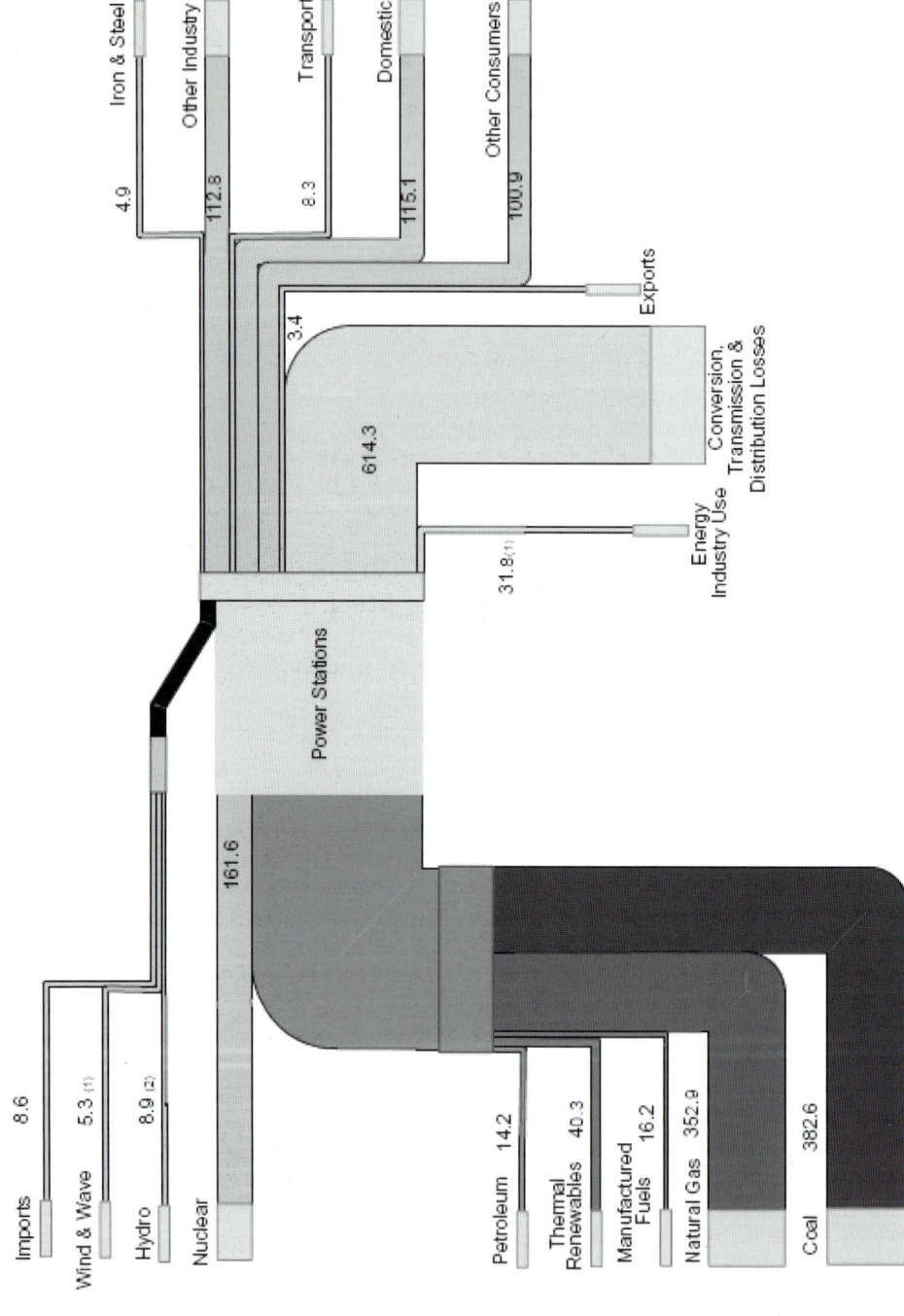

Imports 8.6

Wind & Wave 5.3 (1)

Hydro 8.9 (2)

Nuclear 161.6

Petroleum 14.2

Thermal Renewables 40.3

Manufactured Fuels 16.2

Natural Gas 352.9

Coal 382.6

Power Stations

Energy Industry Use 31.8 (1)

Conversion, Transmission & Distribution Losses

614.3

3.4

Exports

Iron & Steel 4.9

Other Industry 112.8

Transport 8.3

Domestic 115.1

Other Consumers 100.9

Notes:

This flow chart is based on the data in Tables 5.1 (for imports, exports, use, losses and consumption) and 5.6 (fuel used).

(1) Includes solar photovoltaics.

(2) Hydro includes generation from pumped storage while electricity used in pumping is included under Energy Industry Use.

(c)   Between September 1998 and May 1999, the remaining part of the electricity market (ie below 100 kW peak load) was opened up to competition. Paragraph 5.9 and Table 5A give more details of the opening up of the domestic gas and electricity markets to competition.

5.5   Since the late 1990s, there have been commercial moves toward vertical re-integration between generating, electricity distribution and/or electricity supply businesses. Those mergers that have taken place were approved by the relevant competition authority. Initially the National Grid Company was owned by the 12 privatised regional electricity companies, but was floated on the Stock Exchange in 1995. National Grid (and its predecessors since 1990) has owned and operated the high voltage transmission system in England and Wales linking generators to distributors and some large customers. This transmission system is linked to the transmission system of continental Europe via an interconnector to France under the English Channel (see Table 5.11). Up to March 2005, the Scottish transmission system was regarded as being linked to that in England and Wales by two interconnectors but under BETTA National Grid took on responsibility for operating the transmission system in Scotland as well as England and Wales. Thus a single Great Britain market has been created and the transmission network is regarded as a single system.

5.6   In Scotland, until the end of March 2005, the two main companies, Scottish Power and Scottish and Southern Energy, covered the full range of electricity provision. They operated generation, transmission, distribution and supply businesses. In addition, there were a number of small independent hydro stations and some independent generators operating fossil-fuelled stations, which sold their output to Scottish Power and Scottish and Southern Energy.

5.7   The electricity supply industry in Northern Ireland has been in private ownership since 1993 with Northern Ireland Electricity plc (NIE) (part of the Viridian Group) responsible for power procurement, transmission, distribution and supply in the Province. Generation is provided by three private sector companies who own the four major power stations. In December 2001, the link between Northern Ireland's grid and that of Scotland was inaugurated. A link between the Northern Ireland grid and that of the Irish Republic was re-established in 1996, along which electricity is both imported and

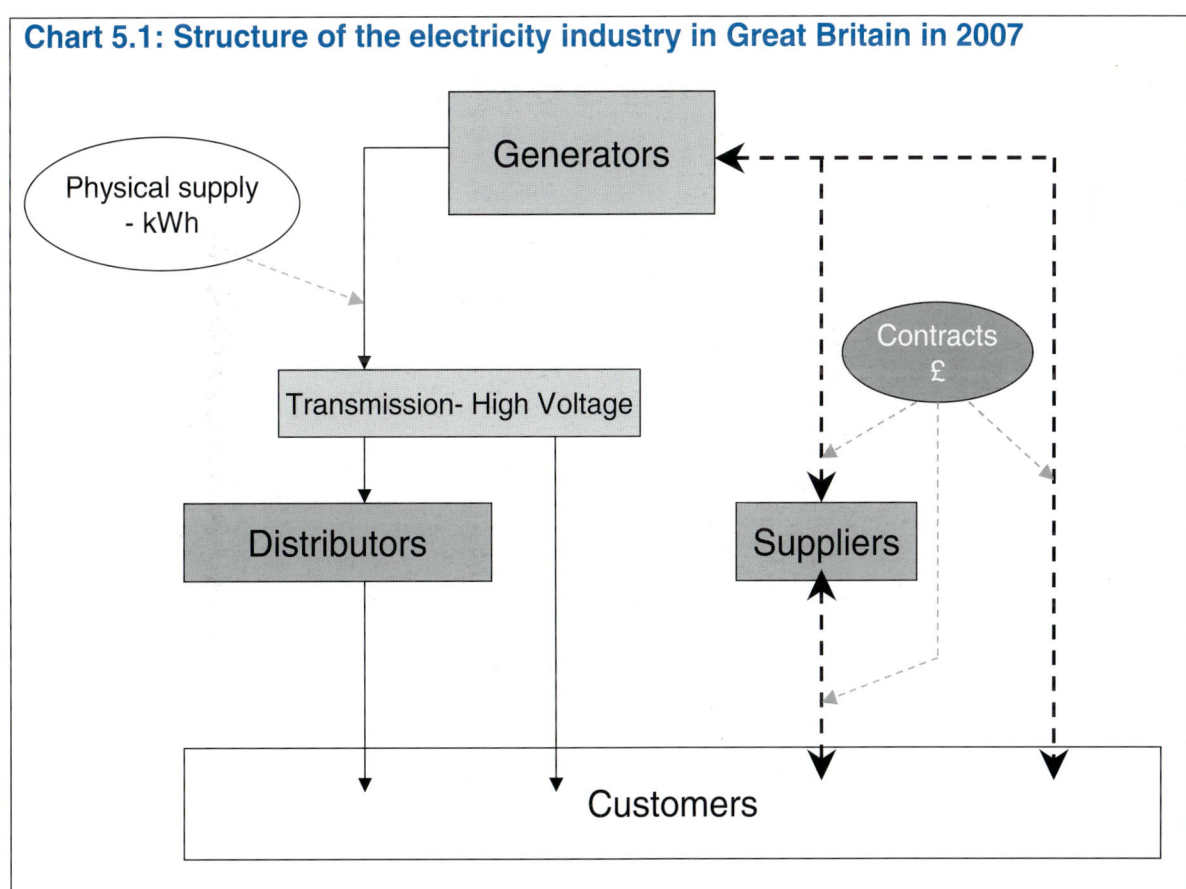

Chart 5.1: Structure of the electricity industry in Great Britain in 2007

Physical supply - kWh

Generators

Contracts £

Transmission- High Voltage

Distributors

Suppliers

Customers

Note: Distribution of electricity is also covered by contracts regulated by Ofgem exported. However, on 1 November 2007 the two grids were fully integrated and a joint body SEMO (Single Electricity Market Operator) was set up by SONI (System Operator for Northern Ireland) and Eirgrid from the Republic to oversee the new single market.

5.8   In March 2001, the means of trading electricity changed with the introduction in England and Wales of the New Electricity Trading Arrangements (NETA). This replaced the Electricity Pool of England and Wales. These arrangements were based on bi-lateral trading between generators, suppliers, traders and customers. They were designed to be more efficient and provide greater choice for market participants, whilst maintaining the operation of a secure and reliable electricity system. The system included forwards and futures markets, a balancing mechanism to enable National Grid, as system operator, to balance the system, and a settlement process. In April 2005 this system was extended to Scotland under BETTA. The system is shown in simplified form in Chart 5.1.

5.9   By December 2007, 13 million electricity consumers (50 per cent) were no longer with their home supplier. Table 5A gives market penetration in the fourth quarter of 2007. By the end of 2007, the home suppliers (i.e. the former regional electricity companies) had lost 46 per cent of the credit, 60 per cent of the direct debit, and 47 per cent of the prepayment market. However, as Table 5A shows there is considerable regional variation with much higher retention in Northern Scotland.

Table 5A: Domestic electricity market penetration (in terms of percentage of customers supplied) by Public Electricity Supply area and payment type, fourth quarter of 2007

Region	Home Supplier			Non-Home Supplier		
	Credit	Direct Debit	Prepayment	Credit	Direct Debit	Prepayment
North West	46	30	34	54	70	66
East Midlands	48	37	42	52	63	58
West Midlands	49	34	45	51	66	55
Merseyside and North Wales	52	42	51	48	58	49
Eastern	53	39	41	47	61	59
Yorkshire	55	39	48	45	61	52
North East	58	31	34	42	60	66
South East	58	38	57	42	62	43
London	58	41	60	42	59	40
Southern Scotland	58	50	62	42	50	38
South West	63	36	63	37	64	37
Southern	68	54	62	32	46	38
South Wales	71	64	86	29	36	14
Northern Scotland	83	71	84	17	29	16
Great Britain	54	40	53	46	60	47

Commodity balances for electricity (Table 5.1)

5.10   The first page of this two page balance table shows that just under 99 per cent of UK electricity supply in 2007 was home produced and just over 1 per cent was from imports net of exports. Just 0.9 per cent of home produced electricity was exported. The second page of the balance shows that of the 393,000 GWh produced (excluding pumped storage production), 90 per cent was from major power producers and 10 per cent from other generators, 19 per cent was from primary sources and 81 per cent from secondary sources.

5.11   Electricity generated by each type of fuel is shown on the second page of the commodity balance table. The link between electricity generated and electricity supplied is made in Table 5.6 and electricity supplied by each type of fuel is illustrated in Chart 5.3. Paragraph 5.27 examines further the ways of presenting each fuel's contribution to electricity production.

5.12   Demand for electricity is predominantly from final consumers, who accounted for just over 85 per cent in 2007. The remainder is split 8 per cent to energy industries' use and just under 7 per cent to losses. The electricity industry itself uses 57 per cent of the energy industries' total use of electricity. This does not include the 16 per cent of energy industry use accounted for by pumping at pumped storage stations. Petroleum refineries are the next most significant consumer with almost 15 per cent of energy industry use. The losses item has three components:

- transmission losses from the high voltage transmission system, which represented about 24 per cent of the figure in 2007,
- distribution losses, which occur between the gateways to the public supply system's network and the customers' meters, and accounted for about 72 per cent of losses,
- theft or meter fraud (around 4 per cent) (see paragraph 5.66).

5.13 Industrial consumption was just over 34 per cent of final consumption in 2007 (29 per cent of total demand for electricity which includes losses and fuel industries' use), marginally more than the consumption by households (just under 34 per cent), with transport, storage and communications and the services sector accounting for the remaining 32 per cent. Within the industrial sector the three largest consuming industries are chemicals, paper and food, which together account for 40 per cent of industrial consumption. Taken together, the engineering industries accounted for a further 19 per cent of industrial consumption of electricity. The iron and steel sector is also a substantial user of electricity but part of its consumption is included against blast furnaces and coke ovens under energy industry uses. This is because electricity is used by coke ovens and blast furnaces in the transformation of solid fuels into coke, coke oven gas and blast furnace gas. A note on the estimates included within these figures is to be found at paragraph 5.67. Chart 5.2 shows diagrammatically the total demand for electricity in 2007.

5.14 The transport sector covers electricity consumed by companies involved in transport, storage and communications. Within the overall total of 8,250 GWh, it is known that national railways consumed about 2,900 GWh each year for traction purposes, and this figure has been shown separately in the balances.

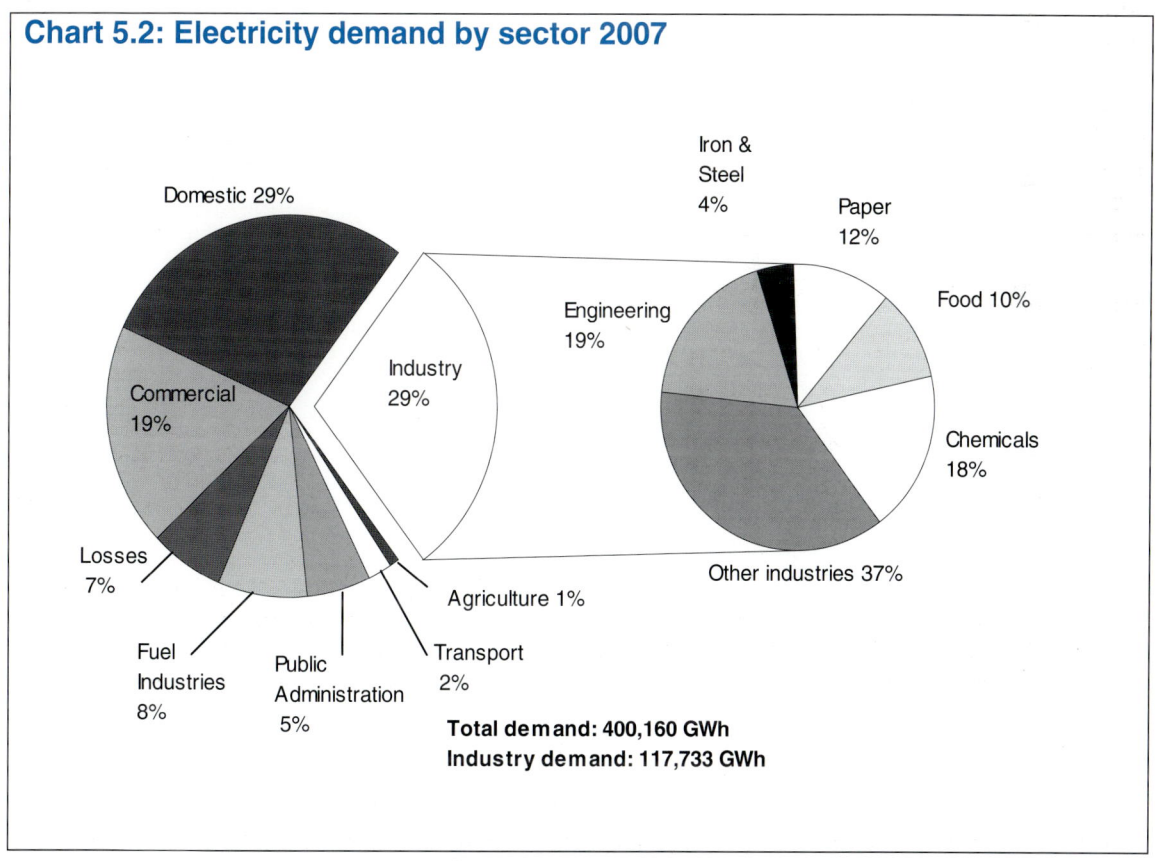

Chart 5.2: Electricity demand by sector 2007

Domestic 29%
Commercial 19%
Losses 7%
Fuel Industries 8%
Public Administration 5%
Transport 2%
Agriculture 1%
Industry 29%

Iron & Steel 4%
Paper 12%
Food 10%
Chemicals 18%
Other industries 37%
Engineering 19%

Total demand: 400,160 GWh
Industry demand: 117,733 GWh

Supply and consumption of electricity (Table 5.2)

5.15 There was a 1.1 per cent decrease in the supply of electricity in 2007. This follows on from a marginal fall in 2006 which was the first year on year fall in supply since 1997. From these two years' data it is difficult to determine the significance of this combined fall of 1.2 per cent since 2005. 2006 was a year of both high electricity prices and mild weather (see paragraph 5.16) and while electricity prices fell back a little in 2007, in both years there was an increased adoption of energy efficiency measures through the climate change agenda. Production (including pumped storage production) fell by 0.6 per cent, while imports of electricity net of exports were 31 per cent lower than in 2006. In 2003 high prices in continental Europe fostered a growth in electricity exports, which were nearly four times their level a year earlier in 2002. After that exports to continental Europe returned to their 2002 level but exports remained high because of the increase in exports from Northern Ireland to the Irish Republic. In 2007 exports to continental Europe were again high so while Northern Ireland exports accounted for about 65 per cent of the total UK electricity exports in 2006, they accounted for only around 40 per cent of the total in 2007. An analysis of electricity flows across Europe was carried out by BERR in 2007 using data published by the International Energy Agency and Eurostat. This was published in *Energy Trends*, March 2008[2]. (See Annex C for more information about *Energy Trends*).

5.16   Losses as a proportion of electricity demand in 2007, at 6.6 per cent, were slightly lower than in 2006 (6.8 per cent). Energy industry use as a proportion of total demand, at 8.0 per cent, was also slightly lower than in 2006 (8.1 per cent). Industrial consumption of electricity was 0.5 per cent lower than in 2006, and consumption in the services sector was 0.8 per cent lower. Domestic sector consumption of electricity was 1.2 per cent lower than in 2006. Consumption by transport, storage and communications was 0.5 per cent higher than in 2006. Temperatures influence the actual level of consumption in any one year in the winter months, as customers adjust heating levels in their homes and businesses. In addition, the hot summers of 2003 and 2006 led to an increased use of electricity for air conditioning and cooling. Temperatures in 2005 were mild in the first quarter, but there was a cold end to the year; thus on average during the year temperatures were similar to those of 2004 and 2003. 2006 had a cold first quarter, a hot third quarter and a mild fourth quarter which combined to give an average temperature during the year that was the warmest since 2002. 2007 had particularly mild first and second quarters but a cool third quarter so on average the year was cooler than 2006.

Table 5B: Electricity sales 2006

	Domestic sector sales (GWh)	Number of domestic customers (thousand) (1)	Industrial and commercial sector sales (GWh)	Number of I & C customers (thousand) (1)	All consumers sales (GWh)
Wales	5,600	1,328	11,794	125	17,394
Scotland	12,117	2,696	17,311	225	29,427
North East	4,494	1,174	9,315	82	13,809
North West	13,111	3,064	23,355	243	36,465
Yorkshire and the Humber	9,564	2,273	17,157	180	26,721
East Midlands	8,510	1,928	14,989	161	23,499
West Midlands	10,436	2,325	17,300	201	27,736
East of England	11,973	2,457	16,827	217	28,801
Greater London	13,701	3,267	29,143	415	42,843
South East	17,041	3,565	24,632	337	41,673
South West	11,181	2,337	15,783	244	26,964
Unallocated Consumption	88	20	2,410	10	2,499
Sales direct from high voltage lines (2)					6,600
Great Britain	117,817	26,434	200,015	2,441	324,432
Northern Ireland (3)					8,076
Total (4)					332,508

(1)   Figures are the number of Meter Point Administration Numbers (MPANs); every metering point has this unique reference number.
(2)   Based on estimate provided by Ofgem.
(3)   Northern Ireland data are based on data for electricity distributed provided by Northern Ireland Electricity.
(4)   This is close to the figure for UK electricity sales in 2006 of 328,300 GWh shown in Table 5.5; see paragraph 5.18.

Regional electricity data
5.17   The collection of data relating to regional and local consumption of electricity began in autumn 2004 and regional and local data on electricity consumption were published on an experimental basis

2 www.berr.gov.uk/files/file45397.pdf

in December of that year. The exercise was repeated in the autumns of 2005, 2006 and 2007 building on lessons learned from the 2004 exercise and the results were published in the December 2005, 2006 and 2007 *Energy Trends*. (See Annex C for more information about *Energy Trends*). For details of the availability of local level electricity (and gas) data see Chapter 4, paragraph 4.11 and the regional statistics pages of the BERR energy statistics web site:
www.berr.gov.uk/energy/statistics/regional/index.html .
From 2005, these statistics are no longer experimental and have been granted "National Statistics" status. A summary of electricity consumption at regional level is given in Table 5B and relates to 2006.

5.18   The difference between total UK electricity sales, shown in Table 5B, and total UK electricity sales shown in Table 5.5 is small (4,208 GWh or 1.3 per cent). This mainly arises from the fact that the regional data are not based exactly on a calendar year.

Commodity balances for the public distribution system and for other generators (Table 5.3)

5.19   Table 5.3 expands on the commodity balance format to show consumption divided between electricity distributed over the public distribution system and electricity provided by other generators (largely autogeneration and generation from renewable sources). Autogeneration is the generation of electricity wholly or partly for a company's own use as an activity which supplements its primary activity. However, most generators of electricity from renewable sources (apart from large scale hydro and some biofuels) are included as other generators because of their comparatively small size, even though their main activity is electricity generation. For a full list of companies included as major power producers see paragraph 5.51.

5.20   Table 5.3 also expands the domestic sector to show consumption by payment type and the commercial sector is expanded to show detailed data beyond that presented in Tables 5.1 and 5.2.

5.21   The proportion of electricity supplied by generators other than major power producers rose slightly in 2007 to 9.7 per cent. This is just below the previous peak share it reached in 2000, but is a new peak level in volume terms. The proportion of this electricity transferred to the public distribution system in 2007 was 34 per cent; in the previous four years this proportion has varied between 23 and 33 per cent according to market conditions. In three of the previous four years high gas prices and low electricity prices made it more difficult for electricity produced by other generators to compete in the electricity market, but in 2006 and 2007 the high level of electricity prices promoted greater sales by other generators to electricity suppliers. During the last 5 years there has also been greater generation from renewables and wastes which are included in the "Other generators" category (see Chapter 7).

5.22   In 2007, 6 per cent of final consumption of electricity was by other generators and did not pass over the public distribution system. This was the same proportion as in 2006. A substantial proportion of electricity used in the energy industries is self-generated (above 16 per cent in all three years shown in the table). At petroleum refineries the proportion is even higher and in 2007, 69 per cent of electricity was self-generated.

5.23   In 2007, 14 per cent of the industrial demand for electricity was met by autogeneration. There was also a lesser proportion (3.0 per cent) from autogeneration within the public administration and transport sectors. Table 1.9 in Chapter 1 shows the fuels used by autogenerators to generate this electricity within each major sector and also the quantities of electricity generated and consumed.

5.24   Within the domestic sector, the amount of electricity consumed reported as being purchased under some form of off-peak pricing structure was 31 per cent which is a reduction from the 32 per cent reported for the two previous years. Sixteen per cent of consumption was through prepayment systems, about the same proportion as in the two previous years.

Fuel used in generation (Table 5.4)

5.25   In this table fuel used by electricity generators is measured in both original units and for comparative purposes, in the common unit of million tonnes of oil equivalent. In Table 5.6 figures are quoted in a third unit, namely GWh, in order to show the link between fuel use and electricity generated.

5.26   The energy supplied basis defines the primary input (in million tonnes of oil equivalent, Mtoe) needed to produce 1 TWh of hydro, wind, or imported electricity as:

$$\text{Electricity generated (TWh)} \times 0.085985$$

The primary input (in Mtoe) needed to produce 1 TWh of nuclear electricity is similarly

$$\frac{\text{Electricity generated (TWh)} \times 0.085985}{\text{Thermal efficiency of nuclear stations}}$$

In the United Kingdom the thermal efficiency of nuclear stations has risen in stages from 32 per cent in 1982 to nearly 39 per cent in 2007 (see Table 5.10 and paragraph 5.58 for the definition)[3]. The factor of 0.085985 is the energy content of one TWh divided by the energy content of one million tonnes of oil equivalent (see page 209 and inside back cover flap).

5.27   Figures on fuel use for electricity generation can be compared in two ways. Table 5.4 illustrates one way by using the volumes of **fuel input** to power stations (after conversion of inputs to an oil equivalent basis), but this takes no account of how efficiently that fuel is converted into electricity. The fuel input basis is the most appropriate to use for analysis of the quantities of particular fuels used in electricity generation (eg to determine the additional amount of gas or other fuels required as coal use declines under tighter emissions restrictions). A second way uses the amount of electricity generated and supplied by each fuel. This **output** basis is appropriate for comparing how much, and what percentage, of electricity generation comes from a particular fuel. It is the most appropriate method to use to examine the dominance of any fuel and for diversity issues. Percentage shares based on fuel outputs reduce the contribution of coal and nuclear, and increase the contribution of gas (by over 5 percentage points in 2007) compared with the fuel input basis. This is because of the higher conversion efficiency of gas. This output basis is used in Chart 5.3, taking electricity supplied (gross) figures from Table 5.6. Trends in fuel used on this electricity supplied basis are described in the section on Table 5.6, in paragraphs 5.30 to 5.33, below.

5.28   A historical series of fuel used in generation on a consistent, energy supplied, fuel input basis is available at Table 5.1.1 on BERR's energy statistics web site and accessible from the Digest of UK Energy Statistics home page: www.berr.gov.uk/energy/statistics/publications/dukes/page45537.html

Relating measurements of supply, consumption and availability (Table 5.5)

5.29   The balance methodology uses terms that cannot be readily employed for years before 1998 because statistics were not available in sufficient detail. Table 5.5 shows the relationship between these terms for the latest five years. For the full definitions of the terms used in the commodity balances see the Annex A, paragraphs A.7 to A.42.

Electricity generated and supplied (Table 5.6)

5.30   The main data on generation and supply in Table 5.6 are presented by type of fuel. However, there remains an interest in the type of station and so the final part of the table shows generation from conventional steam stations and from combined cycle gas turbine stations over the most recent five years.

5.31   Total electricity generated in the United Kingdom in 2007 was 0.6 per cent lower than generation in 2006. Major power producers (as defined in paragraph 5.50) accounted for 90 per cent of electricity generation in 2007. Generation by other generators was 4.5 per cent up on a year earlier mainly because most (66 per cent in 2007) of the generation from thermal and non-thermal renewables (which has increased by 8.9 per cent) is included in the "other generators" category.

5.32   Generation from coal was 9.2 per cent lower in 2007 than in 2006. Generation from gas in 2007 was 16 per cent higher than in 2006 and 4.7 per cent higher than the previous record level in 2004. Generation from nuclear sources fell by 16 per cent with the nuclear sector again affected by a high

---

[3] *Note that the International Energy Agency uses 0.33 in its calculations, which is the European average thermal efficiency of nuclear stations in 1989, measured in net terms rather than the UK's gross terms.*

level of outages for repairs and maintenance as well as the closure at the end of 2006 of the two oldest Magnox stations at Dungeness and Sizewell.

5.33   Table 5.6 also shows electricity supplied data. These data take into account the fact that some stations use relatively more electricity than others in the generation process itself. In total, electricity supplied (gross) was 0.3 per cent less than the volume generated in 2006. For gas-fired stations it was 17 per cent more, while for nuclear stations it was 17 per cent less. Chart 5.3 shows how shares of the generation market in terms of electricity output have changed since 2006. Gas' share of electricity supplied (net) plus imports in 2007 at 43 per cent was 6 percentage points higher than in 2006 and 3 percentage points higher than the previous record high in 2004. Coal's share at 34 per cent was 3 percentage points less than in 2006 and 4 percentage points lower than the 40 per cent record share in 2004. Nuclear's 15 per cent share in 2007 was 3 percentage points lower than in 2006, and at its lowest since 1987. Oil's share remained at the 1 per cent it has recorded in each of the last seven years and imports' share has fallen from 3 per cent in 2001 to 1½ per cent in 2007.

Chart 5.3: Fuel used in electricity generation, on an output basis[1]

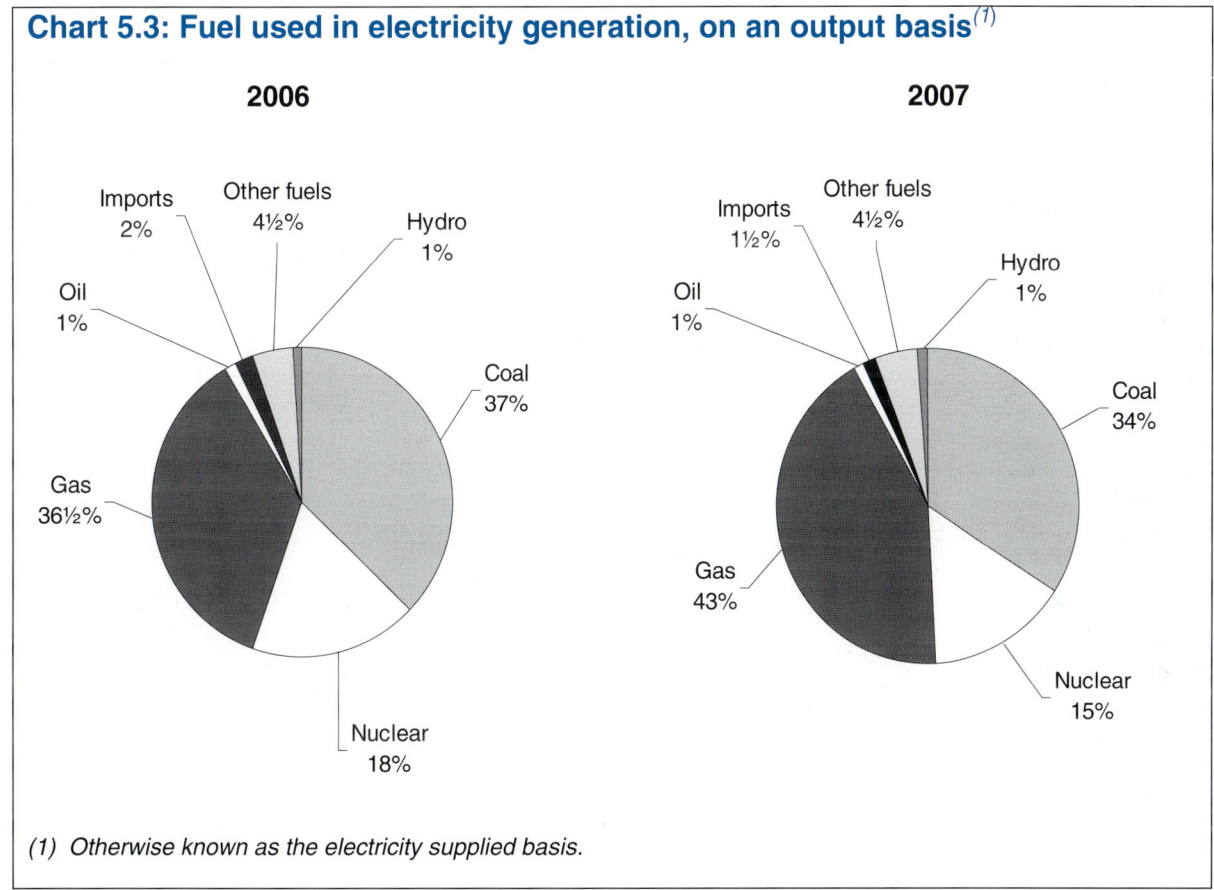

2006

Imports 2%
Other fuels 4½%
Hydro 1%
Oil 1%
Coal 37%
Gas 36½%
Nuclear 18%

2007

Imports 1½%
Other fuels 4½%
Hydro 1%
Oil 1%
Coal 34%
Gas 43%
Nuclear 15%

(1) Otherwise known as the electricity supplied basis.

Plant capacity (Tables 5.7, 5.8 and 5.9)

5.34   Table 5.7 shows capacity, ie the maximum power available at any one time, for major power producers and other generators by type of plant.

5.35   As in last year's Digest, for major power producers capacities at the end of December 2007 and at the end of December 2006 are measured in Transmission Entry Capacity (TEC) terms. A full definition is given in the Technical Notes section at paragraph 5.55. The effect of this change has been to increase the capacity of major power producers by about 2,000 GWh in total with the majority of fossil fuel stations increasing their capacity under the TEC measurement but some decreasing.

5.36   In 2007, there was an increase of 194 MW in the capacity of major power producers. The three main contributory factors were a new biomass fired power station, a new gas fired power station and an uprating following the installation of new steam turbines at an existing station. In addition there was a net increase from more minor up-ratings and down-ratings. In December 2007, major power producers accounted for 91 per cent of the total generating capacity, the same proportion as at the ends of December 2006. The capacity of other generators rose by 284 MW (+3.8 per cent). This included a 220 MW (27 per cent) increase in the capacity of wind turbines (all of which are currently included under Other generators) and a 147 MW (12 per cent) increase in the capacity of other renewables (excluding hydro) (see Chapter 7). Other generators' conventional steam plant capacity was 59 MW lower (-1.9 per cent).

5.37   A breakdown of the capacity of the major power producers' plants at the end of March each year from 1993 to 1996 and at the end of December for 1996 to 2007 is shown in Chart 5.4.

5.38   Table 5.8 separates the capacities of major power producers geographically to show England and Wales, Scotland and Northern Ireland. So as not to disclose data for individual stations that have been provided in confidence, the breakdowns by type of station cannot be given in as much detail as in Table 5.7. In 2007, nearly 85 per cent of the generating capacity in the UK owned by major power producers was in England and Wales, 13 per cent was in Scotland and just under 3 per cent in Northern Ireland. Out of the net increase in UK capacity of 194 MW in 2007, 149 MW was in England and Wales, 45 MW was in Scotland and there was no change in Northern Ireland.

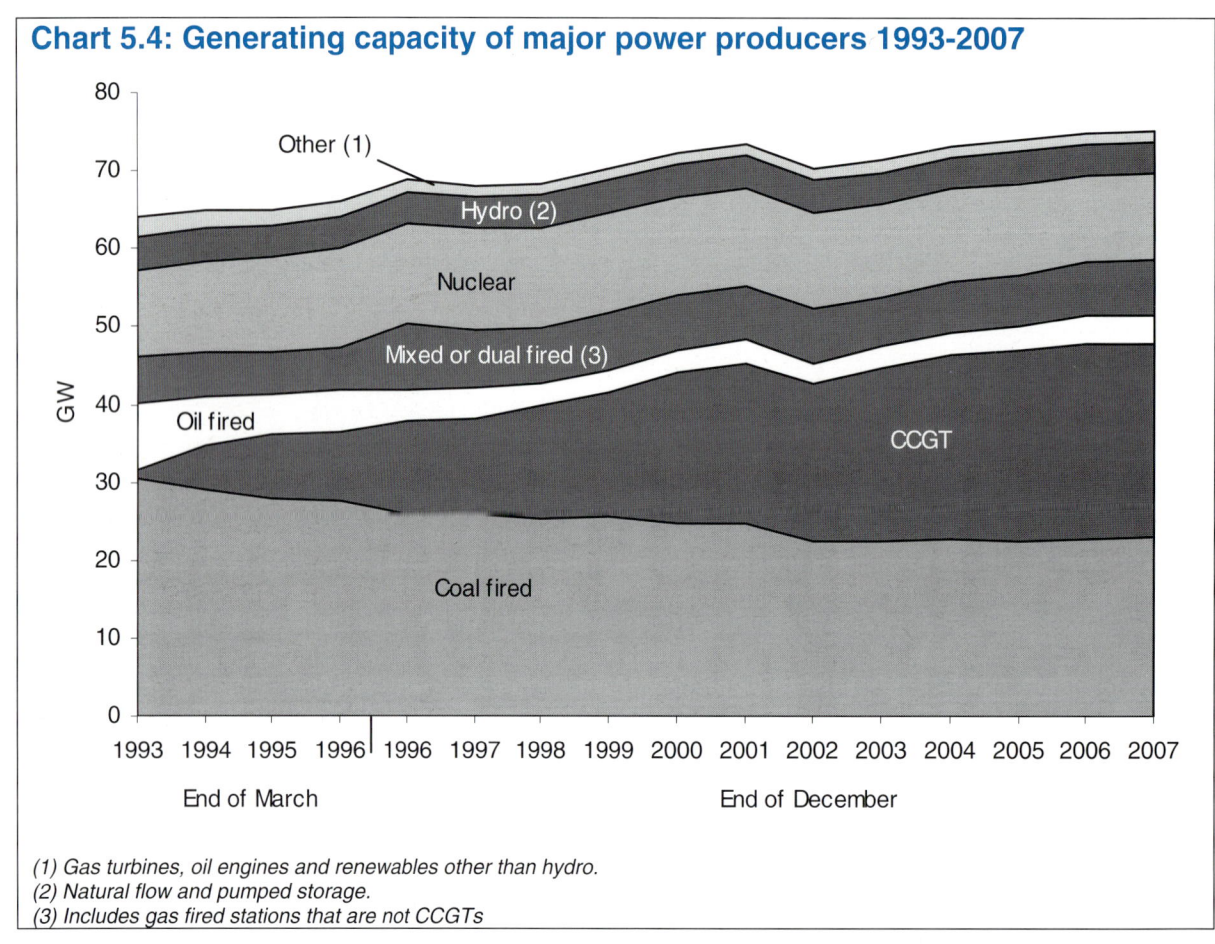

Chart 5.4: Generating capacity of major power producers 1993-2007

(1) Gas turbines, oil engines and renewables other than hydro.
(2) Natural flow and pumped storage.
(3) Includes gas fired stations that are not CCGTs

5.39   In Table 5.9, data for the generating capacity of industrial, commercial and transport undertakings are shown, according to the industrial classification of the generator. Because CHP schemes have now been classified according to the sector that receives the majority of the heat there are some changes to the allocation between sectors this year. 2005, 2006 and 2007 are all on the new basis but there are some discontinuities between 2004 and 2005 (see Chapter 6 paragraphs 6.19, 6.33 and Table 6G). In 2007, 15 per cent of the capacity was in the chemicals sector. Oil and gas terminals and oil refineries had just under 13 per cent of capacity, Paper, printing and publishing had a 10 per cent share, and engineering and other metal trades had an 8 per cent share. However, the major share (45 per cent) was outside the industrial sector.

Plant loads, demand and efficiency (Table 5.10)

5.40   Table 5.10 shows the maximum load met each year, load factors (by type of plant and for the system in total) and indicators of thermal efficiency. Maximum demand figures cover the winter period ending the following March. Until 2004, maximum demand figures for England and Wales, Scotland, and Northern Ireland are shown separately as well as total UK maximum demand. With the advent of BETTA (see paragraph 5.3), England, Wales and Scotland are covered by a single network and a single maximum load is shown for Great Britain in 2005, 2006 and 2007.

5.41   Maximum demand in the UK during the winter of 2007/2008 occurred in December 2007. This was 4.2 per cent higher than the previous winter's maximum in January 2007. Maximum demand in 2007/2008 was 82 per cent of the UK capacity of major power producers (as shown in Table 5.7) as measured at the end of December 2007, up from the 2006/07 percentage of 79.   Both these percentages are lower than the percentages for earlier years, in part due to the definitional change to TEC explained in paragraph 5.35.   On an unchanged basis the percentages would have been 84 (2007/08) and 81 (2006/07) compared with 83 in 2005/06.   In Great Britain maximum demand in December 2007 was 82 per cent of the England, Wales and Scotland capacity of major power producers (Table 5.8) compared with 79 per cent for winter 2006/07.   For Northern Ireland the proportion was 81 per cent (78 per cent in 2006/07).   These percentages do not include the capacities available via the interconnectors with neighbouring grid systems nor demand for electricity via these interconnectors.

5.42   Plant load factors measure how intensively each type of plant has been used. The trend up to the end of the 1990s had been for conventional thermal plant to be used less intensively and CCGT stations more intensively.   From 2000 increased maintenance and repair at nuclear stations (and initially at CCGT stations), and a rising trend in gas prices led to a departure from this trend. The load factor of nuclear stations in 2007 at 59.6 per cent was 20.5 percentage points below the recent peak load factor of 80.1 per cent in 1998. Very high gas prices in 2006 resulted in a particularly low CCGT load factor but this recovered in 2007 to close to the levels seen 3 or 4 years earlier. More intensive use of coal fired stations saw their plant load factor rise to 69.4 per cent in 2006, but fall back to 62.5 per cent in 2007.

5.43   2003 and to a lesser extent 2006 were particularly dry years, especially in some areas where hydro electricity is produced. As a result the load factors for natural flow hydro show falls in those years (a substantial fall in 2003). Conversely the wetter year of 2007 saw a greater use of large scale hydro plant. Pumped storage use was less affected by the dry weather and high electricity prices encouraged its use in 2006 and 2007. The lower availability of hydro was another contributory factor to the high load factors for conventional thermal in 2003 and 2006.

5.44   Thermal efficiency measures the efficiency with which the heat energy in fuel is converted into electrical energy. The efficiency of coal-fired stations had been on a downward trend as coal became the marginal fuel for generation, but coal's increased role since 2000 has halted the decline in the thermal efficiency of coal-fired generation. CCGT efficiency fell back slightly in 2006, but only to its 2004 level because it has tended to be the older, less efficient CCGT stations that have not been used so intensively because of the high gas prices. The efficiency of nuclear stations has been on a rising trend in recent years as older, less efficient stations have closed and this trend continued into 2007. However, outages have tended to counteract these efficiency gains in some years, including 2006. The efficiencies presented in this table are calculated using **gross** calorific values to obtain the energy content of the fuel inputs. If **net** calorific values are used, efficiencies are higher, for example CCGT efficiencies rise by about 5 percentage points.

Power stations in the United Kingdom (Tables 5.11 and 5.12)

5.45   Table 5.11 lists the operational power stations in the United Kingdom as at the end of May 2008 along with their installed capacity and the year they began to generate electricity. Where a company operates several stations the stations are grouped together. In general the table aims to list all stations of 1 MW installed capacity or over. Digest Table 5.12 adds to the information in Table 5.11. It shows CHP schemes of 1 MW and over for which the information is publicly available, but it is the total power output of these stations that is given, not just that which is classed as good quality CHP under the CHP Quality Assurance programme (CHPQA, see Chapter 6), since CHPQA information for individual sites is not publicly available. In Table 5.11 generating stations using renewable sources are also listed in aggregate form in the "Other power stations" section apart from hydro stations operated by the major power producers and selected major wind farms, which appear in the main table. For completeness CHP stations not appearing in the main table are also listed in aggregate in this section. Details of the interconnectors between England and France, Scotland and Northern Ireland and Northern Ireland and the Irish Republic are also given in this table. The total installed capacity of all the power stations individually listed in Table 5.11 is 78,024 MW.

Carbon dioxide emissions from power stations.

5.46   It is estimated that carbon dioxide emissions from power stations accounted for 33 per cent of the UK's total carbon dioxide emissions in 2007. Emissions vary by type of fuel used to generate the electricity and emission estimates for all electricity generation for 2005 to 2007 are shown in Table 5C below:

Table 5C: Estimated carbon dioxide emissions from electricity generation 2005 to 2007

Fuel	Emissions (tonnes of carbon dioxide per GWh electricity supplied)		
	2005	2006	2007
Coal	932	928	939
Oil	675	606	658
Gas	408	415	405
All fossil fuels	651	674	643
All fuels (including nuclear and renewables)	483	506	501

The Electricity Supply System in Great Britain in 2007

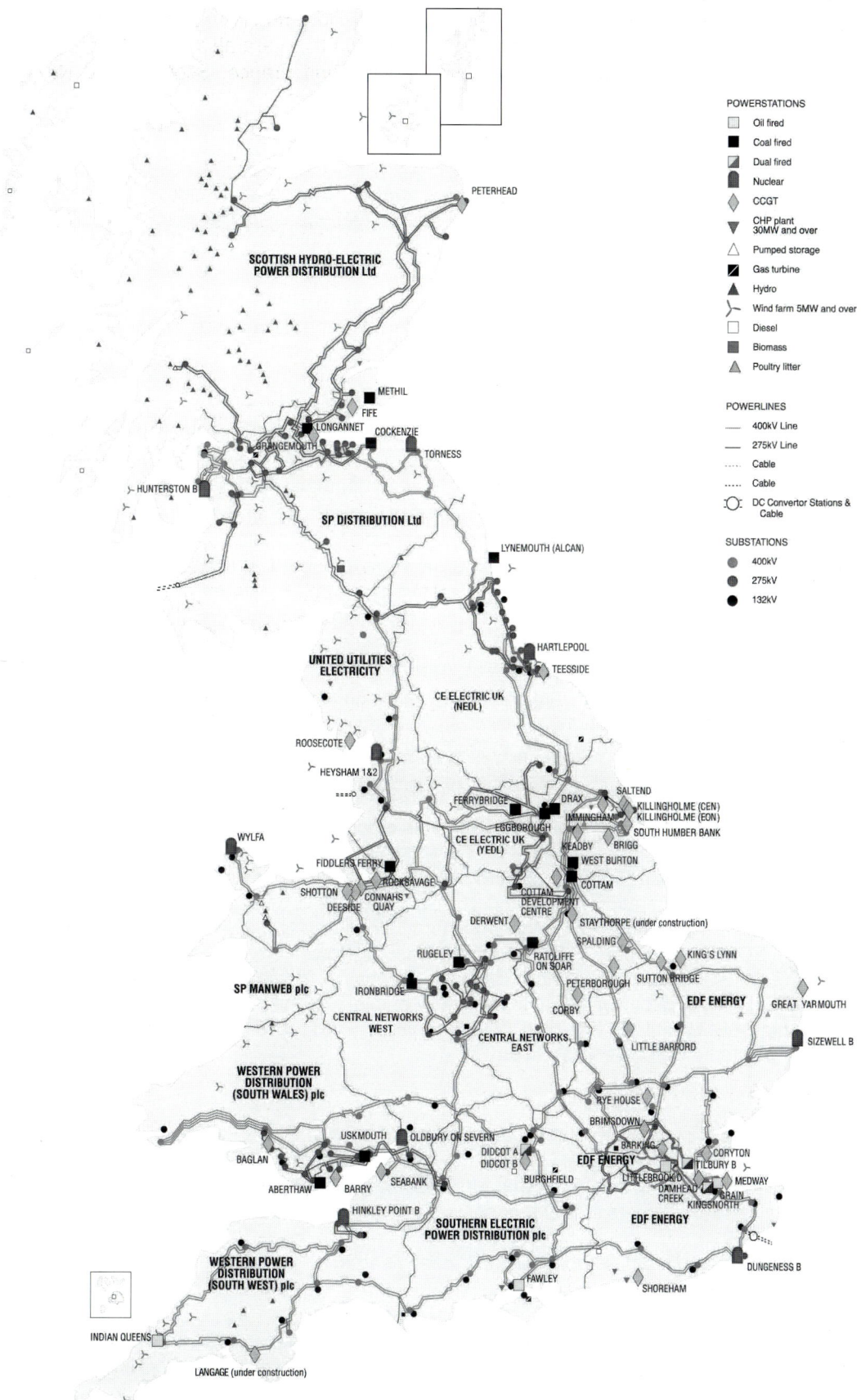

This map has been adapted from a map provided by Reed Business Publishing and National Grid; it is available in colour on the BERR energy website.

Technical notes and definitions

5.47 These notes and definitions are in addition to the technical notes and definitions covering all fuels and energy as a whole in Chapter 1, paragraphs 1.26 to 1.58. For notes on the commodity balances and definitions of the terms used in the row headings see Annex A, paragraphs A.7 to A.42. While the data in the printed and bound copy of this Digest cover only the most recent 5 years, these notes also cover data for earlier years that are available on the BERR web site.

Electricity generation from renewable sources
5.48 Figures on electricity generation from renewable energy sources are included in the tables in this section. Further detailed information on renewable energy sources is included in Chapter 7.

Combined heat and power
5.49 Electricity generated from combined heat and power (CHP) schemes, CHP generating capacities and fuel used for electricity generation are included in the tables in this chapter. However, more detailed analyses of CHP schemes are set out in Chapter 6.

Generating companies
5.50 Following the restructuring of the electricity supply industry in 1990, the term "Major generating companies" was introduced into the electricity tables to describe the activities of the former nationalised industries and distinguish them from those of autogenerators and new independent companies set up to generate electricity. The activities of the autogenerators and the independent companies were classified under the heading "Other generating companies". In the 1994 Digest, a new terminology was adopted to encompass the new independent producers, who were then beginning to make a significant contribution to electricity supply. Under this terminology, all companies whose prime purpose is the generation of electricity are included under the heading "Major power producers" (or MPPs). The term "Other generators" ("Autogenerators" in the balance tables) is restricted to companies who produce electricity as part of their manufacturing or other commercial activities, but whose main business is not electricity generation. "Other generators" also covers generation by energy services companies at power stations on an industrial or commercial site where the main purpose is the supply of electricity to that site, even if the energy service company is a subsidiary of a major power producer. Most generators of electricity from renewable sources (apart from large scale hydro and some biofuels) are also included as "Other generators" because of their comparatively small size, even though their main activity is electricity generation.

5.51 **Major power producers at the end of 2007 were:**
AES Electric Ltd., Baglan Generation Ltd., Barking Power Ltd., British Energy plc., Centrica Energy, Coolkeeragh ESB Ltd., Corby Power Ltd., Coryton Energy Company Ltd., Derwent Cogeneration Ltd., Drax Power Ltd., EDF Energy plc., E.On UK plc., Energy Power Resources, Immingham CHP, International Power Mitsui, Magnox Electric Ltd., Premier Power Ltd., RGS Energy Ltd, Rocksavage Power Company Ltd., RWE Npower plc., Scottish Power plc., Scottish and Southern Energy plc., Seabank Power Ltd., SELCHP Ltd., Spalding Energy Company Ltd., Teesside Power Ltd., Uskmouth Power Company, Western Power Generation Ltd.

Types of station
5.52 The various types of station identified in the tables of this chapter are as follows:

Conventional steam stations are stations that generate electricity by burning fossil fuels to convert water into steam, which then powers steam turbines.

Nuclear stations are also steam stations but the heat needed to produce the steam comes from nuclear fission.

Gas turbines use pressurised combustion gases from fuel burned in one or more combustion chambers to turn a series of bladed fan wheels and rotate the shaft on which they are mounted. This then drives the generator. The fuel burnt is usually natural gas or gas oil.

Combined cycle gas turbine (CCGT) stations combine in the same plant gas turbines and steam turbines connected to one or more electrical generators. This enables electricity to be produced at higher efficiencies than is otherwise possible when either gas or steam turbines are used in isolation. The gas turbine (usually fuelled by natural gas or oil) produces mechanical power (to drive the generator) and waste heat. The hot exhaust gases (waste heat) are fed to a boiler, where steam is raised at pressure to drive a conventional steam turbine that is also connected to an electrical generator.

Natural flow hydro-electric stations use natural water flows to turn turbines.

Pumped storage hydro-electric stations use electricity to pump water into a high level reservoir. This water is then released to generate electricity at peak times. Where the reservoir is open, the stations also generate some natural flow electricity; this is included with natural flow generation. As electricity is used in the pumping process, pumped storage stations are net consumers of electricity.

Other stations include wind turbines and stations burning fuels such as landfill gas, sewage sludge, biomass and waste.

Public distribution system

5.53　This comprises the grid systems in England and Wales, Scotland and Northern Ireland. In April 2005 the Scotland and England and Wales systems were combined into a single grid.

Sectors used for sales/consumption

5.54　The various sectors used for sales and consumption analyses are standardised across all chapters of the 2008 Digest. For definitions of the sectors see Chapter 1 paragraphs 1.54 to 1.58 and Annex A paragraphs A.31 to A.42.

Transmission Entry Capacity, Declared Net Capacity and Installed Capacity

5.55　Transmission Entry Capacity (TEC) is a Connection and Use of System Code term that defines a generator's maximum allowed export capacity onto the transmission system. In the generating capacity statistics of the 2007 Digest it replaced Declared Net Capacity (DNC) as the basis of measurement of the capacity of Major Power Producers from 2006. Declared Net Capacity is the maximum power available for export from a power station on a continuous basis minus any power generated or imported by the station from the network to run its own plant. It represents the nominal maximum capability of a generating set to supply electricity to consumers. The maximum rated output of a generator (usually under specific conditions designated by the manufacturer) is referred to as its Installed Capacity. For the nuclear industry, the World Association of Nuclear Operators (WANO) recommends that capacity of its reactors is measured in terms of Reference Unit Power (RUP) and it is the RUP figure that is given for both as equivalent to the transmission entry capacity, the installed capacity and the declared net capacity of nuclear stations.

5.56　DNC is used to measure the maximum power available from generating stations that use renewable resources. For wind and tidal power and small scale hydro a factor is applied to declared net capability to take account of the intermittent nature of the energy source (eg 0.43 for wind and 0.365 for small scale hydro).

Load factors

5.57　The following definitions are used in Table 5.10:

Maximum load – Twice the largest number of units supplied in any consecutive thirty minutes commencing or terminating at the hour.

Simultaneous maximum load met – The maximum load on the grid at any one time. From 2005 (following the introduction of BETTA – see paragraph 5.5) it is measured by the sum of the maximum load met in Great Britain and the load met at the same time in Northern Ireland. Prior to 2005 it was measured by the sum of the maximum load met in England and Wales and the loads met at the same time by companies in other parts of the United Kingdom. In 2007/08 the maximum load in Great Britain occurred on 17 December 2007 at the half hour period ending 18.00 (59,880 MW). However, in Northern Ireland the maximum load occurred on 9 January 2008 at the period ending 17.30 (1,710 MW), which was 1.7 per cent above the province's previous record on 20 December 2004. In Great Britain the highest ever load met was 60,118 MW on 10 December 2002.

Plant load factor – The average hourly quantity of electricity supplied during the year, expressed as a percentage of the average output capability at the beginning and the end of year.

System load factor – The average hourly quantity of electricity available during the year expressed as a percentage of the maximum demand nearest the end of the year or early the following year.

Thermal efficiency

5.58 Thermal efficiency is the efficiency with which heat energy contained in fuel is converted into electrical energy. It is calculated for fossil fuel burning stations by expressing electricity generated as a percentage of the total energy content of the fuel consumed (based on average gross calorific values). For nuclear stations it is calculated using the quantity of heat released as a result of fission of the nuclear fuel inside the reactor. The efficiency of CHP systems is discussed separately in Chapter 6, paragraph 6.22 and 6.23 and Table 6D. Efficiencies based on gross calorific value of the fuel (sometimes referred to as higher heating values or HHV) are lower than the efficiencies based on net calorific value (or lower heating value LHV). The difference between HHV and LHV is due to the energy associated with the latent heat of the evaporation of water products from the steam cycle which cannot be recovered and put to economic use.

Period covered

5.59 Until 2004 figures for the major power producers relate to periods of 52 weeks as listed below (although some data provided by electricity supply companies related to calendar months and were adjusted to the statistical calendar). In 2004 a change was made to a calendar year basis. This change was made in the middle of the year and the data are largely based on information collected monthly. The January to May 2004 data are therefore based on the 21 weeks ended 29 May 2004 and the calendar months June to December 2004, making a total of 361 days. In terms of days 2004 is therefore 1.1 per cent shorter than 2005:

Year	52 weeks ended
2002	29 December 2002
2003	28 December 2003
2004	21 weeks ended 29 May 2004 and 7 months ended 31 December 2004
2005	12 months ended 31 December 2005
2006	12 months ended 31 December 2006
2007	12 months ended 31 December 2007

5.60 Figures for industrial, commercial and transport undertakings relate to calendar years ending on 31 December, except for the iron and steel industry where figures relate to the following 52 or 53 week periods:

Year	52 weeks ended
2002	28 December 2002
	53 weeks ended
2003	3 January 2004
	52 weeks ended
2004	1 January 2005
2005	31 December 2005
2006	30 December 2006
2007	29 December 2007

Monthly and quarterly data

5.61 Monthly and quarterly data on fuel use, electricity generation and supply and electricity availability and consumption are available on BERR's Energy Statistics web site www.berr.gov.uk/energy/statistics/source/index.html . Monthly data on fuel used in electricity generation by major power producers are given in Monthly Table 5.3 and monthly data on supplies by type of plant and type of fuel are given in Monthly Table 5.4. Monthly data on availability and consumption of electricity by the main sectors of the economy are given in Monthly Table 5.5. A quarterly commodity balance for electricity is published in BERR's quarterly statistical bulletin *Energy Trends* (Quarterly Table 5.2) along with a quarterly table of fuel use for generation by all generators and electricity supplied by major power producers (Quarterly Table 5.1). Both these quarterly tables are also available from BERR's Energy Statistics web site. See Annex C for more information about *Energy Trends*.

Data collection

5.62   For Major Power Producers, as defined in paragraph 5.50, the data for the tables in this Digest are obtained from the results of an annual BERR inquiry, sent to each company, covering generating capacity, fuel use, generation, sales and distribution of electricity.

5.63   Another annual inquiry is sent to electricity distributors to establish electricity distributed by these companies. Similarly an annual inquiry is sent to licensed suppliers of electricity to establish electricity sales by these companies. Electricity consumption for the iron and steel sector is based on data provided by the Iron and Steel Statistics Bureau (ISSB) rather than electricity suppliers since electricity suppliers were over-estimating their sales to this sector. The difference between the ISSB and electricity suppliers' figures has been re-allocated to other sectors based on the results of the latest Office for National Statistics' Purchases Inquiry. A further means of checking electricity consumption data has been employed this year on the 2006 and 2007 figures. The data were validated using information on sectors from EU Emissions Trading Scheme (EU-ETS) sources. The figures could not be used directly in the allocation because not all electricity use is recorded by the EU-ETS as some companies are not signed up to the scheme. The EU-ETS was used to check minimum consumption by sectors against other data collected by BERR.

5.64   A sample of companies that generate electricity mainly for their own use (known as autogenerators or autoproducers – see paragraph 5.50, above) is covered by a quarterly inquiry commissioned by BERR but carried out by the Office for National Statistics (ONS). Where autogenerators operate a combined heat and power (CHP) plant, this survey is supplemented (on an annual basis) by information from the CHP Quality Assessment scheme (for autogenerators who have registered under the scheme – see Chapter 6 on CHP). There are two areas of autogeneration that are covered by direct data collection by BERR, mainly because the return contains additional energy information needed by the Department. These are the Iron and Steel industry, and generation on behalf of London Underground.

Losses and statistical differences

5.65   Statistical differences are included in Tables 5.1, 5.2 and 5.3. These arise because data collected on production and supply do not match exactly with data collected on sales or consumption. One of the reasons for this is that some of the data are based on different calendars as described in paragraphs 5.59 and 5.60, above. Sales data based on calendar years will always have included more electricity consumption than the slightly shorter statistical year of exactly 52 weeks.

5.66   Of the losses shown in the commodity balance for electricity of just over 26,400 GWh in 2007, it is estimated that about 6,200 GWh (around 1.6 per cent of electricity available) were lost from the high voltage transmission system of the National Grid and 19,000 GWh (5.4 per cent) between the grid supply points (the gateways to the public supply system's distribution network) and customers' meters. The balance (0.3 per cent of electricity available) is accounted for by theft and meter fraud, accounting differences and calendar differences (as described in paragraph 5.65, above).

5.67   Care should be exercised in interpreting the figures for individual industries in the commodity balance tables. As new suppliers have entered the market and companies have moved between suppliers, it has not been possible to ensure consistent classification between and within industry sectors and across years. The breakdown of final consumption includes some estimated data. In 2007, for about 3 per cent of consumption of electricity supplied by the public distribution system, the sector figures are partially estimated.

Contact: *Mike Janes (Statistician)*   *Joe Ewins*
           Energy Markets Unit   *Energy Markets Unit*
           mike.janes@berr.gsi.gov.uk   *joe.ewins@berr.gsi.gov.uk*
           020 7215 5186   *020 7215 5190*

5.1 Commodity balances

Electricity

	2005	2006	GWh 2007
Total electricity			
Supply			
Production	395,383r	394,971r	392,597
Other sources (1)	2,930	3,853	3,859
Imports	11,160	10,282	8,613
Exports	-2,839	-2,765	-3,398
Marine bunkers	-	-	-
Stock change	-	-	-
Transfers	-	-	-
Total supply	**406,633r**	**406,341r**	**401,671**
Statistical difference (2)	**+2,382r**	**+1,504r**	**+1,511**
Total demand	**404,251r**	**404,837r**	**400,160**
Transformation	-	-	-
Electricity generation	-	-	-
Major power producers	-	-	-
Other generators	-	-	-
Heat generation	-	-	-
Petroleum refineries	-	-	-
Coke manufacture	-	-	-
Blast furnaces	-	-	-
Patent fuel manufacture	-	-	-
Other	-	-	-
Energy industry use	**30,102r**	**32,677r**	**31,813**
Electricity generation	17,871r	19,210r	18,087
Oil and gas extraction	505	546	560
Petroleum refineries	4,459r	4,559r	4,662
Coal extraction and coke manufacture	1,165	1,133r	1,073
Blast furnaces	515	497r	479
Patent fuel manufacture	-	-	-
Pumped storage	3,707	4,918	5,071
Other	**1,881r**	**1,815r**	1,881
Losses	**27,674r**	**27,470r**	**26,401**
Final consumption	**346,475r**	**344,690r**	**341,945**
Industry	**120,524r**	**118,297r**	**117,733**
Unclassified	-	-	-
Iron and steel	5,020r	5,860r	4,924
Non-ferrous metals	7,693	7,682r	7,827
Mineral products	7,978	7,963r	8,204
Chemicals	24,125r	22,909r	21,266
Mechanical engineering, etc	8,633r	8,491r	8,720
Electrical engineering, etc	7,420r	7,325r	7,271
Vehicles	5,841r	5,847r	5,937
Food, beverages, etc	12,773r	12,318r	12,308
Textiles, leather, etc	3,393r	3,382r	3,349
Paper, printing, etc	13,725r	13,421r	13,703
Other industries	21,995r	21,460r	22,625
Construction	1,929	1,640	1,599
Transport	**8,816r**	**8,211r**	**8,254**
Air	-	-	-
Rail (3)	2,800	2,900	2,900
Road	-	-	-
National navigation	-	-	-
Pipelines	-	-	-
Other	**217,135r**	**218,181r**	**215,958**
Domestic	116,811	116,449	115,050
Public administration	20,878r	22,227r	21,848
Commercial	75,294	75,376	75,235
Agriculture	4,152	4,130	3,825
Miscellaneous	-	-	-
Non energy use	**-**	**-**	**-**

5.1 Commodity balances (continued)

Electricity

GWh

	2005	2006	2007
Electricity production			
Total production (4)	**395,383r**	**394,971r**	**392,597**
Primary electricity			
Major power producers	**85,444**	**79,144r**	**67,172**
Nuclear	81,618	75,451	63,028
Large scale hydro (4)	3,637	3,481r	3,906
Small scale hydro	189	212	238
Wind	-	-	-
Other generators	**4,008**	**5,136r**	**6,229**
Nuclear	-	-	-
Large scale hydro	841	634r	648
Small scale hydro	254	266r	297
Wind	2,912	4,236r	5,285
Secondary electricity			
Major power producers	**273,838**	**279,263r**	**287,221**
Coal	130,894	146,356r	132,535
Oil	2,716	2,883	2,349
Gas	137,483	126,637	149,346
Renewables	2,746	3,386r	2,991
Other	-	-	-
Other generators	**32,093r**	**31,428r**	**31,975**
Coal	3,947r	4,036r	4,010
Oil	2,417r	2,374r	2,293
Gas	15,159r	14,587r	15,127
Renewables	6,894r	6,951r	7,514
Other	3,676r	3,481r	3,032
Primary and secondary production (5)			
Nuclear	81,618	75,451	63,028
Hydro	4,921	4,593r	5,088
Wind	2,912	4,236r	5,285
Coal	134,841r	150,392r	136,545
Oil	5,133r	5,257r	4,641
Gas	152,642r	141,225r	164,473
Other renewables	9,639r	10,337r	10,505
Other	3,676r	3,481r	3,032
Total production	**395,383r**	**394,971r**	**392,597**

(1) Pumped storage production.
(2) Total supply minus total demand.
(3) See paragraph 5.14.
(4) Excludes pumped storage production.
(5) These figures are the same as the electricity generated figures in Table 5.6 except that they exclude
    pumped storage production. Table 5.6 shows that electricity used on works is deducted to obtain
    electricity supplied. It is electricity supplied that is used to produce Chart 5.3 showing each fuel's share
    of electricity output (see paragraph 5.27).

5.2 Electricity supply and consumption

GWh

	2003	2004	2005	2006	2007
Supply					
Production	395,475	391,219	395,383r	394,971r	392,597
Other sources (1)	2,734	2,649	2,930	3,853	3,859
Imports	5,119	9,784	11,160	10,282	8,613
Exports	-2,959	-2,294	-2,839	-2,765	-3,398
Total supply	**400,369**	**401,357**	**406,633r**	**406,341r**	**401,671**
Statistical difference (2)	+2,208	+2,387	+2,382r	+1,504r	+1,511
Total demand	**398,161**	**398,970**	**404,251r**	**404,837r**	**400,160**
Transformation	-	-	-	-	-
Energy industry use	**32,081**	**29,294**	**30,102r**	**32,677r**	**31,813**
Electricity generation	18,136	17,030	17,871r	19,210r	18,087
Oil and gas extraction	551	558	505	546	560
Petroleum refineries	5,769	4,681	4,459r	4,559r	4,662
Coal and coke	1,190	1,118	1,165	1,133r	1,073
Blast furnaces	492	468	515	497r	479
Pumped storage	3,546	3,497	3,707	4,918	5,071
Other	2,398	1,942	1,881r	1,815r	1,881
Losses	**29,862**	**30,728**	**27,674r**	**27,470r**	**26,401**
Final consumption	**336,218**	**338,949**	**346,475r**	**344,690r**	**341,945**
Industry	**113,358**	**115,842**	**120,524r**	**118,297r**	**117,733**
Unclassified	-	-	-	-	-
Iron and steel	5,434	5,412	5,020r	5,860r	4,924
Non-ferrous metals	7,284	7,518	7,693	7,682r	7,827
Mineral products	7,651	7,835	7,978	7,963r	8,204
Chemicals	20,941	21,128	24,125r	22,909r	21,266
Mechanical engineering. etc	8,839	8,510	8,633r	8,491r	8,720
Electrical engineering, etc	6,019	6,809	7,420r	7,325r	7,271
Vehicles	5,660	5,682	5,841r	5,847r	5,937
Food, beverages, etc	11,949	12,348	12,773r	12,318r	12,308
Textiles, leather, etc	3,443	3,407	3,393r	3,382r	3,349
Paper, printing, etc	12,750	13,671	13,725r	13,421r	13,703
Other industries	21,686	21,716	21,995r	21,460r	22,625
Construction	1,701	1,804	1,929	1,640	1,599
Transport	**8,212**	**8,463**	**8,816r**	**8,211r**	**8,254**
Other	**214,648**	**214,644**	**217,135r**	**218,181r**	**215,958**
Domestic	115,761	115,526	116,811	116,449	115,050
Public administration	20,623	20,708	20,878r	22,227r	21,848
Commercial	74,238	74,215	75,294	75,376	75,235
Agriculture	4,025	4,194	4,152	4,130	3,825
Miscellaneous	-	-	-	-	-
Non energy use	-	-	-	-	-

(1) Pumped storage production.
(2) Total supply minus total demand.

5.3 Commodity balances
Public distribution system and other generators

GWh

	2005			2006			2007		
	Public distribution system	Other generators	Total	Public distribution system	Other generators	Total	Public distribution system	Other generators	Total
Supply									
Major power producers	359,282	-	359,282	358,407r	-	358,407r	354,393	-	354,393
Other generators	-	36,100r	36,100r	-	36,564r	36,564r	-	38,204	38,204
Other sources (1)	2,930	-	2,930	3,853	-	3,853	3,859	-	3,859
Imports	11,160	-	11,160	10,282	-	10,282	8,613	-	8,613
Exports	-2,839	-	-2,839	-2,765	-	-2,765	-3,398	-	-3,398
Transfers	+8,537r	-8,537r	-	+10,112r	-10,112r	-	12,876	-12,876	-
Total supply	**379,070r**	**27,564r**	**406,633r**	**379,889r**	**26,451r**	**406,341r**	**376,343**	**25,328**	**401,671**
Statistical difference (2)	+2,382r	-r	+2,382r	+1,524r	-20r	+1,504r	+1,446	+65	+1,511
Total demand	**376,688r**	**27,564r**	**404,251r**	**378,365r**	**26,471r**	**404,837r**	**374,896**	**25,264**	**400,160**
Transformation	-	-	-	-	-	-	-	-	-
Energy industry use	**24,815r**	**5,288r**	**30,102r**	**27,321r**	**5,356r**	**32,677r**	**26,348**	**5,466**	**31,813**
Electricity generation	16,265	1,606r	17,871r	17,706r	1,504r	19,210r	16,510	1,577	18,087
Oil and gas extraction	505	-	505	546	-	546	560	-	560
Petroleum refineries	1,593	2,866r	4,459r	1,501	3,058r	4,559r	1,461	3,201	4,662
Coal and coke	1,066	98	1,165	1,037	96r	1,133r	983	90	1,073
Blast furnaces	-	515	515	-	497r	497r	-	479	479
Pumped storage	3,707	-	3,707	4,918	-	4,918	5,071	-	5,071
Other fuel industries	1,679	203r	1,881r	1,614	201r	1,815r	1,763	119	1,881
Losses	**27,643r**	**31**	**27,674r**	**27,443r**	**27r**	**27,470r**	**26,375**	**26**	**26,401**
Final consumption	**324,230**	**22,245r**	**346,475r**	**323,602**	**21,089r**	**344,690r**	**322,173**	**19,772**	**341,945**
Industry	**101,035**	**19,489r**	**120,524r**	**100,731**	**17,566r**	**118,297r**	**101,188**	**16,545**	**117,733**
Iron and steel	4,033	987r	5,020r	4,871r	989r	5,860r	3,983	941	4,924
Non-ferrous metals	4,355	3,338	7,693	4,429r	3,253r	7,682r	4,733	3,094	7,827
Mineral products	7,772	206	7,978	7,795r	167r	7,963r	8,052	151	8,204
Chemicals	15,190	8,935r	24,125r	15,102r	7,807r	22,909r	15,012	6,254	21,266
Mechanical engineering, etc	8,531	102r	8,633r	8,401	90r	8,491r	8,397	323	8,720
Electrical engineering, etc	7,415	5r	7,420r	7,322	3r	7,325r	7,268	3	7,271
Vehicles	5,742	98r	5,841r	5,775	72r	5,847r	5,872	65	5,937
Food, beverages, etc	10,887	1,886r	12,773r	10,821r	1,498r	12,318r	10,818	1,491	12,308
Textiles, leather, etc	3,387	6r	3,393r	3,377	5r	3,382r	3,344	5	3,349
Paper, printing, etc	10,235	3,490r	13,725r	10,161r	3,260r	13,421r	9,975	3,728	13,703
Other industries	21,574	421r	21,995r	21,052r	408r	21,460r	22,151	474	22,625
Construction	1,914	15	1,929	1,625	15	1,640	1,584	15	1,599
Transport	**7,464**	**1,352r**	**8,816r**	**7,520**	**691r**	**8,211r**	**7,515**	**739**	**8,254**
Of which National Rail (3)	2,800	-	2,800	2,900	-	2,900	2,900	-	2,900
Other	**215,730**	**1,404r**	**217,135r**	**215,350**	**2,831r**	**218,181r**	**213,470**	**2,488**	**215,958**
Domestic	116,811	-	116,811	116,449	-	116,449	115,050	-	115,050
Standard	64,676	-	64,676	65,668r	-	65,668r	66,408	-	66,408
Economy 7 and other off-peak	32,344	-	32,344	31,321r	-	31,321r	29,625	-	29,625
Prepayment (standard)	13,091	-	13,091	12,934r	-	12,934r	12,921	-	12,921
Prepayment (off-peak)	6,202	-	6,202	6,098r	-	6,098r	5,791	-	5,791
Sales under any other arrangement	498	-	498	428r	-	428r	306	-	306
Public administration	19,474	1,404r	20,878r	19,395	2,831r	22,227r	19,360	2,488	21,848
Public lighting (4)	2,095	-	2,095	2,147	-	2,147	2,223	-	2,223
Other public sector	17,379	1,404r	18,783r	17,248	2,831r	20,080r	17,137	2,488	19,625
Commercial	75,294	-	75,294r	75,376	-	75,376	75,235	-	75,235
Shops	34,258	-	34,258	34,348	-	34,348	30,674	-	30,674
Offices	22,422	-	22,422	22,811	-	22,811	25,557	-	25,557
Hotels	8,705	-	8,705	8,360	-	8,360	8,921	-	8,921
Combined domestic/ commercial premises	2,230	-	2,230	2,078	-	2,078	2,196	-	2,196
Post and telecommunications	5,880	-	5,880	5,979	-	5,979	6,086	-	6,086
Unclassified	1,800	-	1,800	1,800	-	1,800	1,800	-	1,800
Agriculture	4,152	-	4,152	4,130	-	4,130	3,825	-	3,825

(1) Pumped storage production.
(2) Total supply minus total demand.
(3) See paragraph 5.14.
(4) Sales for public lighting purposes are increasingly covered by wider contracts that cannot distinguish the public lighting element.

5.4 Fuel used in generation[1]

	Unit	2003	2004	2005	2006	2007
		Original units of measurement				
Major power producers (2)						
Coal	M tonnes	50.90	48.97	50.58	55.81	51.02
Oil (3)	"	0.63	0.55	0.79	0.81	0.54
Gas	GWh	284,662	304,497	295,643	278,127	319,836
Other generators (2)						
Transport undertakings:						
Gas	GWh	93	27	38	24	21
Undertakings in industrial and commercial sectors:						
Coal	M tonnes	1.57	1.49	1.48	1.56r	1.54
Oil (4)	"	0.48	0.48	0.42	0.43r	0.42
Gas	GWh	39,825	35,706	32,691r	31,643r	33,072
		Million tonnes of oil equivalent				
Major power producers (2)						
Coal		31.570	30.375	31.654	34.998r	31.934
Oil (4)		0.654	0.577	0.883	1.010r	0.750
Gas		24.476	26.182	25.421	23.915	27.501
Nuclear		20.041	18.164	18.372	17.131	14.036
Hydro (natural flow) (5)		0.221	0.337	0.329	0.318r	0.356
Other renewables (5)		0.381	0.540	0.818	0.731	0.625
Net imports		0.186	0.644	0.715	0.646	0.448
Total major power producers (2)		**77.530**	**76.819**	**78.192**	**78.749r**	**75.650**
Of which: conventional thermal and other stations (6)		34.631	33.667	35.824	39.821r	36.221
combined cycle gas turbine stations		22.451	24.007	22.952	20.833r	24.589
Other generators (2)						
Transport undertakings:						
Gas		0.008	0.002	0.003	0.002	0.002
Undertakings in industrial and commercial sectors:						
Coal		0.972	0.937	0.921r	0.949	0.966
Oil (4)		0.539	0.523	0.469	0.483r	0.469
Gas		3.424	3.070	2.811r	2.721r	2.844
Hydro (natural flow) (5)		0.057	0.080	0.094	0.077r	0.081
Wind (5)		0.111	0.166	0.250	0.363	0.453
Other renewables (5)		2.027	2.228	2.535	2.720	2.862
Other fuels (7)		1.522	1.387	1.834r	1.578r	1.393
Total other generators (2)		**8.659**	**8.394**	**8.917r**	**8.894r**	**0.070**
All generating companies						
Coal		32.542	31.312	32.575r	35.947r	32.900
Oil (3)(4)		1.192	1.100	1.352	1.493r	1.219
Gas		27.909	29.254	28.235r	26.639r	30.347
Nuclear		20.041	18.164	18.372	17.131	14.036
Hydro (natural flow) (5)		0.278	0.418	0.423	0.395r	0.437
Wind (5)		0.111	0.166	0.250	0.363	0.453
Other renewables (5)		2.408	2.768	3.353	3.451	3.487
Other fuels (7)		1.522	1.387	1.834r	1.578r	1.393
Net imports		0.186	0.644	0.715	0.646	0.448
Total all generating companies		**86.189**	**85.214**	**87.108r**	**87.645r**	**84.720**

(1) For details of where to find monthly updates of fuel used in electricity generation by major power producers and quarterly updates of fuel used in electricity generation by all generating companies see paragraph 5.61.

(2) See paragraphs 5.50 and 5.51 for information on companies covered.

(3) Includes oil used in gas turbine and diesel plants, and oil used for lighting up coal fired boilers. Other fossil fuels such as petcoke are included with oil where the figures shown in million tonnes of oil equivalent.

(4) Includes refinery gas.

(5) Renewable sources, which are included under hydro and other renewables in this table, are shown separately in Table 7.6 of Chapter 7.

(6) Includes gas turbines, oil engines and plants producing electricity from renewable sources other than hydro.

(7) Main fuels included are coke oven gas, blast furnace gas, colliery methane and waste products from chemical processes.

5.5 Electricity supply, electricity supplied (net), electricity available, electricity consumption and electricity sales

GWh

	2003	2004	2005	2006	2007
Total supply					
(as given in Tables 5.1 and 5.2)	400,369	401,357	406,633r	406,341r	401,671
less imports of electricity	-5,119	-9,784	-11,160	-10,282	-8,613
plus exports of electricity	+2,959	+2,294	+2,839	+2,765	+3,398
less electricity used in pumped storage	-3,546	-3,497	-3,707	-4,918	-5,071
less electricity used on works	-18,136	-17,030	-17,871r	-19,210r	-18,087
equals					
Electricity supplied (net)	376,527	373,340	376,734r	374,693r	373,298
(as given in Tables 5.6, 5.1.2 and 5.1.3)					
Total supply					
(as given in Tables 5.1 and 5.2)	400,369	401,357	406,633r	406,341r	401,671
less electricity used in pumped storage	-3,546	-3,497	-3,707	-4,918	-5,071
less electricity used on works	-18,136	-17,030	-17,871r	-19,210r	-18,087
equals					
Electricity available	378,687	380,830	385,055r	382,213r	378,513
(as given in Table 5.1.2)					
Final consumption					
(as given in Tables 5.2 and 5.3)	336,218	338,949	346,475r	344,690r	341,945
plus Iron and steel consumption counted as energy industry use	+648	+625	+675r	+637r	+607
equals					
Final users	336,866	339,574	347,150r	345,327r	342,552
(as given in Table 5.1.2)					
Final consumption					
Public distribution system					
(as given in Table 5.3)	319,492	319,066	324,230	323,602	322,173
plus Oil and gas extraction use	+551	+558	+505	+546	+560
plus Petroleum refineries use	+1,550	+1,478	+1,593	+1,501	+1,461
plus Coal and coke use	+1,091	+1,027	+1,066	+1,037	+983
plus Other fuel industries use	+1,649	+1,585	+1,679	+1,614	+1,763
equals					
UK Electricity sales (1)	324,333	323,714	329,073	328,300	326,940

(1) The renewables obligation percentage is calculated using total renewables generation on an obligation basis from Table 7.5 (x 100) as the numerator, and this figure as the denominator. Separate electricity sales data for public electricity suppliers are given for England and Wales, Scotland and Northern Ireland in Table 5.5 of Energy Trends on the BERR web site at www.berr.gov.uk/energy/statistics/source/electricity/page18527.html (and scroll to the Monthly Tables section).

5.6 Electricity fuel use, generation and supply

GWh

	Coal	Oil	Gas	Nuclear	Renew-ables (1)	Other (3)	Total	Hydro-natural flow	Hydro-pumped storage	Other (4)	Total All sources
	Thermal sources							**Non-thermal sources**			
2003											
Major power producers (2)											
Fuel used	367,162	7,604	284,662	233,080	4,434	-	896,941	2,568	2,734	-	902,243
Generation	134,023	2,197	131,238	88,686	1,154	-	357,299	2,568	2,734	-	362,600
Used on works	6,325	249	3,201	6,775	95	-	16,645	9	92	-	16,747
Supplied (gross)	127,698	1,948	128,037	81,911	1,059	-	340,654	2,559	2,641	-	345,854
Used in pumping											3,546
Supplied (net)											342,308
Other generators (2)											
Fuel used	11,301	6,263	39,265	-	23,574	17,703	98,106	660	-	1,288	100,054
Generation	4,282	2,397	17,643	-	5,537	3,800	33,660	660	-	1,288	35,609
Used on works	220	174	547	-	306	135	1,382	7	-	-	1,389
Supplied	4,062	2,223	17,097	-	5,231	3,665	32,278	653	-	1,288	34,220
All generating companies											
Fuel used	378,463	13,867	323,926	233,080	28,008	17,703	995,047	3,228	2,734	1,288	1,002,297
Generation	138,305	4,594	148,881	88,686	6,692	3,800	390,959	3,228	2,734	1,288	398,209
Used on works	6,545	424	3,747	6,775	401	135	18,027	16	92	-	18,136
Supplied (gross)	131,760	4,171	145,134	81,911	6,290	3,665	372,932	3,212	2,641	1,288	380,074
Used in pumping											3,546
Supplied (net)											376,528
2004											
Major power producers (2)											
Fuel used	353,256	6,743	304,495	211,248	6,282	-	882,024	3,908	2,649	-	888,581
Generation	127,827	1,883	140,577	79,999	1,471	-	351,757	3,908	2,649	-	358,313
Used on works	5,890	354	2,819	6,317	104	-	15,486	7	90	-	15,582
Supplied (gross)	121,937	1,528	137,758	73,682	1,367	-	336,271	3,901	2,559	-	342,731
Used in pumping											3,497
Supplied (net)											339,234
Other generators (2)											
Fuel used	10,902	6,083	35,733	-	25,902	16,132	94,751	936	-	1,939	97,627
Generation	3,961	2,761	16,487	-	6,407	3,062	32,679	936	-	1,939	35,554
Used on works	208	196	511	-	405	111	1,431	17	-	-	1,448
Supplied	3,753	2,565	15,976	-	6,002	2,951	31,248	919	-	1,939	34,106
All generating companies											
Fuel used	364,158	12,827	340,228	211,248	32,184	16,132	976,776	4,844	2,649	1,939	986,207
Generation	101,700	4,044	157,064	79,999	7,878	3,062	384,436	4,844	2,649	1,939	393,867
Used on works	6,098	550	3,330	6,317	509	111	16,917	24	90	-	17,030
Supplied (gross)	125,689	4,094	153,734	73,682	7,369	2,951	367,519	4,820	2,559	1,939	376,837
Used in pumping											3,497
Supplied (net)											373,340
2005											
Major power producers (2)											
Fuel used	368,134	10,268	295,643	213,661	9,515	-	897,222	3,826	2,930	-	903,978
Generation	130,894	2,716	137,483	81,618	2,746	-	355,456	3,826	2,930	-	362,212
Used on works	6,038	511	2,959	6,445	154	-	16,106	5	154	-	16,265
Supplied (gross)	124,857	2,205	134,524	75,173	2,592	-	339,350	3,821	2,776	-	345,947
Used in pumping											3,707
Supplied (net)											342,240r
Other generators (2)											
Fuel used	10,712r	5,454r	32,729r	-	29,475r	21,327r	99,696r	1,096	-	2,912	103,704r
Generation	3,947r	2,417r	15,159r	-	6,894r	3,676r	32,093r	1,096	-	2,912	36,101r
Used on works	210	157r	470r	-	470r	133	1,440r	166	-	-	1,606r
Supplied	3,737r	2,260r	14,689r	-	6,424r	3,543r	30,653r	930r	-	2,912	34,495r
All generating companies											
Fuel used	378,846r	15,722r	328,372r	213,661	38,990r	21,327r	996,918r	4,922	2,930	2,912	1,007,682r
Generation	134,841r	5,133r	152,642r	81,618	9,639r	3,676r	387,549r	4,922	2,930	2,912	398,313r
Used on works	6,247r	668r	3,428r	6,445	624r	133	17,546r	171	154	-	17,871r
Supplied (gross)	128,594r	4,465r	149,214r	75,173	9,016r	3,543r	370,003r	4,750	2,776	2,912	380,442r
Used in pumping											3,707
Supplied (net)											376,735r

5.6 Electricity fuel use, generation and supply (continued)

GWh

		Thermal sources						Non-thermal sources			
	Coal	Oil	Gas	Nuclear	Renew-ables (1)	Other (3)	Total	Hydro-natural flow	Hydro-pumped storage	Other (4)	Total All sources
2006											
Major power producers (2)											
Fuel used	407,027r	11,751r	278,149	199,237r	8,504	-	904,668r	3,693r	3,853	-	912,214r
Generation	146,356	2,883	126,637	75,451	3,386r	-	354,714r	3,693r	3,853	-	362,260r
Used on works	7,937r	567r	2,634	6,214	211r	-	17,562r	13	130	-	17,706r
Supplied (gross)	138,419r	2,316r	124,003	69,237	3,175r	-	337,151r	3,680r	3,722r	-	344,554r
Used in pumping											4,918
Supplied (net)											339,636r
Other generators (2)											
Fuel used	11,038	5,614r	31,667r	-	31,628r	18,356r	98,302r	899r	-	4,236r	103,437r
Generation	4,036r	2,374r	14,587r	-	6,951r	3,481r	31,428r	899r	-	4,236r	36,563r
Used on works	216r	167r	452r	-	531r	124r	1,490r	14r	-	-	1,504r
Supplied	3,820r	2,207r	14,135r	-	6,420r	3,356r	29,938r	885r	-	4,236r	35,059r
All generating companies											
Fuel used	418,065r	17,365r	309,816r	199,237r	40,131r	18,356r	1,002,970r	4,593r	3,853	4,236r	1,015,651r
Generation	150,392r	5,257r	141,225r	75,451	10,337r	3,481r	386,142r	4,593r	3,853	4,236r	398,823r
Used on works	8,153r	734r	3,086r	6,214	742r	124r	19,052r	27r	130	-	19,210r
Supplied (gross)	142,239r	4,523r	138,139r	69,237	9,595r	3,356r	367,090r	4,565r	3,722	4,236r	379,613r
Used in pumping											4,918
Supplied (net)											374,695r
2007											
Major power producers (2)											
Fuel used	371,396	8,718	319,836	161,582	7,271	-	868,804	4,144	3,859	-	876,807
Generation	132,535	2,349	149,346	63,028	2,991	-	350,249	4,144	3,859	-	358,252
Used on works	6,976	400	2,894	5,779	417	-	16,466	30	13	-	16,510
Supplied (gross)	125,559	1,949	146,452	57,249	2,574	-	333,783	4,114	3,846	-	341,742
Used in pumping											5,071
Supplied (net)											336,671
Other generators (2)											
Fuel used	11,234	5,452	33,093	-	33,045	16,198	99,022	946	-	5,285	105,252
Generation	4,010	2,293	15,127	-	7,514	3,032	31,975	946	-	5,285	38,205
Used on works	214	161	469	-	611	106	1,561	16	-	-	1,577
Supplied	3,796	2,131	14,658	-	6,903	2,926	30,414	930	-	5,285	36,628
All generating companies											
Fuel used	382,630	14,171	352,929	161,582	40,316	16,198	967,827	5,089	3,859	5,285	982,060
Generation	136,545	4,641	164,473	63,028	10,505	3,032	382,224	5,089	3,859	5,285	396,457
Used on works	7,190	561	3,363	5,779	1,029	106	18,028	46	13	-	18,087
Supplied (gross)	129,355	4,080	161,110	57,249	9,477	2,926	364,197	5,043	3,846	5,285	378,370
Used in pumping											5,071
Supplied (net)											373,299

	2003		2004		2005		2006		2007	
	Conv-entional thermal (5)	CCGT	Conv-entional thermal (5)	CCGT	Conv-entional thermal (5)	CCGT	Conv-entional thermal (5)	CCGT	Conv-entional thermal (5)	CCGT
Major power producers (2)										
Generated	147,536	121,076	140,576	131,182	143,149	130,689	160,768r	118,495	147,395	139,826
Supplied (gross)	140,196	118,546	133,607	128,983	135,999	128,179	151,516r	116,398	138,973	137,561
Other generators										
Generated	22,781	10,879	20,827	11,852	20,301r	11,792r	19,867r	11,561r	20,459	11,516
Supplied (gross)	21,942	10,336	19,988	11,260	19,449	11,204r	18,594r	10,984r	19,473	10,941
All generating companies										
Generated	170,318	131,955	161,403	143,034	163,450r	142,481r	180,636r	130,056r	167,854	151,342
Supplied (gross)	162,139	128,882	153,595	140,243	155,448	139,383r	170,471r	127,382r	158,446	148,502

(1) Thermal renewable sources are those included under biofuels and non-biodegradable wastes in Chapter 7.
(2) See paragraphs 5.50 and 5.51 on companies covered.
(3) Other thermal sources include coke oven gas, blast furnace gas and waste products from chemical processes.
(4) Other non-thermal sources include wind, wave and solar photovoltaics.
(5) Includes gas turbines, oil engines and plants producing electricity from thermal renewable sources; also stations with some CCGT capacity but mainly operate in conventional thermal mode.

5.7 Plant capacity - United Kingdom

MW

	2003	2004	2005	2006	2007
				end December	
Major power producers (1)					
Total transmission entry capacity (2)	**71,471**	**73,293**	**73,941**	**74,996r**	**75,190**
Of which:					
Conventional steam stations:	31,867	31,982	32,292	33,608r	33,776
Coal fired	22,524	22,639	22,627	22,882r	23,008
Oil fired	2,930	2,930	3,262	3,778	3,778
Mixed or dual fired (3)	6,413	6,413	6,403	6,948r	6,990
Combined cycle gas turbine stations	22,037	23,783	24,263	24,859	24,854
Nuclear stations	11,852	11,852	11,852	10,969	10,979
Gas turbines and oil engines	1,537	1,495	1,356	1,444	1,404
Hydro-electric stations:					
Natural flow	1,273	1,276	1,273	1,294	1,293
Pumped storage	2,788	2,788	2,788	2,726	2,744
Renewables other than hydro	117	117	117	96	140
Other generators (1)					
Total capacity of own generating plant (6)	**6,793**	**6,829**	**7,422r**	**7,477r**	**7,761**
Of which:					
Conventional steam stations (4)	3,480	3,275	3,269r	3,106r	3,047
Combined cycle gas turbine stations	1,927	1,968	2,182	2,147r	2,119
Hydro-electric stations (natural flow)	129	132	120r	123r	127
Wind (5)	312	393	658	822	1,042
Renewables other than hydro and wind	945	1,061	1,194	1,278	1,425
All generating companies					
Total capacity	**78,264**	**80,122**	**81,363r**	**82,473r**	**82,951**
Of which:					
Conventional steam stations (4)	35,347	35,257	35,561r	36,714r	36,823
Combined cycle gas turbine stations	23,964	25,751	26,445	27,006r	26,973
Nuclear stations	11,852	11,852	11,852	10,969	10,979
Gas turbines and oil engines	1,537	1,495	1,356	1,444	1,404
Hydro-electric stations:					
Natural flow	1,402	1,408	1,393r	1,417r	1,420
Pumped storage	2,788	2,788	2,788	2,726	2,744
Wind (5)	312	393	658	822	1,042
Renewables other than hydro and wind	1,062	1,178	1,311	1,374	1,565

(1) See paragraphs 5.50 and 5.51 for information on companies covered.
(2) See paragraphs 5.55 for definition. Data before 2006 are based on declared net capability.
(3) Includes gas fired stations that are not Combined Cycle Gas Turbines, or have some CCGT capability but mainly operate as conventional thermal stations.
(4) For other generators, conventional steam stations include combined heat and power plants (electrical capacity only) but exclude combined cycle gas turbine plants, hydro-electric stations and plants using renewable sources.
(5) At present all wind generation is allocated to "Other generators"
(6) "Other generators" capacities are given in declared net capacity terms, see paragraph 5.55

5.8 Plant capacity - England and Wales, Scotland, and Northern Ireland

MW

	2003	2004	2005	2006	2007
				end December	
Major power producers in England and Wales (1)					
Total transmission entry capacity (2)	**60,056**	**61,875**	**62,343**	**63,390r**	**63,540**
Of which:					
Conventional steam stations:	26,171	26,286	26,792	28,132r	28,300
Coal fired	19,068	19,183	19,171	19,426r	19,552
Oil fired	2,750	2,750	3,262	3,778	3,778
Mixed or dual fired (3)	4,353	4,353	4,359	4,928	4,970
Combined cycle gas turbine stations	20,967	22,664	22,765	23,358	23,353
Nuclear stations	9,412	9,412	9,412	8,559	8,569
Gas turbines and oil engines	1,168	1,177	1,038	1,124r	1,086
Hydro-electric stations:					
Natural flow	133	131	131	136	133
Pumped storage	2,088	2,088	2,088	1,986	2,004
Renewables other than hydro	117	117	117	96	96
Major power producers in Scotland (1)					
Total transmission entry capacity (2)	**9,500**	**9,555**	**9,537**	**9,582r**	**9,627**
Of which:					
Conventional steam and combined cycle gas turbine stations	5,070	5,119	5,103	5,119	5,119
Nuclear stations	2,440	2,440	2,440	2,410	2,410
Gas turbines and oil engines	150	152	152	155r	153
Hydro-electric stations:					
Natural flow	1,140	1,144	1,142	1,158	1,161
Pumped storage	700	700	700	740	740
Renewables other than hydro	-	-	-	-	44
Major power producers in Northern Ireland (1)					
Total transmission entry capacity (2)	**1,915**	**1,862**	**2,061**	**2,023**	**2,023**

(1) See paragraphs 5.50 and 5.51 for information on companies covered.
(2) See paragraph 5.55 for definition. Data before 2006 are based on declared net capability.
(3) Includes gas fired stations that are not Combined Cycle Gas Turbines.

5.9 Capacity of other generators

MW

	2003	2004	2005	2006	2007
				end December	
Capacity of own generating plant (1)					
Undertakings in industrial and commercial sector:					
Oil and gas terminals and oil refineries	955	868	923r	919r	974
Iron and steel	312	313	313	314	315
Chemicals	1,324	1,176	1,265r	1,226r	1,174
Engineering and other metal trades	670	620	616r	588r	616
Food, drink and tobacco	379	375	411r	412r	410
Paper, printing and publishing	804	788	779r	759r	760
Other (2)	2,246	2,586	3,011r	3,156r	3,410
Total industrial and commercial sector	6,690	6,726	7,319r	7,374r	7,658
Undertakings in transport sector	103	103	103	103	103
Total other generators	**6,793**	**6,829**	**7,422r**	**7,477r**	**7,761**

(1) For combined heat and power plants the electrical capacity only is included. Further CHP capacity is included under major power producers in Table 5.7. A detailed analysis of CHP capacity is given in the tables of Chapter 6.
It has not been possible to include revisions to CHP capacity for years before 2005 in this table see paragraph 5.39.
(2) Includes companies in the commercial sector.

5.10 Plant loads, demand and efficiency

Major power producers [1]

	Unit	2003	2004	2005	2006	2007
Simultaneous maximum load met [2][3]	MW	60,501	61,013	61,697	59,071	61,527
of which England and Wales	MW	52,965	53,795	..	..	..
Scotland	MW	5,909	5,579	..	..	..
Great Britain	MW	58,874	59,374	60,100	57,490	59,880
Northern Ireland	MW	1,627	1,639	1,597	1,581	1,647
Maximum demand as a percentage of UK capacity	Per cent	84.7	83.3	83.4	78.7	81.8
Plant load factor						
Combined cycle gas turbine stations	Per cent	64.2	65.0	60.9	54.1	63.2
Nuclear stations	"	77.8	71.8	72.4	69.3	59.6
Hydro-electric stations:						
Natural flow	"	22.7	35.3	34.2	32.7r	36.3
Pumped storage	"	10.8	10.6	11.4	15.4	16.1
Conventional thermal and other stations [4]	"	47.7	46.0	46.1	50.2	45.0
of which coal-fired stations	"	65.0	62.3	63.0	69.4r	62.5
All plant	"	**55.8**	**54.7**	**53.6**	**52.8**	**52.0**
System load factor	"	**67.2**	**67.2**	**66.3r**	**68.8r**	**65.6**
Thermal efficiency						
(gross calorific value basis)						
Combined cycle gas turbine stations	"	46.4	47.0	49.0	48.9	48.9
Coal fired stations	"	36.5	36.2	35.6	36.0r	35.7
Nuclear stations	"	38.1	37.9	38.2	37.9	38.6

(1) See paragraphs 5.50 and 5.51 for information on companies covered.

(2) Data cover the 12 months ending March of the following year, e.g. 2007 data are for the year ending March 2008.

(3) Prior to 2005 the demands shown are those that occurred in Scotland and Northern Ireland at the same time as England and Wales had their maximum demand. In 2005, 2006 and 2007 the Northern Ireland demand shown is that which occurred at the time as in Great Britain. See paragraph 5.57 for further details.

(4) Conventional steam plants, gas turbines and oil engines and plants producing electricity from renewable sources other than hydro.

5.11 Power Stations in the United Kingdom
(operational at the end of May 2008)[1]

Company Name	Station Name	Fuel	Installed Capacity (MW)	Year of commission or year generation began	Location Scotland, Wales Northern Ireland, or English region
AES	Kilroot	coal/oil	520	1981	Northern Ireland
Alcan	Lynemouth	coal	420	1995	North East
	Fort William	hydro	72	1929	Scotland
	Kinlochleven	hydro	19.5	1907	Scotland
Ardrossan Windfarm (2)	Ardrossan	wind	24	2004	Scotland
Baglan Generation Ltd	Baglan Bay	gas turbine	575	2002	Wales
Barking Power	Barking	CCGT	1,000	1994	London
Beaufort Wind Ltd (3)	Bears Down	wind	10	2001	South West
	Bein Ghlas	wind	8	1999	Scotland
	Bryn Titli	wind	10	1994	Wales
	Carno	wind	34	1996	Wales
	Causeymire	wind	48	2004	Scotland
	Kirkby Moor	wind	5	1993	North West
	Lambrigg	wind	7	2000	North West
	Llyn Alaw	wind	20	1997	Wales
	Mynydd Gorddu	wind	10	1996	Wales
	Novar	wind	17	1997	Scotland
	Taff Ely	wind	9	1993	Wales
	Tow Law	wind	2	2001	North East
	Trysglwyn	wind	6	1996	Wales
	Windy Standard	wind	22	1996	Scotland
	North Hoyle	wind (offshore)	60	2003	Wales
	Farr	wind	92	2006	Scotland
	Ffynnon Oer	wind	32	2006	Wales
Braes of Doune Windfarm (2)	Braes of Doune	wind	72	2006	Scotland
British Energy	Dungeness B	nuclear	1,040	1983	South East
	Hartlepool	nuclear	1,190	1984	North East
	Heysham1	nuclear	1,160	1984	North West
	Heysham 2	nuclear	1,235	1988	North West
	Hinkley Point B	nuclear	820	1976	South West
	Sizewell B	nuclear	1,188	1995	East
	Hunterston B	nuclear	820	1976	Scotland
	Torness	nuclear	1,230	1988	Scotland
	Eggborough	coal	1,960	1967	Yorkshire and the Humber
	Aberdare District Energy	gas	10	2002	Wales
	Bridgewater District Energy	gas	10	2000	South West
	Sevington District Energy	gas	10	2000	South East
	Solutia District Energy	gas	10	2000	Wales
Cemmaes Windfarm Ltd (5)	Cemmaes	wind	15	2002 (6)	Wales
Centrica	Barry	CCGT	230	1998	Wales
	Glanford Brigg	CCGT	260	1993	Yorkshire and the Humber
	Killingholme	CCGT	665	1994	the Humber
	Kings Lynn	CCGT	340	1996	East
	Peterborough	CCGT	405	1993	East
	Roosecote	CCGT	229	1991	North West
	South Humber Bank	CCGT	1,285	1996	Yorkshire and the Humber
	Glens of Foudland	wind	26	2005	Scotland
	Barrow Offshore Windfarm	wind (offshore)	90	2006	North West

For footnotes see page 147

5.11 Power Stations in the United Kingdom
(operational at the end of May 2008)[1] (continued)

Company Name	Station Name	Fuel	Installed Capacity (MW)	Year of commission or year generation began	Location Scotland, Wales Northern Ireland, or English region
Citigen (London) UK Ltd	Charterhouse St, London	gas/gas oil CHP	16	1995	London
Cold Northcott Windfarm Ltd(5)	Cold Northcott	wind	7	1993	South West
Coolkeeragh ESB Ltd	Coolkeeragh	CCGT	408	2005	Northern Ireland
Corby Power Ltd	Corby	CCGT	401	1993	East Midlands
Coryton Energy Company Ltd	Coryton	CCGT	732	2001	East
Derwent Cogeneration	Derwent	gas CHP	236	1994	East Midlands
Drax Power Ltd	Drax	coal	3,870	1974	Yorkshire and
	Drax GT	gas oil	75	1971	the Humber
EDF Energy	Sutton Bridge	CCGT	800	1999	East
	Cottam	coal	2,008	1969	East Midlands
	West Burton	coal	1,972	1967	East Midlands
	West Burton GT	gas oil	40	1967	East Midlands
EPR Ely Limited	Elean	straw/gas	38	2001	East
EPR Glanford Ltd	Glanford	meat & bone meal	13	1993	East
EPR Eye Ltd	Eye, Suffolk	AWDF (7)	13	1992	East
EPR Thetford Ltd	Thetford	poultry litter	39	1998	East
E.On UK	Kingsnorth	coal/oil	1,940	1970	South East
	Ironbridge	coal	970	1970	West Midlands
	Ratcliffe	coal	2,000	1968	East Midlands
	Grain	oil	1,300	1979	South East
	Grain GT	gas oil	55	1978	South East
	Kingsnorth GT	gas oil	34	1967	South East
	Ratcliffe GT	gas oil	34	1966	East Midlands
	Taylor's Lane GT	gas oil	132	1979	London
	Connahs Quay	CCGT	1,380	1996	Wales
	Cottam Development Centre	CCGT	400	1999	East Midlands
	Enfield	CCGT	392	1999	London
	Killingholme	CCGT	900	1993	Yorkshire and the Humber
	Steven's Croft	biomass	44	2007	Scotland
	Rheidol	hydro	49	1961	Wales
	Askam	wind	5	1999	North West
	Bessy Bell	wind	5	1995	Northern Ireland
	Blood Hill	wind	2	1992	East
	Bowbeat	wind	31	2002	Scotland
	Deucheran Hill	wind	16	2001	Scotland
	Hare Hill	wind	5	2004	North East
	High Volts	wind	8	2004	North East
	Holmside	wind	5	2004	North East
	Lowca	wind	5	2000	North West
	Oldside	wind	5	1996	North West
	Out Newton	wind	9	2002	Yorkshire and the Humber
	Rheidol	wind	2	1997	Wales
	Scroby Sands	wind (offshore)	60	2005	East
	Siddick	wind	4	1996	North West
	St Breock	wind	5	1994	South West

For footnotes see page 147

5.11 Power Stations in the United Kingdom

(operational at the end of May 2008)[1] (continued)

Company Name	Station Name	Fuel	Installed Capacity (MW)	Year of commission or year generation began	Location Scotland, Wales Northern Ireland, or English region
E.On UK (continued)	Stags Holt	wind	18	2007	East
	Rhyd-y-Groes	wind	7	1992	Wales
	Blyth Offshore	wind (offshore)	4	2000	North East
Fenland Windfarms Ltd (5)	Deeping	wind	16	2006	East Midlands
	Glass Moor	wind	16	2006	East Midlands
	Red House	wind	12	2006	East Midlands
	Red Tile	wind	24	2007	East Midlands
Fred Olsen	Crystal Rig Windfarm	wind	50	2003	Scotland
	Haverigg III	wind	3	2005	North West
	Paul's Hill	wind	64	2005	Scotland
	Rothes	wind	51	2004	Scotland
Gaz de France	Shotton	gas CHP	180	2001	Wales
Great Orton Windfarm Ltd (5)	Great Orton	wind	4	1999 (6)	North West
HG Capital	Tyr Mostyn & Foel Goch	wind	21	2005	Wales
Immingham CHP LLP	Immingham CHP	gas CHP	741	2004	Yorkshire and the Humber
International Power / Mitsui	Indian Queens	gas oil/kerosene	140	1996	South West
	Dinorwig	pumped storage	1,728	1983	Wales
	Ffestiniog	pumped storage	360	1961	Wales
	Rugeley	coal	1,006	1972	West Midlands
	Rugeley GT	gas oil	50	1972	West Midlands
	Deeside	CCGT	500	1994	Wales
	Saltend	CCGT	1,200	2000	Yorkshire and the Humber
K/S Winscales (5)	Winscales 1	wind	2	1999	North West
	Winscales 2	wind	7	2005	North West
Llangwyryfon Windfarm Ltd (5)	Llangwyryfon	wind	9	2003 (2)	Wales
Magnox Electric Ltd (4)	Oldbury	nuclear	434	1967	South West
	Wylfa	nuclear	980	1971	Wales
	Fellside CHP	gas CHP	180	1995	North West
	Maentwrog	hydro	28	1928	Wales
Premier Power Ltd	Ballylumford B	gas/oil	360	1968	Northern Ireland
	Ballylumford C	CCGT	616	2003	Northern Ireland
RES-Gen Ltd	Dyffryn Brodyn	wind	6	1994	Wales
	Four Burrows	wind	5	1995	South West
	Forss	wind	2	2003	Scotland
	Forss2	wind	5	2007	Scotland
	Lendrum's Bridge	wind	13	2000	Northern Ireland
	Altahullion	wind	26	2003	Northern Ireland
	Altahullion2	wind	12	2007	Northern Ireland
	Black Hill	wind	29	2006	Scotland
	Lough Hill	wind	8	2007	Northern Ireland
RGS Energy Ltd	Knapton	gas	40	1994	Yorkshire and the Humber
Rocksavage Power Co. Ltd	Rocksavage	CCGT	748	1998	North West
RWE Npower Plc	Aberthaw B	coal	1,586	1971	Wales
	Tilbury B	coal	1,063	1968	East

For footnotes see page 147

5.11 Power Stations in the United Kingdom
(operational at the end of May 2008)[1] (continued)

Company Name	Station Name	Fuel	Installed Capacity (MW)	Year of commission or year generation began	Location Scotland, Wales Northern Ireland, or English region
RWE Npower Plc (continued)	Didcot A	coal/gas	1,958	1972	South East
	Aberthaw GT	gas oil	51	1971	Wales
	Cowes	gas oil	140	1982	South East
	Didcot GT	gas oil	100	1972	South East
	Fawley GT	gas oil	34	1969	South East
	Littlebrook GT	gas oil	105	1982	South East
	Tilbury GT	gas oil	68	1968	East
	Little Barford GT	gas oil	17	2006	East
	Fawley	oil	968	1969	South East
	Littlebrook D	oil	2,055	1982	South East
	Didcot B	CCGT	1,390	1998	South East
	Great Yarmouth	CCGT	420	2001	East
	Little Barford	CCGT	665	1995	East
Npower Renewables Ltd	Braevallich	hydro	2	2005	Scotland
(Part of RWE Npower)	Cwm Dyli	hydro	10	2002 (6)	Wales
	Dolgarrog High Head	hydro	18	2002 (6)	Wales
	Dolgarrog Low Head	hydro	15	1926/2002	Wales
	Garrogie	hydro	2	2005	Scotland
	Inverbain	hydro	1	2006	Scotland
	Kielder	hydro	6	2006 (6)	Yorkshire and the Humber
	Burgar Hill	wind	5	2007	Scotland
	Hameldon Hill	wind	5	2007	Northwest
Scottish & Southern Energy plc					
Hydro Schemes:					
Affric/Beauly	Mullardoch Tunnel	hydro	2.4	1955	Scotland
	Fasnakyle	hydro	69	1951	Scotland
	Fasnakyle Compensation Set	hydro	8	2006	Scotland
	Deanie	hydro	38	1963	Scotland
	Culligran	hydro	17	1962	Scotland
	Culligran Compensation Set	hydro	2	1962	Scotland
	Aigas	hydro	20	1962	Scotland
	Kilmorack	hydro	20	1962	Scotland
Breadalbane	Lubreoch	hydro	4	1958	Scotland
	Cashlie	hydro	11	1959	Scotland
	Lochay	hydro	45	1958	Scotland
	Lochay Compensation Set	hydro	2	1959	Scotland
	Finlarig	hydro	17	1955	Scotland
	Lednock	hydro	3	1961	Scotland
	St. Fillans	hydro	17	1957	Scotland
	Dalchonzie	hydro	4	1958	Scotland
Conon	Achanalt	hydro	3	1956	Scotland
	Grudie Bridge	hydro	19	1950	Scotland
	Mossford	hydro	19	1957	Scotland
	Luichart	hydro	34	1954	Scotland
	Orrin	hydro	18	1959	Scotland
	Torr Achilty	hydro	15	1954	Scotland
Foyers	Foyers	hydro/pumped storage	300	1974	Scotland
Great Glen	Foyers Falls	hydro	5	1968	Scotland
	Mucomir	hydro	2	1962	Scotland
	Ceannacroc	hydro	20	1956	Scotland
	Livishie	hydro	17	1962	Scotland

For footnotes see page 147

5.11 Power Stations in the United Kingdom
(operational at the end of May 2008)[1] (continued)

Company Name	Station Name	Fuel	Installed Capacity (MW)	Year of commission or year generation began	Location Scotland, Wales Northern Ireland, or English region
Scottish & Southern Energy plc					
Hydro Schemes (continued)					
Great Glen (continued)	Glenmoriston	hydro	39	1957	Scotland
	Quoich	hydro	18	1955	Scotland
	Invergarry	hydro	20	1956	Scotland
	Kingairloch	hydro	4	2005	Scotland
Shin	Cassley	hydro	10	1959	Scotland
	Lairg	hydro	4	1959	Scotland
	Shin	hydro	19	1958	Scotland
	Loch Dubh	hydro	1	1954	Scotland
Sloy/Awe	Sloy	hydro	153	1950	Scotland
	Sron Mor	hydro	5	1957	Scotland
	Clachan	hydro	40	1955	Scotland
	Allt-na-Lairige	hydro	6	1956	Scotland
	Nant	hydro	15	1963	Scotland
	Inverawe	hydro	25	1963	Scotland
	Kilmelfort	hydro	2	1956	Scotland
	Loch Gair	hydro	6	1961	Scotland
	Lussa	hydro	2	1952	Scotland
	Striven	hydro	8	1951	Scotland
Tummel	Gaur	hydro	8	1953	Scotland
	Cuaich	hydro	3	1959	Scotland
	Loch Ericht	hydro	2	1962	Scotland
	Rannoch	hydro	45	1930	Scotland
	Tummel	hydro	34	1933	Scotland
	Errochty	hydro	75	1955	Scotland
	Clunie	hydro	61	1950	Scotland
	Pitlochry	hydro	15	1950	Scotland
Wind	Artfield Fell	wind	20	2005	Scotland
	Hadyard Hill	wind	120	2005	Scotland
	Spurness	wind	8	2004	Scotland
	Tangy	wind	19	2002	Scotland
	Dalswinton	wind	30	2008	Scotland
	Drumderg	wind	32	2008	Scotland
	Minsca	wind	37	2008	Scotland
	Bessy Bell	wind	9	2008	N Ireland
	Bin Mountain	wind	9	2008	N Ireland
	Tappaghan	wind	20	2008	N Ireland
	Beatrice	wind (offshore)	10	2007	Scotland
Small Hydros:	Chliostair	hydro	1	1960	Scotland
	Cuileig	hydro	3	2002	Scotland
	Kerry Falls	hydro	1	1951	Scotland
	Loch Dubh	hydro	1	1954	Scotland
	Nostie Bridge	hydro	1	1950	Scotland
	Storr Lochs	hydro	2	1952	Scotland
Thermal:	Peterhead (8)	gas/oil	1,540	1980	Scotland
	Fife Power Station	CCGT	123	2000	Scotland
	Keadby	gas/oil	749	1994	Yorkshire and the Humber
	Medway	CCGT	688	1995	South East
	Ferrybridge C	coal/biomass	1,955	1966	Yorkshire and the Humber
	Fiddler's Ferry	coal/biomass	1,961	1971	North West
	Ferrybridge GT	gas oil	34	1966	Yorkshire and the Humber
	Fiddler's Ferry GT	gas oil	34	1969	North West

For footnotes see page 147

5.11 Power Stations in the United Kingdom
(operational at the end of May 2008)[1] (continued)

Company Name	Station Name	Fuel	Installed Capacity (MW)	Year of commission or year generation began	Location Scotland, Wales Northern Ireland, or English region
Scottish & Southern Energy plc					
Thermal (continued)					
	Chickerell	gas/oil	45	1998	South West
	Burghfield	gas/oil	47	1998	South East
	Thatcham	light oil	10	1994	South East
	Five Oaks	light oil	9	1995	South East
	Chippenham	gas	10	2002	South West
	Wheldale	mines gas	8	2002	Yorkshire and the Humber
Island Generation	Arnish	diesel	3	2001	Scotland
	Barra	diesel	2	1990	Scotland
	Bowmore	diesel	6	1946	Scotland
	Kirkwall	diesel	16	1953	Scotland
	Lerwick	diesel	67	1953	Scotland
	Loch Carnan, South Uist	diesel	10	1971	Scotland
	Stornoway	diesel	26	1950	Scotland
	Tiree	diesel	3	1945	Scotland
Scottish Power					
Hydro schemes:					
Galloway	Carsfad	hydro	12	1936	Scotland
	Drumjohn	hydro	2	1985	Scotland
	Earlstoun	hydro	14	1936	Scotland
	Glenlee	hydro	24	1935	Scotland
	Kendoon	hydro	24	1936	Scotland
	Tongland	hydro	33	1935	Scotland
Lanark	Bonnington	hydro	11	1927	Scotland
	Stonebyres	hydro	6	1927	Scotland
Cruachan	Cruachan	pumped storage	440	1966	Scotland
Thermal:	Cockenzie	coal	1,152	1967	Scotland
	Longannet	coal	2,304	1970	Scotland
	Damhead Creek	CCGT	792	2000	South East
	Pilkington - Greengate	gas	10	1998	North West
	Ravenhead	gas	9	1999	North West
	Rye House	CCGT	715	1993	East
	Shoreham	CCGT	400	2000	South East
Wind:	Beinn an Tuirc	wind	30	2001	Scotland
	Beinn Tharsuinn	wind	30	2007	Scotland
	Black Law	wind	124	2005	Scotland
	Callagheen	wind	17	2006	Northern Ireland
	Carland Cross	wind	6	1992	South West
	Coal Clough	wind	10	1992	North West
	Coldham	wind	16	2006	East
	Corkey	wind	5	1994	Northern Ireland
	Cruach Mhor	wind	30	2004	Scotland
	Dun Law	wind	17	2000	Scotland
	Elliots Hill	wind	5	1995	Northern Ireland
	Hagshaw Hill	wind	16	1995	Scotland
	Hare Hill	wind	13	2000	Scotland
	Penryddian & Llidiartywaun	wind	31	1992	Wales
	Rigged Hill	wind	5	1994	Northern Ireland
	Wether Hill	wind	18	2007	Scotland
	Whitelee	wind	23	2007	Scotland
	Wolf Bog	wind	10	2008	Northern Ireland
Seabank Power Limited	Seabank 1	CCGT	812	1998	South West
	Seabank 2	CCGT	410	2000	South West

For footnotes see page 147

5.11 Power Stations in the United Kingdom
(operational at the end of May 2008)[1] (continued)

Company Name	Station Name	Fuel	Installed Capacity (MW)	Year of commission or year generation began	Location Scotland, Wales Northern Ireland, or English region
South East London Combined Heat & Power Ltd	SELCHP ERF	waste	32	1994	London
Spalding Energy Company Ltd	Spalding	CCGT	860	2004	East Midlands
Teesside Power Ltd	Teesside Power Station	CCGT	1,875	1992	North East
Uskmouth Power Company Ltd	Uskmouth	coal/biomass	363	2000	Wales
Vattenfall Wind Power	Kentish Flats	wind (offshore)	90	2005	South East
Western Power Generation	Lynton	gas oil	2	1961	South West
	Princetown	kerosene	3	1959	South West
	Roseland	kerosene	5	1963	South West
	St Marys	gas oil	6	1958	South West
Yorkshire Windpower Ltd	Ovenden Moor	wind	9	1993	Yorkshire and
	Royd Moor	wind	7	1993	the Humber
Total			**78,024**		

Other power stations[9]

Station type	Fuel	Capacity (MW)
Renewable sources and combustible wastes	wind	433
	landfill gas	901
	sewage gas	152
	hydro	129
	waste	294
	other	216
CHP schemes listed in Table 5.12	various fuels	2,183
CHP schemes other than major power producers and renewables and those listed in Table 5.12	mainly gas	1,562
Other autogenerators	various fuels	985

Interconnectors

	Capacity (MW)
England - France	2,000
Scotland - Northern Ireland	500
Northern Ireland - Irish Republic	600

Footnotes

(1) This list covers stations of more than 1 MW capacity, but excludes some renewables stations of over 1 MW which are
     included in the sub table on page147.
(2) Joint venture with Scottish and Southern Energy
(3) Managed by RWE
(4) Owned by NDA but operated by Magnox Electric Ltd
(5) Managed by Cumbria Wind Farms Ltd
(6) Recommissioning dates.
(7) Animal Waste Derived Fuel, i.e. meat and bone meal, poultry litter, feathers and small quantities of other material such as wood chip
(8) Total capacity is 2,370 MW but because of transmission constraints only 1,540 MW can be used at any one time.
(9) As at end December 2007.

5.12 Large scale CHP schemes in the United Kingdom
(operational at the end of December 2007)[1]

Company Name	Scheme Location	Installed Capacity (MWe) [2]
ABB CHP Ltd	Royal Devon and Exeter NHS Trust Wonsford Hospital	1.0
Airbus UK	Broughton, Cheshire	6.3
Alta Estate Services Limited	The University of Birmingham	5.9
Archer Daniels Midland Ltd (ADM Ltd)	Erith, Kent	14.0
Arjo Wiggins Ltd	Dartford, Kent	10.0
Atkins Power	Arreton Valley Nurseries, Isle of Wight	15.5
Atkins Power	Hern Hill Nursery, Kent	9.0
Atkins Power	Waltham Abbey, Essex	3.1
Atkins Power	Cleveland Nurseries, Cleveland	4.5
Atkins Power	Poolbank Salads, Brough, North Humberside	3.9
Atkins Power	Abbey View Nurseries, Waltham Abbey, Essex	3.1
Atkins Power	Tower Nursery, Roydon, Essex	3.1
Atkins Power	Anchor Nurseries Ltd, Beverley, East Yorks	3.1
Atkins Power	Villa Nurseries, Roydon, Essex	3.1
Atkins Power	Stubbins Marketing, Fen Drayton, Cambs	3.1
Atkins Power	Glen Avon Growers, Cottingham, North Humberside	3.0
Atkins Power	Park Lane Nursery, East Yorkshire	2.0
Balcas Ltd	Laragh, Ballycassidy, Enniskillen	2.7
Banham PoultryLtd	Banham, Norfolk	2.4
BHP Billiton	Point of Ayr Terminal, Flintshire	9.1
Bloomsbury CHP	SOAS, London	1.5
BP CHP (UK) Ltd	Polimerieuropa, Hythe	53.0
British Salt Ltd	Middlewich, Cheshire	10.0
British Sugar plc	Wissington, Norfolk	93.6
British Sugar plc	Bury St Edmunds, Suffolk	90.1
British Sugar plc	York	10.0
British Sugar plc	Newark, Nottinghamshire	10.0
British Sugar plc	Allscott, Shropshire	9.0
CIBA Speciality Chemicals plc	Low Moor, Bradford	16.5
Crisp Maltings Ltd	Fakenham, Norfolk	1.2
Cyclerval UK Ltd	Stallingbrough, Grimsby	3.2
Dalkia Clean Power Ltd	Sonoco, Stainland Board Mills, Halifax	7.1
Dalkia Clean Power Ltd	Fribo Foods, Wrexham	1.4
Dalkia Utility Services	Astra Zonooa, Macclesfield	22.8
Dalkia Utility Services	Leeds General Infirmary	18.9
Dalkia Utility Services	Queen Elizabeth Hospital, Edgbaston, Warwickshire	3.5
Dalkia Utility Services	Lincoln County Hospital , Lincoln	1.4
Diageo Distilling Ltd	Port Dundas, Glasgow	5.9
DSM Nutritional Products UK Ltd	Dalry, Ayrshire	45.8
E.On UK (CHP) Ltd	Grovehurst Energy Ltd, Sittingbourne, Kent	80.6
E.On UK (CHP) Ltd	Iggesund Paperboard, Workington	47.8
E.On UK (Cogeneration) Ltd	Port of Liverpool	31.0
E.On UK (Cogeneration) Ltd	A H Marks, Bradford	4.6
Elyo Ltd	Hillhouse International, Lancashire	5.1
Enviroenergy Ltd	Nottingham District Heating Scheme	14.4
Fortum O&M UK Ltd	Sullom Voe, Shetland	88.9
Genzyme td	Haverhill, Suffolk	1.4
GlaxoSmithKline Ltd	Wellcome Foundation, Dartford	6.1
GlaxoSmithKline Ltd	Ulverston, Cumbia	2.3
Haden Building Management	Royal Infirmary of Edinburgh	1.0
Humber Energy	Grimsby, South Humberside	48.4
Hydro Polymers Ltd	Newton Aycliffe, Durham	9.8
Imperial College of Science, Medicine and Technology	Kensington, London	9.0
Ineos Chlor	Runcorn, Cheshire	38.4
Ineos Manufacturing Scotland Ltd	BP Oil Grangemouth	294.9
Innovia Films	Wigton, Cumbria	5.3
INBEV UK Ltd	Magor Brewery, South Wales	6.7
INBEV UK Ltd	Salmesbury Brewery, Preston	6.7

For footnotes see page 150

5.12 Large scale CHP schemes in the United Kingdom
(operational at the end of December 2007)[1] (continued)

Company Name	Scheme Location	Installed Capacity (MWe) [2]
Jaguar Cars Ltd	Castle Bromwich, West Midlands	6.2
Jaguar Cars Ltd	Coventry, West Midlands	3.1
James Cropper Ltd	Kendal, Cumbria	7.0
John Dickens BD	Becton Dickinson, Plymouth	3.0
John Heathcoat and Company Ltd	Devon	1.0
John Thompson and Company	Belfast	2.7
Johnson Matthey	Enfield, Greater London	2.9
Johnson Matthey	Royston, Herts	5.8
Kraft Foods UK Ltd	Banbury, Oxon	7.5
Laporte Industries	Fine Organics Ltd, Seal Sands, Teesside	4.1
Carron Engineering and Construction Ltd	Mill Nurseries, Keyingham, Hull	15.0
North Tees and Hartlepool NHS Trust	North Tees General Hospital, Hartlepool	1.6
Novartis	Grimsby, South Humberside	8.4
Npower Cogen Trading Ltd	Fawley, Hampshire	316.0
Npower Cogen Ltd	Aylesford Newsprint, Kent	99.8
Npower Cogen Ltd	Conoco Phillips Teesside Operations	97.3
Npower Cogen Ltd	Bridgewater Paper, Ellesmere Port, South Wirral	58.0
Npower Cogen Ltd	Dow Corning, Barry	27.2
Npower Cogen Ltd	Huntsman Tioxide, Grimsby	20.3
Npower Cogen Ltd	Millennium Inorganic Chemicals, Stallingborough	15.8
Npower Cogen Ltd	Georgia Pacific, Bridgend Paper Mills, Llangynwyd,	9.0
Npower Cogen Ltd	Lancaster University	1.4
Portals Ltd	Overton Mill, Hampshire	7.4
Prime Energy (MK) Ltd	Ordnance Survey, Southampton	1.7
Prosper De Mulder	Hartshill, Warwickshire	2.9
Rigid Paper Ltd	Selby, East Yorks	5.1
Roquette UK Ltd	Corby, Northants	14.5
Royal Mail Group Property	Slough, Berkshire	3.1
Ryobi Aluminium Castings (UK) Ltd	Carrickfergus, Co Antrim	1.4
Scottish and Southern Energy plc	Hedon Salads, Burstwick, East Yorks	10.0
Scottish and Southern Energy plc	Bradon Farm, Taunton, Somerset	9.7
Scottish and Southern Energy plc	Runcton Nursery, Chichester, West Sussex	4.0
Scottish and Southern Energy plc	Hedon Salads, Newport, East Yorks	3.9
Scottish and Southern Energy plc	Red Roofs, Cottingham, Yorks	3.4
Scottish and Southern Energy plc	West End Nursery, Woking	2.0
Scottish and Southern Energy plc	Koppers UK Ltd, Port Clarence, Teesside	1.9
Scottish and Southern Energy plc	Slough Nurseries, Slough	1.9
Scottish and Southern Energy plc	Western General Hospital, Edinburgh	1.0
Scottish Power Generation	Buckland Nurseries, Reigate, Surrey	1.9
Shell Oil Products Ltd	Stanlow Manufacturing Complex, Cheshire	109.3
Smurfit Kappa SSK Ltd	Nechells, Birmingham	8.7
South West Water	All sites	1.4
Southampton Geothermal	Southampton	6.4
Southern Water	All sites	4.0
St Georges's Healthcare NHS Trust	St George's Hospital, Tooting, London	4.4
St Regis Paper Company	Sudbrook Mill, Caldicot, Monmouthshire	5.0
St Regis Paper Company	Hollins Mill, Darwen, Lancashire	2.0
Sustainable and Renewable Energy	Woking, Surrey	1.4
Tangmere Airfield Nurseries Ltd	Tangmere, Sussex	8.8
Tate and Lyle Europe	Thames Refinery, Silvertown, London	19.5
Thames Valley Power Ltd	Heathrow Airport, London	15.0
Thames Water Utilities	Mogden Treatment Works, London	8.3
Thames Water Utilities	Maple Lodge Sewage Treatment Works, Hertfordshire	3.8
Thames Water Utilities	Deephams Sewage Treatment Works, Edmonton, London	3.3
Thames Water Utilities	Beddington Lane Sewage Treatment Works, Croydon,	3.3
Thames Water Utilities	Long Reach Sewage Treatment Works, Kent	2.3
Thames Water Utilities	Rye Meads Sewage Treatment Works, Hertfordshire	1.5
Thames Water Utilities	Reading (Island Road) Sewage Treatment Works, Reading	1.0
Thameswey Energy Ltd	Midsummer Boulevard, Milton Keynes	3.0
The Boots Company plc	Beeston, Nottingham	14.1

For footnotes see page 150

5.12 Large scale CHP schemes in the United Kingdom
(operational at the end of December 2007)[1] *(continued)*

Company Name	Scheme Location	Installed Capacity (MWe) [2]
UEA Utilities Ltd	University of East Anglia, Norwich	3.1
University of Bristol	Bristol	1.2
University of Dundee	Dundee	3.0
University of Edinburgh	Kings Building, Edinburgh	2.7
University of Edinburgh	George Square Energy Centre, Edinburgh	1.6
University of Southampton	Southampton	2.8
University of Surrey	Guildford, Surrey	1.0
University of Warwick	Warwickshire	4.2
Utilicom Ltd	University College, London	2.9
Weetabix Ltd	Burton Latimer, Northants	6.1
Wessex Water	Bristol Waste Water Treatment Works	5.8
Total [2]		2,265.0
Electrical capacity of good quality CHP for these sites in total		2,183.4

(1) These are sites of 1 MW installed electrical capacity or more that either have agreed to be listed in the Ofgem register of CHP plants or whose details are publicly available elsewhere, or who have provided the information directly to BERR. It excludes CHP sites that have been listed as major power producers in Table 5.11.

(2) This is the total power capacity from these sites and includes all the capacity at that site, not just that classed as good quality CHP under CHPQA.

Chapter 6
Combined heat and power

Introduction

6.1   This chapter sets out the contribution made by Combined Heat and Power (CHP) to the United Kingdom's energy requirements. The data presented in this Chapter have been derived from information submitted to the CHP Quality Assurance programme (CHPQA) or by following the same procedures where no information has been provided directly. The CHPQA programme was introduced by the Government to provide the methods and procedures to assess and certify the quality of the full range of CHP schemes. It is a rigorous system for the Government to ensure that the incentives on offer are targeted fairly and benefit schemes in relation to their environmental performance.

6.2   CHP is the simultaneous generation of usable heat and power (usually electricity) in a single process. The term CHP is synonymous with cogeneration and total energy, which are terms often used in other Member States of the European Community and the United States, respectively. CHP uses a variety of fuels and technologies across a wide range of sites and scheme sizes. The basic elements of a CHP plant comprise one or more prime movers (a reciprocating engine, gas turbine, or steam turbine) driving electrical generators, where the steam or hot water generated in the process is utilised via suitable heat recovery equipment for use either in industrial processes or in community heating and space heating.

6.3   A CHP plant provides primary energy savings compared to separate generation of heat and power. In addition, CHP is typically sized to make use of the available heat[1], and connected to the lower voltage distribution system (ie embedded). Compared to an electricity-only plant, which is larger and connected at very high voltage to the grid transmission system, CHP provides efficiency gains by avoiding significant transmission and distribution losses. CHP can also provide important network services such as black start, improvements to power quality, and some have the ability to operate in island mode if the grid goes down.

6.4   There are four principal types of CHP system: steam turbine, gas turbine, combined cycle systems and reciprocating engines. Each of these is defined in paragraph 6.34 later in this chapter.

Government policy towards CHP

6.5   To reduce carbon emissions and help deliver the UK's Climate Change Programme, the Government has a target of achieving at least 10,000 MWe of Good Quality CHP capacity (GQCHP) by the end of 2010.

6.6   However, in recent years the CHP industry has faced serious economic difficulties. In order to overcome these difficulties, and to help promote the security and diversity of energy supply that accompanies CHP, the Government has made a number of interventions in the market to help support CHP. These interventions include:

- Exemption from the Climate Change Levy (CCL) of all fuel inputs to, and electricity outputs from, Good Quality CHP.
- Eligibility for Enhanced Capital Allowances of Good Quality CHP.
- Grants for public sector led district heating schemes using CHP.
- Business Rates exemption for CHP power generation plant and machinery.
- Reduction of VAT (from 17.5 to 5 per cent) on domestic micro-CHP installations.
- Extension of the eligibility for Renewable Obligation Certificates (ROCs) to CHP plant using mixed wastes.
- The granting, from April 2009, of 'double ROCs' for the Good Quality electricity output of CHP fuelled by biomass.

---

[1] But not always, see paragraph 6.10.

- In April 2010 the Carbon Reduction Commitment (CRC) will come into force. The CRC is a mandatory emissions trading scheme that will cover large, non-energy intensive business, currently not covered under other policy measures like Climate Change Agreements (CCAs) and the EU Emissions Trading Scheme (EU ETS). In the CRC, organisations covered will be required to purchase allowances to cover the CO_2 emissions from all fixed-point energy sources. This means that allowances must be purchased to cover the use of electricity, gas and all other fuel types such as Liquified Petroleum Gas (LPG) and diesel. However, under CRC heat is zero rated, meaning that allowances will not have to be purchased by a site to cover any imported heat. It is expected that this treatment will stimulate a growth in the heat market, and this will in turn incentivise the use of CHP. From 2013, the first capped phase will commence and allowances will be sold to participants by auction.

International context

6.7   The EU Emissions Trading Scheme (EU-ETS) commenced on 1st January 2005 and involves the trading of emissions allowances. The purpose of the EU-ETS is to reduce emissions by a fixed amount at least cost to the regulated sources. Each year participants in the scheme are allocated a set number of allowances. At the end of each trading year allowances equal to the reported emissions must be given up. In the EU-ETS Phase I National Allocation Plan (NAP), the sectoral classification of CHP plant depended on the sector in which it was modelled in DTI's (now BERR's) Updated Energy Projections (UEP) and the presence of CHP at an installation was not considered explicitly in their allocation calculations. The sector in which an installation is classified has an effect on the level of its allocation, because allocations are calculated on the basis of sectoral growth projections. It was argued that this method of allocation would have an impact on CHP because its future growth and emissions are different to those of non-CHP installations in Phase I sectors. For this reason the Government decided to create a specific sector for GQCHP in Phase II, to ensure that incumbent CHP plant would not be disincentivised and to ensure that investment in GQCHP would be encouraged by the implementation of Phase II.

6.8   All sites wishing to be included in the CHP sector were required to submit to CHPQA in 2006. This was to enable the Qualifying Power Capacity (QPC) to be determined for all schemes, and was necessary to ensure that the correct number of allocations was made to individual sites in the CHP sector. A consequence of this was that a number of sites that had previously not submitted to CHPQA, on the grounds that there were no fiscal benefits to accrue to them from the CHPQA process, submitted to CHPQA for the first time in 2006. This made available for the first time accurate data on capacities and energy inputs and outputs for these schemes. Where corrections were necessary, they were made back in time and revised historical data incorporating these corrections were presented in the 2007 Digest. The details of these corrections were provided in paragraph 6.31 of the 2007 Digest which is available on the BERR Energy web site at
www.berr.gov.uk/energy/statistics/publications/dukes/page39771.html.

UK energy markets, and their effect on CHP[2]

6.9   Two major factors affecting the economics of CHP are the relative cost of fuel (principally natural gas) and the value that can be realised for electricity. Energy price trends that are applicable to CHP schemes differ depending upon the size and sector of the scheme. During 2007 the market improved due to a fall in the price of gas relative to that of electricity. This is known as the spark spread (i.e. the difference between the price of electricity and the price of the gas required to generate that electricity) and it caused a consequential improvement in the viability of CHP. While these movements in the market are good for the economics of operating existing CHP, investment in the installation of new CHP capacity appears to remain subdued and this may be due to uncertainty regarding future movements in the gas and electricity markets.

Use of CHPQA in producing CHP statistics

6.10   The CHPQA programme is the major source for CHP statistics. The following factors need to be kept in mind when using the statistics produced:

---

[2] *Reference source for price trends is the BERR's 'Quarterly Energy Prices March 2008', available at*
www.berr.gov.uk/energy/statistics/publications/prices/index.html

- Through CHPQA, scheme operators have been given guidance on how to determine the boundary of a CHP scheme (what is regarded as part of the CHP installation and what is not). A scheme can include multiple CHP prime movers, along with supplementary boilers and generating plant, subject to appropriate metering being installed to support the CHP scheme boundaries proposed, and subject to appropriate metering and threshold criteria. (See CHPQA Guidance Note 11 available at www.chpqa.com).

- The output of a scheme is based on gross power output, ignoring parasitic loads (i.e. ignoring power used in pumps, fans, etc. within the scheme itself).

- The main purpose of a number of CHP schemes is the generation of electricity including export to others. Such schemes may not be sized to use all of the available heat. In such cases, the schemes' total electrical capacity and electrical output have been scaled back using the methodologies outlined in CHPQA. Only the portion of the electrical capacity and electrical output that qualifies as Good Quality is counted in this chapter. The remaining electrical capacity and electrical output are regarded as power only, and these are reported in Chapter 5. The fuel allocated to the power-only portion of the output is calculated from the power efficiency of the prime mover.

- The load factor presented in Table 6A is based on the Good Quality Power Output and Good Quality Power Capacity reported in this Chapter. For schemes that are scaled back, this load factor is likely to be smaller than the actual load factor (hours run) for the prime mover in these schemes. The load factor of all schemes between 2006 and 2007 was substantially unchanged, with a very slight fall of 0.4 percentage points in 2007 with respect to 2006.

Table 6A: A summary of the recent development of CHP[1]

	Unit	2003	2004	2005	2006	2007
Number of schemes		1,356	1,345	1,378	1,388	1,438
Net number of schemes added during year		*2*	*-11*	*33*	*10*	*50*
Electrical capacity (CHP$_{QPC}$)	MWe	4,495	5,398	5,536	5,484	5,474
Net capacity added during year		*-70*	*902*	*138*	*-52*	*-10*
Capacity added in percentage terms	Per cent	*-1.5*	*20.1*	*2.6*	*-0.9*	*-0.2*
Heat capacity	MWth	10,934	11,775	11,504	11,242	11,261
Heat to power ratio *(2)*		2.27	2.08	1.94	1.83	1.83
Fuel input *(3)*	GWh	113,100	120,201	124,632	122,823	122,742
Electricity generation (CHP$_{QPO}$)	GWh	23,938	26,859	28,836	28,904	28,677
Heat generation (CHP$_{QHO}$)	GWh	54,985	56,531	56,455	53,519	53,050
Overall efficiency *(4)*	Per cent	69.2	68.9	67.9	66.6	66.1
Load factor *(3)*	Per cent	60.8	56.8	59.5	60.2	59.8

(1)   All data in this table for 2003 to 2006 have been revised since last year's Digest – see paragraph 6.11.
(2)   Heat to power ratios are calculated from the qualifying heat output (QHO) and the qualifying power output (QPO).
(3)   The load factor reported in this table is based on the qualifying power generation and capacity and does not correspond exactly to the number of hours run by the prime movers in a year (see paragraph 6.10).
(4)   These are calculated using gross calorific values; overall net efficiencies are some 5 percentage points higher.

Progress towards the Government's targets

6.11   Chart 6.1 shows the change in installed CHP capacity over the last eleven years. Installed capacity at the end of 2007 stood at 5,474 MWe. This represents a net increase of 50 schemes between 2006 and 2007, but there was a net decrease of 10 MWe in installed capacity. During 2007 ongoing improvements to the data set were continued. This resulted in the removal of a number of schemes from the data base which do not supply data regularly and for whom research indicated they no longer exist. These removals must be kept in mind when attempting to reconcile the number of schemes and total capacity reported for a particular year in this year's Digest with that reported for the same year in an earlier year's Digest. The details of the data set improvement and the consequences it has had on the statistics is explained in detail in paragraph 6.32.

Installed capacity and output in 2007

6.12   Since the start of 2006 87 new schemes have come into operation, 37 in 2006 and 50 in 2007. During the same period 37 schemes have ceased to operate. These have been removed from the statistics for 2007, as they did not operate at all in 2007, but remain in the statistics for 2006 as they operated for at least part of that year. Schemes ceasing to operate in 2007 remain in the statistics for that year as they operated for at least part of that year. This activity has resulted in a net increase of 50 schemes between 2006 and 2007.

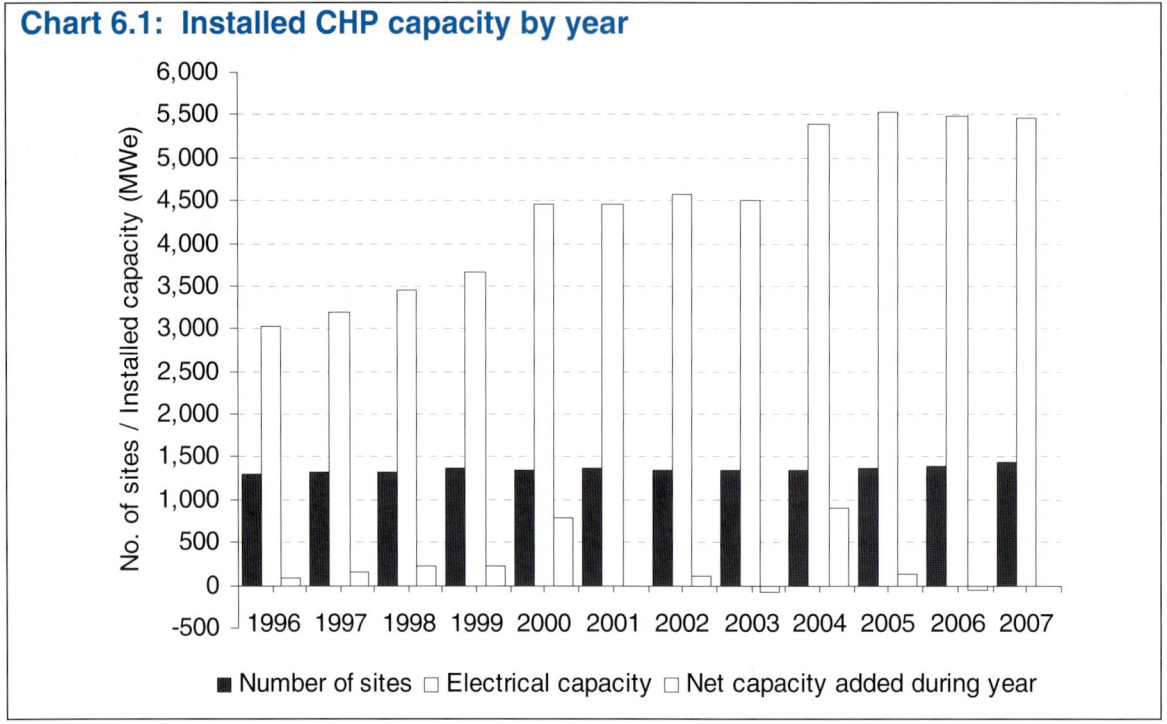

Chart 6.1: Installed CHP capacity by year

Y-axis: No. of sites / Installed capacity (MWe)

Legend: ■ Number of sites □ Electrical capacity □ Net capacity added during year

6.13   In the current market conditions, a number of operators have chosen to mothball their CHP schemes rather than continue to operate. As these schemes are still able to operate they have been included in the capacity figures. At the end of 2007, there were 77 mothballed schemes with a Good Quality capacity of 69 MWe.

6.14   Table 6A gives a summary of the overall CHP market. The electricity generated by CHP schemes in 2007 was 28,677 GWh. This represents a little over 7 per cent of the total electricity generated in the UK. Across the commercial and industrial sectors (including the fuel industries other than electricity generation) electrical output from CHP accounted for around 12 per cent of electricity consumption. CHP schemes in total supplied 53,050 GWh of heat in 2007.

6.15   In terms of electrical capacity by size of scheme, schemes larger than 10 MWe represent over 82 per cent of the total electrical capacity of CHP schemes as shown in Table 6B. However, in terms of number of schemes, the largest share (>80 per cent) is in schemes less than 1 MWe. Schemes of 1 MWe or larger, make up approximately 19 per cent of the total number of schemes. Table 6.5 provides data on electrical capacity for each type of CHP installation.

Table 6B: CHP schemes by capacity size ranges in 2007

Electrical capacity size range	Number of schemes	Share of total (per cent)	Total electricity capacity (MWe)	Share of total (per cent)
Less than 100 kWe	475	33.0	30	0.5
100 kWe - 999 kWe	693	48.2	180	3.3
1 MWe - 9.9 MWe	198	13.8	728	13.3
Greater than 10 MWe	72	5.0	4,536	82.9
Total	**1,438**	**100.0**	**5,474**	**100.0**

6.16   Seventy eight per cent of capacity is now gas turbine based, with about 86 per cent of this (67 per cent in total) in combined cycle mode. Combined Cycle Gas Turbine (CCGT) schemes also account for 47 per cent of total heat capacity. After combined cycle, reciprocating engines represent the second largest technology in terms of installed electrical capacity, closely followed by closed cycle gas turbines, both with very similar individual shares of total installed capacity. Table 6.7 provides data on heat capacity for each type of CHP installation. Over the years there has been a clear downward trend in the capacity of back pressure and pass out condensing steam turbines.

Fuel used by types of CHP installation

6.17   Table 6.2 shows the fuel used to generate electricity and heat in CHP schemes, (see paragraphs 6.35 to 6.37, below for an explanation of the convention for dividing fuel between electricity and heat production). Table 6.3 gives the overall fuel used by types of CHP installation (which are explained in paragraph 6.34). Total fuel use is summarised in Chart 6.2. In 2007, 71 per cent of the total fuel use was natural gas, which is approximately the same proportion as was used in 2006. CHP schemes accounted for 9 per cent of UK gas consumption in 2007 (see Table 4.3). Over the last year the refineries sector has seen an increase in the use of refinery gas, this is a result of a new scheme becoming fully operational in 2007 which uses refinery gases as the main fuel. In addition, Table 6.9 shows a decrease in the use of fuel oil and an increase in the use of natural gas. This is a result of two sites trading between these two fuels, which may be due to changes in the site specific requirements.

6.18   Non-conventional fuels (liquids, solids or gases which are by-products or waste products from industrial processes, or are renewable fuels) account for 24 per cent of all fuel used in CHP in 2007. These are fuels that are not commonly used by the mainstream electricity generating industry, and some would otherwise be flared or disposed of by some means. These fuels, with the exception of some waste gases, will generally be utilised in steam turbines being fed by boilers. In almost all cases, the technical nature of the combustion process, and the lower fuel quality (lower calorific value of the fuel, high moisture content of the fuel, the need to maintain certain combustion conditions to ensure complete disposal etc.) will generally result in a lower efficiency. However, given that the use of such fuels avoids the use of fossil fuels, and since they need to be disposed of in some way, the use of these fuels in CHP provides environmental benefits.

Chart 6.2: Types of fuel used by CHP schemes in 2007

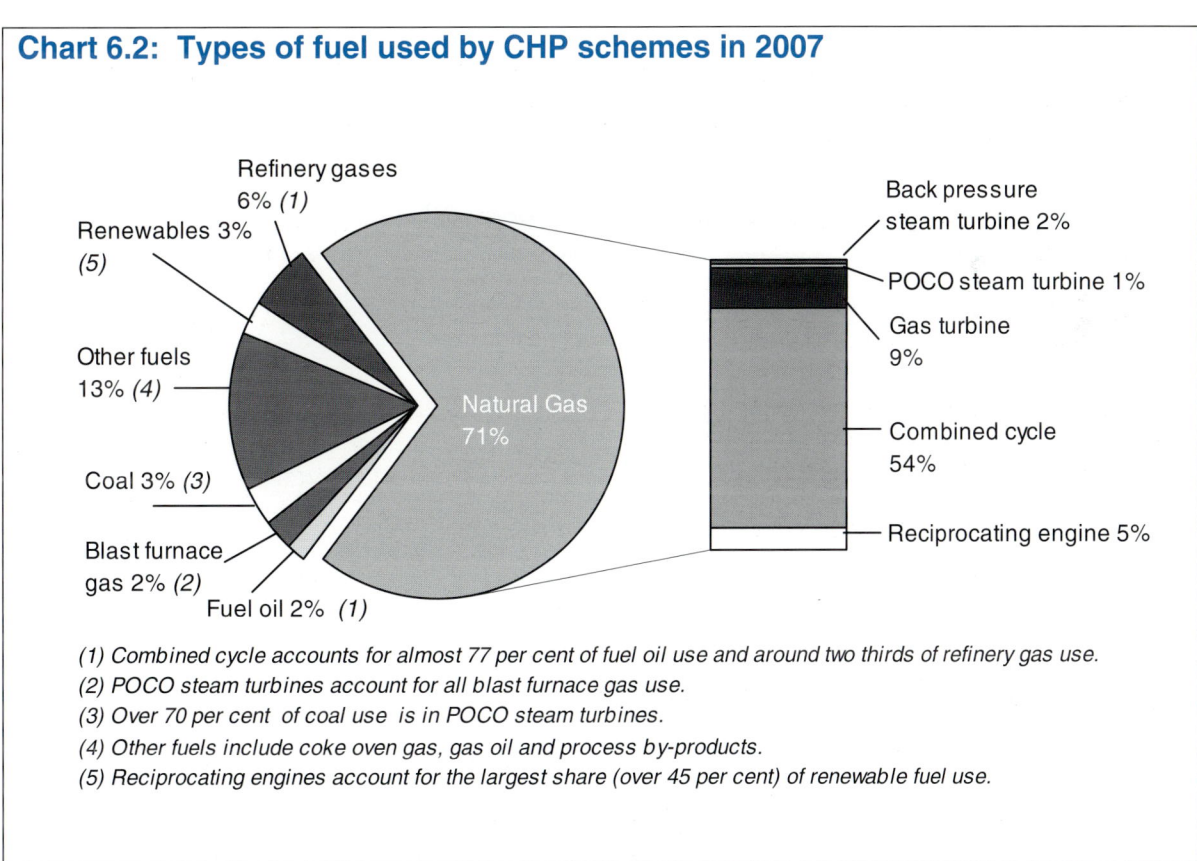

(1) Combined cycle accounts for almost 77 per cent of fuel oil use and around two thirds of refinery gas use.
(2) POCO steam turbines account for all blast furnace gas use.
(3) Over 70 per cent of coal use is in POCO steam turbines.
(4) Other fuels include coke oven gas, gas oil and process by-products.
(5) Reciprocating engines account for the largest share (over 45 per cent) of renewable fuel use.

CHP capacity, output and fuel use by sector

6.19   For this edition of the Digest the results of a review of the sectors to which the CHP schemes have been allocated have been implemented. CHP is now allocated to the sector using the heat, or, where the heat is sent to users in more than one sector, to the sector taking the majority of the heat. This means that all the CHP schemes previously allocated to "electricity supply" now appear in other sectors. There has been no change to total capacity, generation or fuel use as a result of this re-allocation. For further details see paragraph 6.33 and Table 6J.

6.20   Table 6.8 gives data on all operational schemes by economic sector. A definition of the sectors used in this table can be found in Chapter 1, paragraph 1.57 and Table 1F:

- 328 schemes (90 per cent of electrical capacity) are in the industrial sector and 1,110, schemes (10 per cent of capacity) are in the agricultural, commercial, public administration, residential and transport sectors.

- Three industrial sectors account for 69 per cent of the CHP electrical capacity – chemicals (34 per cent), oil refineries (32 per cent of capacity), and paper and publishing and printing (10 per cent). Capacity by sector is shown in Chart 6.3.

Chart 6.3: CHP electrical capacity by sector in 2007

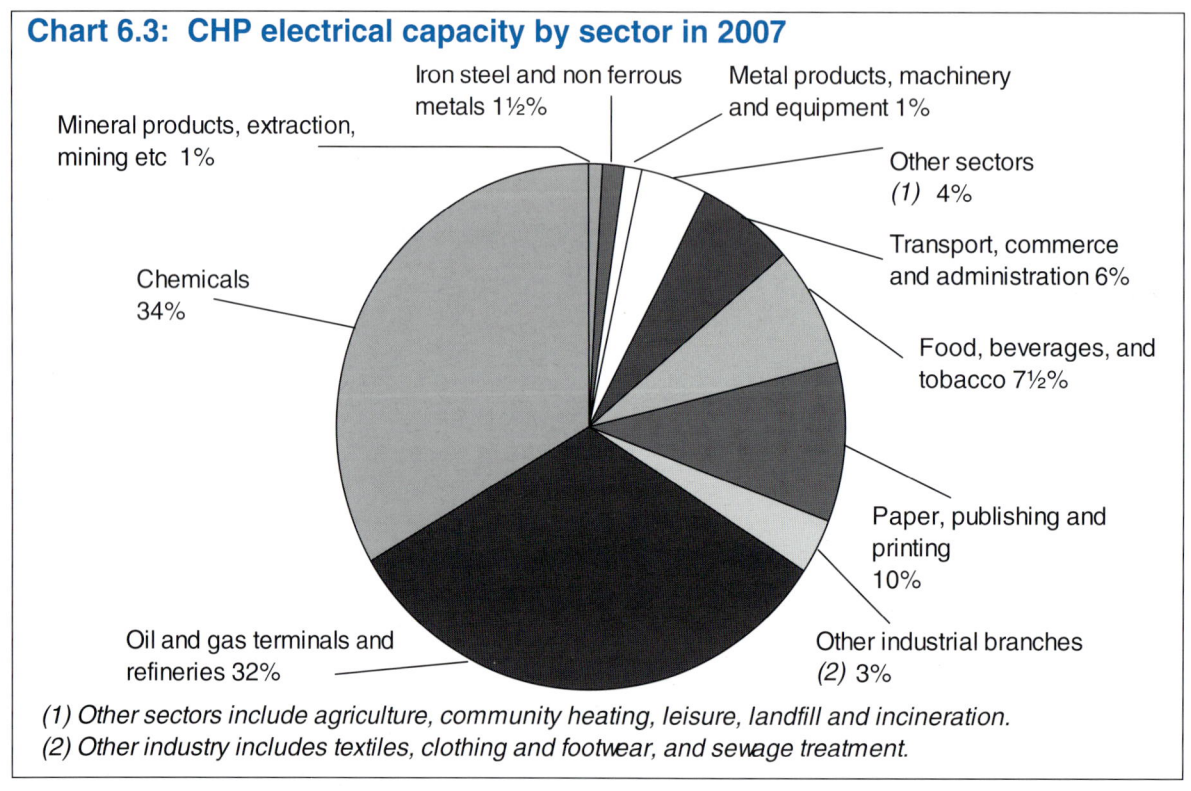

(1) Other sectors include agriculture, community heating, leisure, landfill and incineration.
(2) Other industry includes textiles, clothing and footwear, and sewage treatment.

6.21   Table 6C gives a summary of the 979 schemes installed in the commercial, public sector and residential buildings. These schemes form a major part of the "Transport, commerce and administration" and "Other" sectors in Tables 6.8 and 6.9. The vast majority of these schemes are

Table 6C: Number and capacity of CHP schemes installed in buildings by sector in 2007

	Number of schemes	Electrical capacity (MWe)	Heat capacity (MWth)
Leisure	391	47.1	56.2
Hotels	254	35.6	47.3
Health	194	122.6	196.7
Residential Group Heating	36	27.4	60.7
Universities	36	44.7	83.8
Offices	19	17.7	18.5
Education	17	10.0	17.7
Government Estate	17	15.9	18.6
Retail	11	2.6	3.4
Other *(1)*	4	16.0	21.9
Total	**979**	**339.5**	**524.8**

(1) Other includes: agriculture, airports, and domestic buildings

based on spark ignition reciprocating engines fuelled with natural gas, though the larger schemes use compression ignition reciprocating engines or gas turbines. The largest proportion of the capacity is in the health sector, mainly hospitals. Leisure and hotels account for nearly two-thirds of the total number of schemes but only about 25 per cent of the electrical capacity. Table 6.9 gives details of the quantities of fuels used in each sector.

CHP performance by main prime mover

6.22   Table 6D gives a summary of the performance of schemes in 2007 by main prime mover type. Combined cycle gas turbines have the highest average operating hours at 5,745 hours. The average for all schemes of 5,239 hours is higher than in 2006 (5,041 hours).

6.23   The average electrical efficiency is 23 per cent and heat efficiency 43 per cent, giving an overall average (because of rounding) of 67 per cent (all measured on a gross calorific value (GCV) basis).

Table 6D: A summary of scheme performance in 2007

	Average operating hours per annum (Full load equivalent)	Average electrical efficiency (% GCV)	Average heat efficiency (% GCV)	Average overall efficiency (% GCV)	Average heat to power ratio
Main prime mover in CHP plant					
Back pressure steam turbine	4,420	10	62	72	5.9
Pass out condensing steam turbine	3,730	13	46	59	3.6
Gas turbine	5,300	20	48	68	2.4
Combined cycle	5,745	26	40	67	1.5
Reciprocating engine	3,357	26	41	68	1.3
All schemes	**5,239**	**23**	**43**	**67**	**1.8**

CHP schemes which export and schemes with mechanical power output

6.24   Table 6E shows the electrical exports from CHP schemes between 2005 and 2007. Where a scheme that exports is Good Quality for only a portion of its capacity and output, the exports have been scaled back in the same way as power output has been scaled back (see paragraph 6.10, above). Exports accounted for about 30 per cent of power generation from CHP in 2007.

Table 6E: Electrical exports from CHP

GWh

	2005	2006	2007
To part of same qualifying group *(1)*	1,522	1,074r	749
To a firm NOT part of same qualifying group	1,289	3,075r	1,881
To an electricity supplier	7,852	4,621r	5,876
Total	**10,663**	**8,770r**	**8,506**

(1) A qualifying group is a group of two or more corporate consumers that are connected or related to each other, for example, as a subsidiary, or via a parent or holding company, or in terms of share capital.

6.25   Twenty five large schemes also export heat, some larger schemes to more than one customer. As Table 6F shows, together they supplied 7,483 GWh of heat in 2007, a decrease of around 1 per cent compared to 2006.

Table 6F: Electrical exports from CHP

GWh

	2005	2006	2007
To part of same qualifying group *(1)*	1,781	2,566	3,001
To a firm NOT part of same qualifying group	6,862	5,815	4,482
Total	**8,643**	**8,381**	**7,483**

(1) A qualifying group is a group of two or more corporate consumers that are connected or related to each other, for example, as a subsidiary, or via a parent or holding company, or in terms of share capital.

6.26   There are an estimated 12 schemes with mechanical power output. For those schemes, mechanical power accounts for around 6 per cent of their capacity (Table 6G). These schemes are predominantly on petro-chemicals or steel sites, using by-product fuels in boilers to drive steam turbines. The steam turbine is used to provide mechanical rather than electrical power, driving compressors, blowers or fans, rather than an alternator.

Table 6G: CHP schemes with mechanical power output in 2007

	Unit	
Number of schemes		12
Total Power Capacity of these schemes (CHP_{TPC})	MWe	3,725
Mechanical power capacity of these schemes	MWe	231

Emissions savings

6.27   The calculation of carbon emissions savings from CHP is important, given the substantial contribution that CHP can make to the Climate Change Programme. However the derivation of the savings is complex because CHP displaces a variety of fuels, technologies and sizes of plant. The methodology and assumptions used for calculating carbon emission savings are outlined in Energy Trends June 2003 www.berr.gov.uk/files/file11869.pdf and the figures compare CHP with the UK fossil fuel basket carbon intensity and the UK total basket carbon intensity which includes nuclear and renewable generation. The carbon emission savings from CHP in 2007 as compared to the fossil fuel basket were 14.3 $MtCO_2$, which equates to 2.61 Mt CO_2 per 1,000 MWe installed capacity. Against the total basket, in 2007 CHP saved 10.2 Mt CO_2, or 1.87 Mt CO_2 per 1,000 MWe installed capacity. Corresponding figures for 2005 and 2006 are shown in Table 6H.

Table 6H: Carbon dioxide savings due to CHP, absolute and per 1,000 MWe of installed good quality CHP capacity

GWh

	2005		2006		2007	
	$MtCO_2$	$MtCO_2$/1000 MWe	$MtCO_2$	$MtCO_2$/1000 MWe	$MtCO_2$	$MtCO_2$/1000 MWe
Carbon savings against all fossil fuels	15.1	2.73	15.3	2.79	14.3	2.61
Carbon savings against all fuels (including nuclear and renewables)	10.3	1.86	10.5	1.91	10.2	1.87

Technical notes and definitions

6.28   These notes and definitions are in addition to the technical notes and definitions covering all fuels and energy as a whole in Chapter 1, paragraphs 1.26 to 1.58.

Data for 2007

6.29   The data are summarised from the results of a long-term project undertaken by AEA on behalf of the Department for Business, Enterprise and Regulatory Reform (BERR) and the Department for Environment, Food and Rural Affairs (Defra). Data are included for CHP schemes installed in all sectors of the UK economy.

6.30   The project continues to be overseen by a Steering Group that comprises officials from BERR Defra, the Office of Gas and Electricity Markets (Ofgem), the Combined Heat and Power Association (CHPA) and the Office for National Statistics (ONS) all of whom have an interest in either the collection of information on CHP schemes or the promotion of the wider use of CHP in the UK.

6.31   Data for 2007 were based largely on data supplied to the CHPQA programme, supplemented by a survey carried out by the Office for National Statistics (ONS), by information from the Iron and Steel Statistics Bureau (ISSB) and from Ofgem "Renewables Obligation Certificates" (ROCs) information. Over 94 per cent of the total capacity is from schemes certified under the CHPQA programme, while about 1.7 per cent is from schemes covered by the ONS and ISSB sources. Data for schemes not applying for CHPQA and not available from other sources (e.g. because they were below the cut off capacity for the ONS survey) were interpolated from historical data. These schemes account for about 3 per cent of total capacity. Since 2005, Sewage Treatment Works that do not provide returns either to CHPQA or ONS in a format that can be used within these statistics, have been included based on ROCs information from Ofgem returns. The sewage treatment works data from this source account for approximately 1.3 per cent of total electrical capacity.

6.32   An ongoing data cleansing exercise continued in 2007. This cleansing is targeted at schemes included in the CHP database that do not supply annual data through the normal channels of CHPQA, ISSB, ROCs information or ONS. In these cases historical data has been interpolated. During 2007 the recorded operators of schemes in this category were contacted directly in order to ascertain the existence, or otherwise, of CHP at the site in question. Where this research indicated that CHP did not exist the CHP scheme was removed from the CHP database not just for 2007 but retrospectively back to 2000 where appropriate. This cleansing action resulted in the removal of 152 schemes with a Good Quality CHP capacity of 24 MWe. This represents 0.4 per cent of the electrical capacity listed prior to their removal. This cleansing will be ongoing and will target schemes about which the information held is most uncertain or out of date. Hitherto unresearched schemes classified in this way represent another 40 MWe or 0.7 per cent of the currently listed electrical capacity.

6.33   During 2007 other work carried out by BERR, for which the CHP database was used, indicated that revisions should be made to the sector classification of some schemes. The revisions made were consistent with assigning the CHP scheme to the sector in which the majority of the CHP heat output is used, rather than the sector in which the CHP operator was considered to operate. This has led to the migration of schemes between sectors, some of which have been significant. A summary of these reclassifications and the impact they have had on the number and capacity of schemes in the different sectors is summarised in Table 6J. For completeness, Table 6H includes all sectors used in this chapter of the Digest, even those unaffected by the reclassification.

Definitions of schemes

6.34   There are four principal types of CHP system:

- **Steam turbine,** where steam at high pressure is generated in a boiler. In **back pressure steam turbine systems**, the steam is wholly or partly used in a turbine before being exhausted from the turbine at the required pressure for the site. In **pass-out condensing steam turbine systems**, a proportion of the steam used by the turbine is extracted at an intermediate pressure from the turbine with the remainder being fully condensed before it is exhausted at the exit. (Condensing

Table 6J: Reclassification of schemes in 2007

Sector	Pre reclassification		Post reclassification	
	Number of schemes	Electrical capacity, QPC (MWe)	Number of schemes	Electrical capacity, QPC (MWe)
Iron and steel and non-ferrous metals	7	79	7	79
Chemicals	42	1,332	52	1,847
Oil and gas terminals and oil refineries	11	1,819	9	1,763
Paper, publishing and printing	28	541	27	536
Food, beverages and tobacco	40	377	40	408
Metal products, machinery and equipment	9	56	17	66
Mineral products, extraction, mining and agglomeration of solid fuels	8	48	7	38
Other industrial branches	145	611	169	173
Transport, commerce and administration	978	350	991	331
Other	83	214	119	231
Unassigned under pre reclassification system	87	47	-	-
Total	**1,438**	**5,474**	**1,438**	**5,474**

steam turbines without passout and which do not utilise steam are not included in these statistics as they are not CHP). The boilers used in such schemes can burn a wide variety of fuels including coal, gas, oil, and waste-derived fuels. With the exception of waste-fired schemes, steam turbine plant has often been in service for several decades. Steam turbine schemes capable of supplying useful steam have electrical efficiencies of between 10 and 20 per cent, depending on size, and thus between 70 per cent and 30 per cent of the fuel input is available as useful heat. Steam turbines used in CHP applications typically range in size from a few MWe to over 100 MWe.

- **Gas turbine systems**, often aero-engine derivatives, where fuel (gas, or gas-oil) is combusted in the gas turbine and the exhaust gases are normally used in a waste heat boiler to produce usable steam, though the exhaust gases may be used directly in some process applications. Gas turbines range from 30 kWe upwards, achieving electrical efficiency of 23 to 30 per cent (depending on size) and with the potential to recover up to 50 per cent of the fuel input as useful heat. They have been common in CHP since the mid 1980s. The waste heat boiler can include supplementary or auxiliary firing using a wide range of fuels, and thus the heat to power ratio of the scheme can vary.

- **Combined cycle systems**, where the plant comprises more than one prime mover. These are usually gas turbines where the exhaust gases are utilised in a steam generator, the steam from which is passed wholly or in part into one or more steam turbines. In rare cases reciprocating engines may be linked with steam turbines. Combined cycle is suited to larger installations of 7 MWe and over. They achieve higher electrical efficiency and a lower heat to power ratio than steam turbines or gas turbines. Recently installed combined cycle gas turbine (CCGT) schemes have achieved an electrical efficiency approaching 50 per cent, with 20 per cent heat recovery, and a heat to power ratio of less than 1:1.

- **Reciprocating engine systems** range from less than 100 kWe up to around 5 MWe, and are found in applications where production of hot water (rather than steam) is the main requirement, for example, on smaller industrial sites as well as in buildings. They are based on auto engine or marine engine derivatives converted to run on gas. Both compression ignition and spark ignition firing is used. Reciprocating engines operate at around 28 to 33 per cent electrical efficiency with around 50 per cent to 33 per cent of the fuel input available as useful heat. Reciprocating engines produce two grades of waste heat: high grade heat from the engine exhaust and low grade heat from the engine cooling circuits.

Determining fuel consumption for heat and electricity

6.35   In order to provide a comprehensive picture of electricity generation in the United Kingdom and the fuels used to generate that electricity, the energy input to CHP schemes has to be allocated between heat and electricity production. This allocation is notional and is not determinate.

6.36   The convention used to allocate the fuels to heat and electricity relates the split of fuels to the relative efficiency of heat and electricity supply. The efficiency of utility plant varies widely: electricity generation from as little as 25 per cent to more than 50 per cent and boilers from as little as 50 per cent to more than 90 per cent. Thus it is around twice as hard to generate a unit of electricity as it is to generate a unit of heat. Accordingly a simple convention can be implemented whereby twice as many units of fuel are allocated to each unit of electricity generated, as to each unit of heat supplied. This approach is consistent with the Defra Guidelines for Company Reporting on greenhouse gas emissions and for Negotiated Agreements on energy efficiency agreed between Government and industry as part of the Climate Change Levy (CCL) package. It recognises that in developing a CHP scheme, both the heat customer(s) and the electricity generator share in the savings, reflecting the fact that more than three-quarters of CHP build in the last few years has been supplied under an energy services arrangement.

6.37   The assumption in this convention that it is twice as hard to generate a unit of electricity as heat, is appropriate for the majority of CHP schemes. However, for some types of scheme (for example in the iron and steel sector) this allocation is less appropriate and can result in very high apparent heat efficiencies. These, however, are only notional efficiencies.

The effects on the statistics of using CHPQA

6.38   Paragraph 6.10 described how schemes were scaled back so that only CHP_{QPC} and CHP_{QPO} were included in the CHP statistics. This is illustrated in Table 6K. In 2007, 176 schemes have been scaled back. In 2006 176 schemes were also scaled back.

6.39   In 2007 the power output from these schemes was scaled back from a total of 31,943 GWh to 7,736 GWh. The total fuel input to these schemes was 91,855 GWh of which 57,258 GWh was regarded as being for power only.

Table 6K: CHP capacity, output and fuel use which has been scaled back in 2007

	Units	
Number of schemes requiring scaling back		176
Total Power Capacity of these schemes (CHP_{TPC})	MWe	5,218
Qualifying Power Capacity of these schemes (CHP_{QPC})	MWe	1,552
Total power output of these schemes (CHP_{TPO})	GWh	31,943*
Qualifying Power Output of these schemes (CHP_{QPO})	GWh	7,736
Electricity regarded as "Power only" not from CHP ($CHP_{TPO} - CHP_{QPO}$)	GWh	24,208
Total Fuel Input of these schemes (CHP_{TFI})	GWh	91,855
Fuel input regarded as being for "Power only" use i.e. not for CHP	GWh	57,258

*This figure includes generation from major power producers

Contacts:
Richard Hodges,
AEA
Richard.hodges@aeat.co.uk
0870 190 6148

Mike Janes (Statistician),
BERR Energy Markets Unit
mike.janes@berr.gsi.gov.uk
020 7215 5186

6.1 CHP installations by capacity and size range

	2003	2004	2005	2006	2007
Number of schemes *(1)*	**1,356r**	**1,345r**	**1,378r**	**1,388r**	**1,438**
Less than 100 kWe	516r	485r	470r	472r	475
100 kWe to 999 kWe	587r	604r	644r	655r	693
1 MWe to 9.9 MWe	179r	183r	189r	189r	198
10.0 MWe and above	74	73	75	72r	72
					MWe
Total capacity	**4,495r**	**5,398r**	**5,536r**	**5,484r**	**5,474**
Less than 100 kWe	32r	31r	30r	30r	30
100 kWe to 999 kWe	147r	154r	166r	168r	180
1 MWe to 9.9 MWe	712r	732r	733r	712r	728
10.0 MWe and above	3,604	4,481	4,607	4,574r	4,536

(1) A site may contain more than one CHP scheme.

6.2 Fuel used to generate electricity and heat in CHP installations

					GWh
	2003	2004	2005	2006	2007
Fuel used to generate electricity *(1)*					
Coal *(2)*	2,198r	1,696r	1,559r	1,797r	1,781
Fuel oil	1,832	1,888	1,617	1,553r	923
Natural gas	40,806r	44,154r	47,511r	47,185r	47,403
Renewable fuels *(3)*	1,198	1,268	1,420	1,810r	1,957
Other fuels *(4)*	6,528r	9,267r	10,411r	10,801r	11,024
Total all fuels	**52,562r**	**58,274r**	**62,518r**	**63,146r**	**63,088**
Fuel used to generate heat					
Coal *(2)*	3,968r	2,816r	2,591r	2,559r	2,503
Fuel oil	2,587	2,763	2,150	2,006r	1,263
Natural gas	40,073r	40,705r	41,149r	39,365r	39,409
Renewable fuels *(3)*	1,412	1,424	1,440	1,423r	1,473
Other fuels *(4)*	12,498r	14,220r	14,784r	14,323r	15,006
Total all fuels	**60,538r**	**61,928r**	**62,114r**	**59,677r**	**59,654**
Overall fuel use					
Coal *(2)*	6,166r	4,512r	4,150r	4,356r	4,284
Fuel oil	4,419	4,651	3,767	3,560r	2,186
Natural gas	80,879r	84,860r	88,660r	86,550r	86,812
Renewable fuels *(3)*	2,610	2,692	2,860	3,233r	3,430
Other fuels *(4)*	19,026r	23,487r	25,196r	25,124r	26,030
Total all fuels	**113,100r**	**120,201r**	**124,632r**	**122,823r**	**122,742**

(1) See paragraphs 6.35 to 6.37 for an explanation of the method used to allocate fuel use between heat generation and electricity generation.
(2) Includes coke and semi-coke.
(3) Renewable fuels include: sewage gas; other biogases; municipal waste and refuse derived fuels.
(4) Other fuels include: process by-products, coke oven gas, blast furnace gas, gas oil and refinery gas.

6.3 Fuel used by types of CHP installation

GWh

	2003	2004	2005	2006	2007
Coal					
Back pressure steam turbine	970r	942r	903r	693r	592
Gas turbine	42	50	44	-	43
Combined cycle	172	118	41	589r	589
Reciprocating engine	-	-	-	-	-
Pass out condensing steam turbine	4,983	3,402	3,162	3,074	3,059
Total coal	**6,166r**	**4,512r**	**4,150r**	**4,356r**	**4,284**
Fuel oil					
Back pressure steam turbine	482	473	463	207r	138
Gas turbine	8	8	8	12r	3
Combined cycle	3,584	3,756	2,953	2,994r	1,682
Reciprocating engine	190	182	162	141r	132
Pass out condensing steam turbine	156	232	182	206	232
Total fuel oil	**4,419**	**4,651**	**3,767**	**3,560r**	**2,186**
Natural gas					
Back pressure steam turbine	4,313	3,092	2,938	2,154r	1,924
Gas turbine	15,236	13,580	12,358r	11,462r	11,614
Combined cycle	52,543	58,959	65,112	65,283r	65,470
Reciprocating engine	7,383r	7,822r	6,998r	6,523r	6,664
Pass out condensing steam turbine	1,405	1,407	1,254	1,128	1,140
Total natural gas	**80,879r**	**84,860r**	**88,660r**	**86,550r**	**86,812**
Renewable fuels (1)					
Back pressure steam turbine	8	416	326	535r	549
Gas turbine	21	21	30	26	26
Combined cycle	263	411	634	654	654
Reciprocating engine	1,427	1,391	1,497	1,367r	1,550
Pass out condensing steam turbine	892	453	374	651r	651
Total renewable fuels	**2,610**	**2,692**	**2,860**	**3,233r**	**3,430**
Other fuels (2)					
Back pressure steam turbine	5,343	5,356	5,930	5,829	5,824
Gas turbine	3,971	4,002	4,040	4,125r	4,187
Combined cycle	6,691	10,510	11,436	10,516r	11,586
Reciprocating engine	51r	76r	58r	57r	49
Pass out condensing steam turbine	2,971r	3,542r	3,731r	4,597r	4,384
Total other fuels	**19,026r**	**23,487r**	**25,196r**	**25,124r**	**26,030**
Total - all fuels					
Back pressure steam turbine	11,115r	10,279r	10,559r	9,418r	9,026
Gas turbine	19,276	17,662	16,480r	15,625r	15,873
Combined cycle	63,252	73,753	80,176	80,036r	79,982
Reciprocating engine	9,050r	9,471r	8,716r	8,088r	8,395
Pass out condensing steam turbine	10,407r	9,036r	8,702r	9,655r	9,466
Total all fuels	**113,100r**	**120,201r**	**124,632r**	**122,823r**	**122,742**

(1) Renewable fuels include: sewage gas, other biogases, municipal solid waste and refuse derived fuels.
(2) Other fuels include: process by-products, coke oven gas, blast furnace gas, gas oil and refinery gas.

6.4 CHP - electricity generated by fuel and type of installation

GWh

	2003	2004	2005	2006	2007
Coal					
Back pressure steam turbine	92r	106r	99r	77r	63
Gas turbine	7	8	7	-	7
Combined cycle	25	15	3	136r	136
Reciprocating engine	-	-	-	-	-
Pass out condensing steam turbine	795	570	541	517	517
Total coal	**919r**	**699r**	**650r**	**730r**	**723**
Fuel oil					
Back pressure steam turbine	53	53	53	25	16
Gas turbine	2	2	2	3	1
Combined cycle	723	776	636	618r	330
Reciprocating engine	60	59	50	47r	45
Pass out condensing steam turbine	23	34	29	34	36
Total fuel oil	**861**	**923**	**769**	**727r**	**428**
Natural gas					
Back pressure steam turbine	533	255	235	172r	142
Gas turbine	3,412	2,939	2,697r	2,464r	2,559
Combined cycle	12,995	15,305	17,381	17,732r	17,562
Reciprocating engine	1,823r	1,950r	1,781r	1,648r	1,683
Pass out condensing steam turbine	112	170	155	135	132
Total natural gas	**18,876r**	**20,619r**	**22,249r**	**22,150r**	**22,077**
Renewable fuels *(1)*					
Back pressure steam turbine	-	51	36	70r	75
Gas turbine	4	4	5	4	4
Combined cycle	16	25	43	60	60
Reciprocating engine	331	379	419	394r	454
Pass out condensing steam turbine	89	37	34	109r	109
Total renewable fuels	**440**	**496**	**537**	**638r**	**703**
Other fuels *(2)*					
Back pressure steam turbine	640	657	684	641r	641
Gas turbine	614	621	643r	571r	603
Combined cycle	1,348	2,516	2,975	2,860r	3,059
Reciprocating engine	13r	19r	12r	14r	11
Pass out condensing steam turbine	228r	309r	316r	573r	432
Total other fuels	**2,843r**	**4,121r**	**4,631r**	**4,659r**	**4,746**
Total - all fuels					
Back pressure steam turbine	1,318r	1,122r	1,107r	987r	938
Gas turbine	4,039	3,574	3,354r	3,041r	3,174
Combined cycle	15,107	18,637	21,037	21,405r	21,145
Reciprocating engine	2,226r	2,406r	2,262r	2,103r	2,194
Pass out condensing steam turbine	1,247r	1,120r	1,075r	1,367r	1,226
Total all fuels	**23,938r**	**26,859r**	**28,836r**	**28,904r**	**28,677**

(1) Renewable fuels include: sewage gas, other biogases, municipal solid waste and refuse derived fuels.
(2) Other fuels include: process by-products, coke oven gas, blast furnace gas, gas oil and refinery gas.

6.5 CHP - electrical capacity by fuel and type of installation

					MWe
	2003	2004	2005	2006	2007
Coal					
Back pressure steam turbine	24r	32r	32r	26r	28
Gas turbine	1	1	1	-	1
Combined cycle	6	3	-	18r	18
Reciprocating engine	-	-	-	-	-
Pass out condensing steam turbine	227	161	159	150	150
Total coal	**258r**	**197r**	**193r**	**195r**	**197**
Fuel oil					
Back pressure steam turbine	14	15	14	9	7
Gas turbine	-	-	-	-	-
Combined cycle	125	132	116	131r	70
Reciprocating engine	21	21	16	16r	15
Pass out condensing steam turbine	6	9	8	8	8
Total fuel oil	**167**	**176**	**154**	**165**	**101**
Natural gas					
Back pressure steam turbine	156	74	74	61	49
Gas turbine	630r	550r	494r	479r	483
Combined cycle	2,138	2,777	3,045	3,056r	3,051
Reciprocating engine	505r	508r	507r	505r	517
Pass out condensing steam turbine	40	55	61	43	43
Total natural gas	**3,469r**	**3,964r**	**4,181r**	**4,145r**	**4,144**
Renewable fuels *(1)*					
Back pressure steam turbine	-	13	12	16r	16
Gas turbine	1	1	1	1	1
Combined cycle	3	6	6	10	10
Reciprocating engine	80	92	102	102r	117
Pass out condensing steam turbine	23	10	10	23r	23
Total renewable fuels	**107**	**122**	**131**	**151r**	**167**
Other fuels *(2)*					
Back pressure steam turbine	109	136	137	112r	112
Gas turbine	94	113	111	113r	114
Combined cycle	232	596	522	494r	532
Reciprocating engine	4r	5r	4r	4r	4
Pass out condensing steam turbine	55r	89r	103r	105r	103
Total other fuels	**494r**	**939r**	**877r**	**828r**	**864**
Total - all fuels					
Back pressure steam turbine	303r	270r	270r	225r	212
Gas turbine	726r	665r	607r	593r	599
Combined cycle	2,504	3,514	3,688	3,709r	3,681
Reciprocating engine	610r	627r	629r	627r	654
Pass out condensing steam turbine	352r	323r	342r	330r	329
Total all fuels	**4,495r**	**5,398r**	**5,536r**	**5,484r**	**5,474**

(1) Renewable fuels include: sewage gas, other biogases, municipal solid waste and refuse derived fuels.
(2) Other fuels include: process by-products, coke oven gas, blast furnace gas, gas oil and refinery gas.

6.6 CHP - heat generated by fuel and type of installation

					GWh
	2003	2004	2005	2006	2007
Coal					
Back pressure steam turbine	734r	749r	713r	504r	442
Gas turbine	31	28	24	-	24
Combined cycle	89	58	7	170r	170
Reciprocating engine	-	-	-	-	-
Pass out condensing steam turbine	2,543	1,593	1,526	1,503	1,495
Total coal	**3,397r**	**2,428r**	**2,270r**	**2,177r**	**2,131**
Fuel oil					
Back pressure steam turbine	348	347	336	181r	122
Gas turbine	4	4	4	5r	1
Combined cycle	1,932	2,168	1,560	1,568	917
Reciprocating engine	57	56	55	43r	39
Pass out condensing steam turbine	83	114	85	90	101
Total fuel oil	**2,424**	**2,690**	**2,040**	**1,887r**	**1,179**
Natural gas					
Back pressure steam turbine	3,049	2,356	2,271	1,672r	1,482
Gas turbine	7,375	6,756	6,143r	5,648r	5,585
Combined cycle	22,714	24,779	26,435	26,027r	26,102
Reciprocating engine	3,362r	3,437r	3,087r	2,886r	2,896
Pass out condensing steam turbine	1,060	860	696	611	613
Total natural gas	**37,560r**	**38,188r**	**38,631r**	**36,845r**	**36,678**
Renewable fuels *(1)*					
Back pressure steam turbine	3	157	144	242r	207
Gas turbine	11	11	16	14	14
Combined cycle	61	83	108	107	107
Reciprocating engine	527	557	560	456r	521
Pass out condensing steam turbine	324	174	146	143r	143
Total renewable fuels	**926**	**983**	**973**	**961r**	**992**
Other fuels *(2)*					
Back pressure steam turbine	3,361	3,285	3,558	3,314r	3,309
Gas turbine	1,983	2,004	1,973r	1,962r	1,950
Combined cycle	3,527	4,937	4,841	4,209r	4,755
Reciprocating engine	25r	33r	24r	23r	20
Pass out condensing steam turbine	1,784r	1,984r	2,145r	2,141r	2,035
Total other fuels	**10,679r**	**12,242r**	**12,541r**	**11,649r**	**12,070**
Total - all fuels					
Back pressure steam turbine	7,496r	6,894r	7,022r	5,912r	5,562
Gas turbine	9,403	8,803	8,160	7,630	7,575
Combined cycle	28,322	32,024	32,950	32,081	32,051
Reciprocating engine	3,971r	4,084r	3,725r	3,408r	3,476
Pass out condensing steam turbine	5,793r	4,726r	4,598r	4,488r	4,387
Total all fuels	**54,985r**	**56,531r**	**56,455r**	**53,519r**	**53,050**

(1) Renewable fuels include: sewage gas, other biogases, municipal solid waste and refuse derived fuels.
(2) Other fuels include: process by-products, coke oven gas, blast furnace gas, gas oil and refinery gas.

166

6.7 CHP - heat capacity by fuel and type of installation

MWth

	2003	2004	2005	2006	2007
Coal					
Back pressure steam turbine	151r	196r	200r	152r	160
Gas turbine	3	4	4	-	3
Combined cycle	19	10	1	14r	14
Reciprocating engine	-	-	-	-	-
Pass out condensing steam turbine	646	445	445	453	453
Total coal	**819r**	**656r**	**649r**	**619r**	**631**
Fuel oil					
Back pressure steam turbine	93	95	94	51r	42
Gas turbine	1	1	1	1	-
Combined cycle	294	292	247	275r	121
Reciprocating engine	21	21	17	18r	17
Pass out condensing steam turbine	19	29	23	23	28
Total fuel oil	**427**	**437**	**382**	**368r**	**207**
Natural gas					
Back pressure steam turbine	541	459	457	364r	366
Gas turbine	1,522r	1,717r	1,496r	1,424r	1,422
Combined cycle	3,798	4,067	4,267	4,269r	4,432
Reciprocating engine	726r	761r	691r	680r	681
Pass out condensing steam turbine	208	226	183	177	176
Total natural gas	**6,795r**	**7,230r**	**7,093r**	**6,915r**	**7,077**
Renewable fuels (1)					
Back pressure steam turbine	5	46	44	47	47
Gas turbine	2	2	2r	2r	2
Combined cycle	9	13	16	17	17
Reciprocating engine	130	123	117	115r	115
Pass out condensing steam turbine	86	43	43	43	43
Total renewable fuels	**232**	**227**	**223r**	**225r**	**225**
Other fuels (2)					
Back pressure steam turbine	381	435	437	397r	396
Gas turbine	1,465	1,512	1,517	1,586r	1,586
Combined cycle	512	897	813	746r	756
Reciprocating engine	5r	7r	5	4r	4
Pass out condensing steam turbine	297r	375r	385r	383r	379
Total other fuels	**2,661r**	**3,225r**	**3,157r**	**3,116r**	**3,121**
Total - all fuels					
Back pressure steam turbine	1,170r	1,231r	1,231r	1,011r	1,011
Gas turbine	2,993r	3,236r	3,020r	3,013r	3,013
Combined cycle	4,632	5,278	5,343	5,322r	5,341
Reciprocating engine	883r	911r	831r	817r	817
Pass out condensing steam turbine	1,256r	1,119r	1,079r	1,079r	1,079
Total all fuels	**10,934r**	**11,775r**	**11,504r**	**11,242r**	**11,261**

(1) Renewable fuels include: sewage gas, other biogases, municipal solid waste and refuse derived fuels.
(2) Other fuels include: process by-products, coke oven gas, blast furnace gas, gas oil and refinery gas.

6.8 CHP capacity, output and total fuel use[1] by sector

	Unit	2003	2004	2005	2006	2007
Iron and steel and non ferrous metals						
Number of sites		6	7	7	7	7
Electrical capacity	MWe	66	67	67	81	79
Heat capacity	MWth	285	285	285	285	285
Electrical output	GWh	253	243	238	520r	366
Heat output	GWh	1,707	1,708	1,765	1,812r	1,717
Fuel use	GWh	3,085	3,244	3,045	3,984r	3,809
of which : for electricity	GWh	654	662	609	1,426r	1,094
for heat	GWh	2,431	2,582	2,436	2,558r	2,715
Chemicals						
Number of sites		53r	52r	53r	52r	52
Electrical capacity	MWe	1,698r	1,792r	1,865r	1,863r	1,847
Heat capacity	MWth	3,873r	4,058r	3,996r	3,939r	3,939
Electrical output	GWh	9,590r	10,102r	9,853r	10,234r	10,087
Heat output	GWh	18,361r	18,373r	18,296r	18,376r	18,073
Fuel use	GWh	42,338r	42,948r	42,709r	43,401r	42,916
of which : for electricity	GWh	21,947r	22,658r	22,237r	22,846r	22,676
for heat	GWh	20,391r	20,290r	20,472r	20,556r	20,239
Oil and gas terminals and oil refineries						
Number of sites		7r	8r	9r	9r	9
Electrical capacity	MWe	859r	1,672r	1,735r	1,731r	1,763
Heat capacity	MWth	3,077r	3,677r	3,677r	3,677r	3,677
Electrical output	GWh	5,192r	7,211r	9,957r	10,040r	10,255
Heat output	GWh	15,309r	17,065r	17,803r	16,779r	17,709
Fuel use	GWh	27,951r	33,809r	40,713r	40,426r	42,034
of which : for electricity	GWh	11,250r	15,513r	21,306r	21,679r	22,219
for heat	GWh	16,701r	18,296r	19,407r	18,747r	19,816
Paper, publishing and printing						
Number of sites		35r	33r	31r	27r	27
Electrical capacity	MWe	687r	670r	656r	621r	536
Heat capacity	MWth	1,440r	1,384r	1,352r	1,220r	1,220
Electrical output	GWh	3,910r	4,221r	3,845r	3,399r	3,058
Heat output	GWh	8,290r	8,312r	7,987r	6,780r	6,220
Fuel use	GWh	16,666r	17,250r	16,137r	14,049r	12,979
of which : for electricity	GWh	8,004r	8,564r	7,805r	6,938r	6,353
for heat	GWh	8,662r	8,686r	8,332r	7,111r	6,626
Food, beverages and tobacco						
Number of sites		40r	41r	42r	40r	40
Electrical capacity	MWe	378r	403r	408r	408r	408
Heat capacity	MWth	962r	1,063r	968r	923r	923
Electrical output	GWh	1,950r	2,026r	2,091r	1,952r	1,910
Heat output	GWh	5,508r	5,242r	5,148r	4,576r	4,114
Fuel use	GWh	9,524r	9,407r	9,381r	8,604r	8,106
of which : for electricity	GWh	3,916r	4,105r	4,223r	3,945r	3,896
for heat	GWh	5,607r	5,303r	5,158r	4,659r	4,209

For footnotes see page 170

6.8 CHP capacity, output and total fuel use[1] by sector (continued)

	Unit	2003	2004	2005	2006	2007
Metal products, machinery and equipment						
Number of sites		18r	19r	18r	17r	17
Electrical capacity	MWe	106r	77r	73r	34r	66
Heat capacity	MWth	85r	85r	55r	36r	55
Electrical output	GWh	321r	210r	159r	129r	169
Heat output	GWh	428r	320r	216r	203r	231
Fuel use	GWh	1,267r	908r	643r	585r	665
of which : for electricity	GWh	726r	488r	371r	310r	376
for heat	GWh	541r	420r	272r	275r	289
Mineral products, extraction, mining and agglomeration of solid fuels						
Number of sites		9r	8r	8r	7r	7
Electrical capacity	MWe	61r	57r	57r	38r	38
Heat capacity	MWth	173r	210r	210r	208r	208
Electrical output	GWh	332r	270r	217r	171r	172
Heat output	GWh	976r	904r	897r	742r	707
Fuel use	GWh	1,756r	1,565r	1,476r	1,217r	1,159
of which : for electricity	GWh	726r	594r	481r	368r	380
for heat	GWh	1,031r	971r	995r	848r	780
Sewage treatment						
Number of sites		104r	111r	127r	127	163
Electrical capacity	MWe	107	119	126	131r	146
Heat capacity	MWth	157r	150	141r	141	141
Electrical output	GWh	393r	439r	469r	455r	514
Heat output	GWh	607r	635r	621r	531r	596
Fuel use	GWh	1,647r	1,603r	1,675r	1,598r	1,780
of which : for electricity	GWh	919r	930r	1,006r	1,010r	1,128
for heat	GWh	728r	673r	669r	588r	653
Other industrial branches (2)						
Number of sites		6r	5r	6r	6r	6
Electrical capacity	MWe	35r	30r	31r	31r	27
Heat capacity	MWth	16r	16r	16r	16r	16
Electrical output	GWh	93r	89r	47r	52r	75
Heat output	GWh	60r	51r	53r	48r	55
Fuel use	GWh	331r	254r	180r	203r	260
of which : for electricity	GWh	252r	198r	115r	139r	192
for heat	GWh	79r	57r	65r	64r	68
Total industry						
Number of sites		278r	284r	301r	292r	328
Electrical capacity	MWe	3,997r	4,887r	5,017r	4,938r	4,911
Heat capacity	MWth	10,068r	10,927r	10,700r	10,445r	10,464
Electrical output	GWh	22,035r	24,812r	26,876r	26,953r	26,607
Heat output	GWh	51,246r	52,610r	52,786r	49,847r	49,422
Fuel use	GWh	104,564r	110,988r	115,960r	114,067r	113,708
of which : for electricity	GWh	48,393r	53,711r	58,153r	58,661r	58,312
for heat	GWh	56,171r	57,277r	57,807r	55,405r	55,396

For footnotes see page 170

6.8 CHP capacity, output and total fuel use[1] by sector (continued)

	Unit	2003	2004	2005	2006	2007
Transport, commerce and administration						
Number of sites		960r	943r	960r	978r	991
Electrical capacity	MWe	291r	301r	315r	317r	331
Heat capacity	MWth	494r	510r	480r	474r	474
Electrical output	GWh	1,178r	1,257r	1,258r	1,147r	1,254
Heat output	GWh	2,245	2,323r	2,345r	2,260r	2,240
Fuel use	GWh	4,814r	5,164r	5,189r	4,844r	5,041
of which : for electricity	GWh	2,460r	2,663r	2,673r	2,410r	2,643
for heat	GWh	2,354r	2,501r	2,516r	2,434r	2,398
Other (3)						
Number of sites		118r	118r	117r	118r	119
Electrical capacity	MWe	208r	210r	204r	229r	231
Heat capacity	MWth	372	338r	324r	323r	323
Electrical output	GWh	725r	790r	701r	804r	816
Heat output	GWh	1,493r	1,598r	1,324r	1,412r	1,389
Fuel use	GWh	3,722r	4,050r	3,483r	3,912r	3,993
of which : for electricity	GWh	1,709r	1,900r	1,693r	2,074r	2,133
for heat	GWh	2,013r	2,150r	1,790r	1,837r	1,860
Total CHP usage by all sectors						
Number of sites		1,356r	1,345r	1,378r	1,388r	1,438
Electrical capacity	MWe	4,495r	5,398r	5,536r	5,484r	5,474
Heat capacity	MWth	10,934r	11,775r	11,504r	11,242r	11,261
Electrical output	GWh	23,938r	26,859r	28,836r	28,904r	28,677
Heat output	GWh	54,985r	56,531r	56,455r	53,519r	53,050
Fuel use	GWh	113,100r	120,201r	124,632r	122,823r	122,742
of which : for electricity	GWh	52,562r	58,274r	62,518r	63,146r	63,088
for heat	GWh	60,538r	61,928r	62,114r	59,677r	59,654

(1) The allocation of fuel use between electricity and heat is largely notional and the methodology is outlined in paragraphs 6.35 to 6.37.

(2) Other industry includes Textiles, clothing and footwear sector.

(3) Sectors included under Other are agriculture, community heating, leisure, landfill and incineration.

6.9 CHP - use of fuels by sector

GWh

	2003	2004	2005	2006	2007
Iron and steel and non ferrous metals					
Coal	-	-	-	-	-
Fuel oil	45	102	55	79	105
Natural gas	217	282	202	181r	191
Blast furnace gas	2,401	2,523	2,313	3,083	2,885
Coke oven gas	422	337	475	641	628
Other fuels (1)	-	-	-	-	-
Total iron and steel and non ferrous metals	**3,085**	**3,244**	**3,045**	**3,984**	**3,809**
Chemicals					
Coal	4,405	2,923	2,804	3,395r	3,394
Fuel oil	136	185	145r	292r	207
Gas oil	32	409r	545r	98r	45
Natural gas	34,909r	35,502r	35,046r	35,691r	35,347
Refinery gas	1,131	1,132r	1,132r	1,181r	1,181
Renewable fuels (2)	21	21	30	26	26
Other fuels (1)	1,703r	2,776r	3,006r	2,719r	2,717
Total chemical industry	**42,338r**	**42,948r**	**42,709r**	**43,401r**	**42,916**
Oil and gas terminals and oil refineries					
Fuel oil	3,557	3,716	2,910	2,844	1,606
Gas oil	50	94	111	80r	80
Natural gas	11,538r	15,043r	20,818r	21,041r	22,526
Refinery gas	4,836	4,444	4,011	4,651r	5,786
Other fuels (1)	7,969	10,512	12,863	11,810r	12,036
Total oil refineries	**27,951r**	**33,809r**	**40,713r**	**40,426r**	**42,034**
Paper, publishing and printing					
Coal	450	636r	683	595	580
Fuel oil	268	266	308	3r	-
Gas oil	22	30	73r	188	86
Natural gas	15,780r	16,188r	14,926r	13,114r	12,163
Renewable fuels (2)	2	1	-	-	-
Other fuels (1)	145	130	147	150	151
Total paper, publishing and printing	**16,666r**	**17,250r**	**16,137r**	**14,049r**	**12,979**
Food, beverages and tobacco					
Coal	970r	577r	578r	338r	238
Fuel oil	214	191	192	199	137
Gas oil	20	767	97r	81r	85
Natural gas	8,311r	7,850r	8,473r	7,965r	7,641
Renewable fuels (2)	-	-	-	1	-
Other fuels (1)	8	23	42r	20	5
Total food, beverages and tobacco	**9,524r**	**9,407r**	**9,381r**	**8,604r**	**8,106**
Metal products, machinery and equipment					
Coal	-	-	-	-	-
Fuel oil	91	92	89	89	89
Gas oil	-	-	-	-	-
Natural gas	1,003r	684r	450r	391r	472
Renewable fuels (2)	172r	131r	105r	104r	104
Other fuels (1)	-r	-r	-r	-r	-
Total metal products, machinery and equipment	**1,266r**	**907r**	**643r**	**585r**	**665**

For footnotes see page 172

6.9 CHP - use of fuels by sector (continued)

	2003	2004	2005	2006	2007
Mineral products, extraction, mining and agglomeration of solid fuels					
Coal	-	-	-	-	-
Fuel oil	2	-	-	-	-
Gas oil	-	-	-	1	-
Natural gas	1,526r	1,342r	1,207r	945r	890
Coke oven gas	228	223	269	271r	269
Total mineral products, extraction, mining and agglomeration of solid fuels	**1,756r**	**1,565r**	**1,476r**	**1,217r**	**1,159**
Sewage treatment					
Fuel oil	60	60	53	41r	40
Gas oil	30	30	15	20	20
Natural gas	132r	124r	112r	145r	145
Renewable fuels (2)	1,425	1,389	1,495	1,392r	1,575
Total sewage treatment	**1,647r**	**1,603r**	**1,675r**	**1,598r**	**1,780**
Other industrial branches					
Coal	-r	-r	-r	-r	-r
Fuel oil	-	-	-	-	-
Gas oil	-	-	-r	-r	-
Natural gas	331r	254r	180r	203r	260
Renewable fuels (2)	-	-	-	-	-
Total other industrial branches	**331r**	**254r**	**180r**	**203r**	**260**
Transport, commerce and administration					
Coal	42r	50r	44r	-r	43
Fuel oil	42	35	7	3r	2
Gas oil	17	19	60	116r	32
Natural gas	4,711r	5,057r	5,075r	4,725r	4,963
Refinery gas	-	-	-	-	-
Renewable fuels (2)	2	2	2	-	-
Other fuels (1)	1	1	1	1	1
Total transport, commerce and administration	**4,814r**	**5,164r**	**5,189r**	**4,844r**	**5,041**
Other (3)					
Coal	300r	326r	41r	28r	28
Fuel oil	3r	4r	8r	10r	1
Gas oil	10r	37r	34r	14r	25
Natural gas	2,421r	2,535r	2,172r	2,149r	2,215
Renewable fuels (2)	989r	1,149r	1,228r	1,711r	1,724
Other fuels (1)	-r	-r	-r	-r	-
Total other	**3,722r**	**4,050r**	**3,483r**	**3,912r**	**3,993**
Total - all sectors					
Coal	6,166r	4,512r	4,150r	4,356r	4,284
Fuel oil	4,419	4,651	3,767	3,560r	2,186
Gas oil	181r	1,385r	934r	599r	373
Natural gas	80,879r	84,860r	88,660r	86,550r	86,812
Blast furnace gas	2,401	2,523	2,313	3,083	2,885
Coke oven gas	650	559	744	911r	897
Refinery gas	5,967	5,576	5,143	5,832r	6,966
Renewable fuels (2)	2,610	2,692	2,860	3,233r	3,430
Other fuels (1)	9,826	13,443r	16,060r	14,699r	14,909
Total CHP fuel use	**113,100r**	**120,201r**	**124,632r**	**122,823r**	**122,742**

(1) Other fuels include: process by-products.
(2) Renewable fuels include: sewage gas, other biogases, municipal solid waste and refuse derived fuels.
(3) Sectors included under Other are agriculture, community heating, leisure, landfill and incineration.

Chapter 7
Renewable sources of energy

Introduction

7.1 This chapter provides information on the contribution of renewable energy sources to the United Kingdom's energy requirements. It includes sources that under international definitions are not counted as renewable sources or are counted only in part. This is to ensure that this Digest covers all sources of energy available in the United Kingdom. However, within this chapter the international definition of total renewables is used and this excludes non-biodegradable wastes. The energy uses of wastes are still shown in the tables of this chapter but as "below the line" items. This chapter covers both the use of renewables to generate electricity and the burning of renewable fuels to produce heat either in boilers (or cookers) or in combined heat and power (CHP) plants. This year the coverage of liquid biofuels for transport has also been extended and data included in the Commodity balances (Tables 7.1 to 7.3) and the use of renewables table (Table 7.6).

7.2 The data summarise the results of an ongoing study undertaken by the AEA on behalf of the Department for Business, Enterprise and Regulatory Reform (BERR) to update a database containing information on all relevant renewable energy sources in the United Kingdom. This database is called RESTATS, the Renewable Energy STATisticS database.

7.3 The study started in 1989, when all relevant renewable energy sources were identified and, where possible, information was collected on the amounts of energy derived from each source. The renewable energy sources identified were the following: active solar heating; photovoltaics; onshore and offshore wind power; wave power; large and small scale hydro; biomass (both plant and animal based); geothermal aquifers. The technical notes at the end of this chapter define each of these renewable energy sources. The database now contains 19 years of data from 1989 to 2007.

7.4 The information contained in the database is collected by a number of methods. For larger projects, an annual survey is carried out in which questionnaires are sent to project managers. For technologies in which there are large numbers of small projects, the values given in this chapter are estimates based on information collected from a sub-sample of the projects. Further details about the data collection methodologies used in RESTATS, including the quality and completeness of the information, are given in the technical notes at the end of this chapter.

7.5 An energy flow chart for 2007, showing the flows of renewables from fuel inputs through to consumption, is included for the first time, overleaf. This is a way of simplifying the figures that can be found in the commodity balance for renewables energy sources in Table 7.1. It illustrates the flow of primary fuels from the point at which they become available from home production or imports (on the left) to their eventual final uses (on the right) as well as the energy lost in conversion.

7.6 Commodity balances for renewable energy sources covering each of the last three years form the first three tables (Tables 7.1 to 7.3). Unlike in the commodity balance tables in other chapters of the Digest, Tables 7.1 to 7.3 have zero statistical differences. This is because the data for each category of fuel are, in the main, taken from a single source where there is less likelihood of differences due to timing or measurement. These balance tables are followed by 5-year tables showing capacity of, and electricity generation from, renewable sources (Table 7.4), and generation from sources eligible for the Renewables Obligation (RO) and sources qualifying under the Renewables Directive (RD) (Table 7.5). Table 7.6 shows renewable sources used to generate electricity and heat in each of the last five years. A long-term trends commentary and table (Table 7.1.1) covering the use of renewables to generate electricity and heat is available on BERR's energy statistics web site and accessible from the Digest of UK Energy Statistics home page: www.berr.gov.uk/energy/statistics/publications/dukes/page45537.html.
Also available on the web site is Table 7.1.2 summarising all the renewable orders made under the Non Fossil Fuels Obligation (NFFO), Northern Ireland Non Fossil Fuels Obligation, and Scottish Renewables Orders (SRO) along with descriptive text.

Renewables flow chart 2007 (thousand tonnes of oil equivalent)

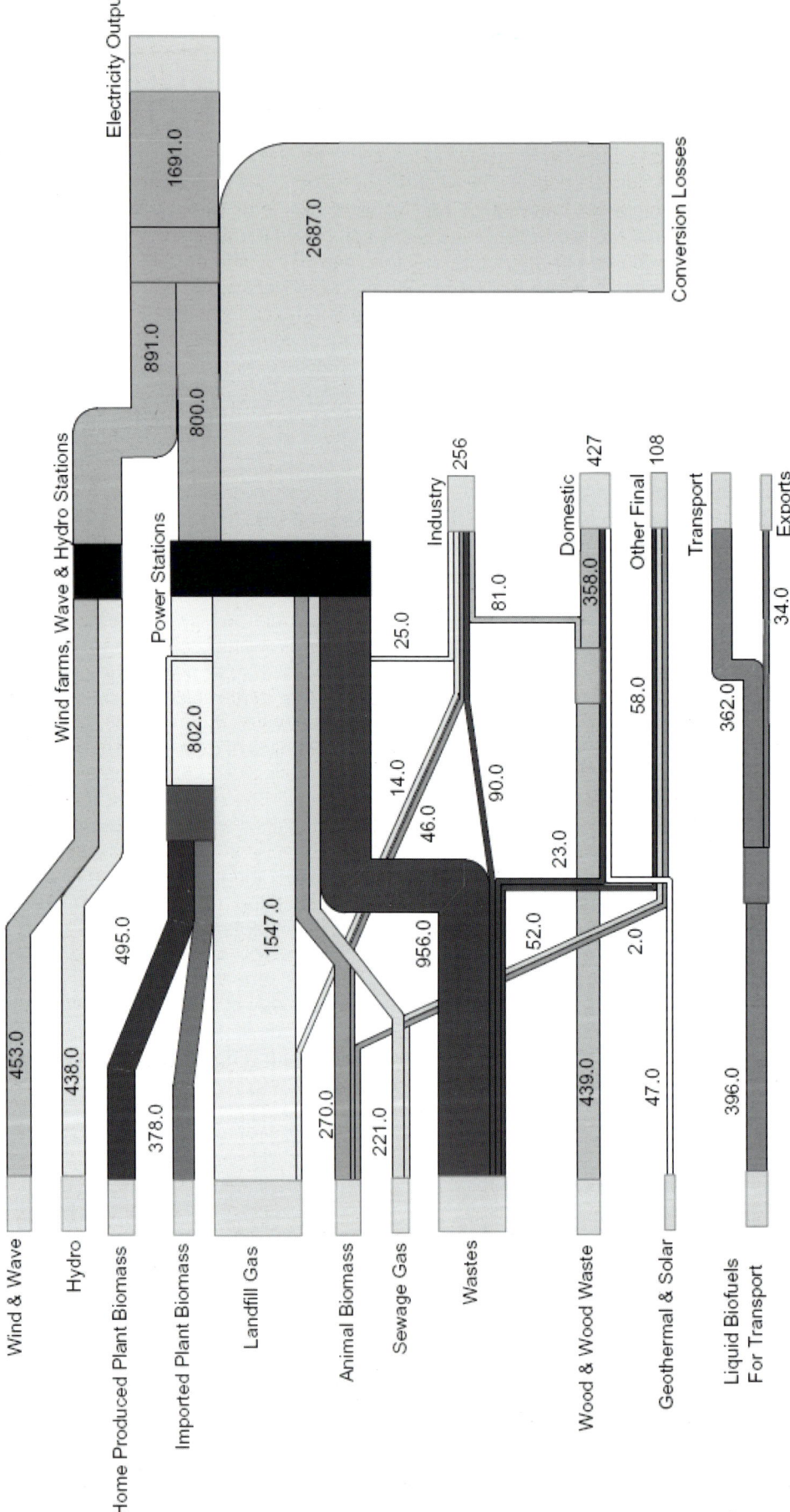

Notes:
This flow chart is based on data that appear in Tables 7.1 and 7.4

Renewables Obligation and Renewables Directive

7.7   In April 2002 the Renewables Obligation (RO) (and the analogous Renewables Obligation (Scotland)) came into effect[1]. It is an obligation on all electricity suppliers to supply a specific proportion of electricity from eligible renewable sources. Eligible sources include all those covered by this chapter but with specific exclusions. These are: existing hydro plant of over 20 MW; all plant using renewable sources built before 1990 (unless re-furbished and less than 20 MW); and energy from mixed waste combustion unless the waste is first converted to fuel using advanced conversion technology. Only the biodegradable fraction of any waste is eligible (in line with the EU Directive, see paragraph 7.8, below). All stations outside the United Kingdom (the UK includes its territorial waters and the continental shelf) are also excluded. The upper part of Table 7.5 shows all the components of total electricity generation on an RO basis. Strictly speaking until 2005, the RO covers only Great Britain, but in these UK based statistics Northern Ireland renewable sources have been treated as if they were also part of the RO.

7.8   The European Union's Renewables Directive (RD) (which came into force in October 2001) uses the same "international definition" as is used elsewhere in this chapter (in that it excludes non-biodegradable wastes). In 2006 the European Commission clarified its definition and confirmed that imports of electricity generated from renewable sources cannot be included, although such imports will be part of the overall consumption of electricity in the UK which forms the denominator in the calculation of the Renewables Directive percentage (see paragraph 7.12, below). AEA has estimated the percentage of municipal solid waste (MSW) that was non-biodegradable for all the years in the RESTATS database. For 2007 the estimate is the same as in earlier years, namely that 37½ per cent of MSW was non-biodegradable as were all of waste tyres (but see paragraph 7.72) and hospital waste. The lower part of Table 7.5 shows the components of total electricity generation on an RD basis.

7.9   Prior to 2002 the main instruments for pursuing the development of renewables capacity were the NFFO Orders for England and Wales and for Northern Ireland, and the Scottish Renewable Orders. In this chapter the term "NFFO Orders" is used to refer to these instruments collectively. For projects contracted under NFFO Orders in England and Wales, the Non Fossil Purchasing Agency (NFPA) provided details of capacity and generation. The Scottish Executive and Northern Ireland Electricity provided information on the Scottish and Northern Ireland NFFO Orders, respectively. Statistics of these Orders can now be found on the BERR energy web site (see paragraph 7.6, above).

Renewables Targets

7.10   Since February 2000, the United Kingdom's renewables policy has consisted of four key strands:

- a new RO on all electricity suppliers in Great Britain to supply a specific proportion of electricity from eligible renewables, introduced in April 2002;
- exemption of electricity from renewable sources[2] from the Climate Change Levy, introduced from April 2001;
- an expanded support programme for new and renewable energy including capital grants and an expanded research and development programme; and
- development of a regional strategic approach to planning and targets for renewables.

The RO is part of the UK's programme to tackle climate change and to encourage a more sustainable approach to energy consumption. Previous policy has been successful in introducing renewables to the UK marketplace and in reducing costs. The focus of current policy is to build on these achievements through the Obligation and a system of capital grants designed to bring forward offshore wind and energy crops, thereby maximising the chances of meeting the Government's targets.

7.11   The EU Directive of October 2001 proposed that Member States adopt national targets for renewables that are consistent with reaching the overall EU target of 12 per cent of energy (22.1 per cent of electricity) from renewables by 2010. The UK "share" of this target is that renewables sources

---

[1] Parliamentary approval of the Renewables Obligation Orders under The Utilities Act 2000 was given in March 2002.
[2] Electricity generated by hydro stations with a declared net capacity of more than 10 MW is not exempt from the Climate Change Levy.

eligible under the RD should account for 10 per cent of UK electricity **consumption** by 2010. In March 2007 the European Council agreed to a common strategy for energy security and tackling climate change. An element of this was establishing a target of 20 per cent of EU's energy to come from renewable sources. In January 2008 the European Commission published proposals for each Member State's contribution to the EU target. The Commission's proposal was that the UK by 2020 15 per cent of final energy consumption should be accounted for by energy from renewable sources (see paragraph 7.13, below).

Table 7A: Percentages of electricity derived from renewable sources

	2005	2006	2007
Overall renewables percentage (international basis)	4.25r	4.54r	4.96
Percentage on a Renewables Obligation basis	4.01r	4.44r	4.88
Percentage on a Renewables Directive basis	4.17r	4.46r	4.90

7.12   Chart 7.3 shows the growth in all sources of renewables generation since 1990 and Table 7A gives renewables shares on three different bases for the three most recent years. They show progress towards the RO and RD targets. Generation from all renewables in the UK (on the international definition basis) accounted for 4.96 per cent of UK electricity generation in 2007. In 2007 the RO percentage rose by 0.44 percentage points to 4.88 per cent of electricity sales by licensed suppliers. On the basis used for the Renewables Directive, the percentage of UK electricity consumption accounted for by RD eligible renewable sources rose from 4.46 per cent in 2005 to 4.90 per cent in 2007. The increases in all three percentages shown in Table 7A are mainly due to growth in the numerators (ie the renewables element) but the small declines in the respective denominators have also played a part. The overall percentage of electricity generation in 2007 fell by 0.6 per cent, while for the RO percentage there was a fall of 0.4 per cent in electricity sales by licensed suppliers. For the RD basis electricity consumption fell by 1.1 per cent in 2007.

7.13   An article published in the March 2008 Energy Trends (see Annex C for further information about Energy Trends) compared current and proposed target levels of the share of renewable energy in total final energy consumption in each of the 27 EU Member States. Total final energy consumption in this article is on the basis favoured by Eurostat. It includes the use of electricity and heat (and other fuels used for heating) by final consumers, and the use of energy for transport purposes. This Eurostat definition of total final energy consumption (which is calculated on a net calorific value basis) also currently includes consumption of electricity by electricity generators, consumption of heat by heat generators, transmission and distribution losses for electricity, and transmission and distribution losses for distributed heat. In the UK, energy balances are usually published on a gross calorific value basis, but in order to facilitate comparisons with EU statistics the balances for 2004, 2005 and 2006 have been calculated on a net calorific value basis and are available at: http://stats.berr.gov.uk/energystats/dukes1_1-1_3net.xls .

Table 7B: Percentages of energy derived from renewable sources

	2005	2006	2007
Percentage of final energy consumption (ie the basis proposed by Eurostat for the new Renewables Directive)	1.32	1.50	1.78
Percentage of primary energy demand (ie the basis previously quoted in this Digest)	1.77	1.88	2.04

7.14   Table 7B shows that overall, renewable sources, excluding wastes and passive solar design (see paragraph 7.35), continues to increase and provided 2.0 per cent of the United Kingdom's total primary energy requirements in 2007. On the basis proposed by Eurostat, which measures renewables contribution relative to final energy consumption, the UK percentage rose by 0.28 percentage points in 2007 to 1.78 per cent. The primary energy demand basis produces higher percentages because thermal renewables are measured including the energy that is lost in transformation. The thermal renewables used in the UK are less efficient in transformation than fossil fuels and currently account for nearly half of electricity generation from renewables. As non-thermal renewables such as wind (which by convention are 100 per cent efficient in transformation) grow as a proportion of UK renewables use, the final energy consumption percentage will overtake the primary consumption percentage.

Commodity balances for renewables in 2007 (Table 7.1), 2006 (Table 7.2) and 2005 (Table 7.3)

7.15   This year eleven different categories of renewable fuels are identified in the commodity balances. Some of these categories are themselves groups of renewables because a more detailed disaggregation could disclose data for individual companies. In the commodity balance tables the distinction between biodegradable and non-biodegradable wastes cannot be maintained for this reason. However, for this Digest the biomass category has been separated into animal based biomass, plant based biomass, and liquid biofuels for transport. To reduce confusion the term "biofuels" is now only used for liquid biofuels used for transport and "biomass" is used as the term to describe all fuels from biological sources. Liquid biofuels for transport were previously not included in these commodity balances. The largest contribution to renewables in **input** terms (82 per cent) is from biomass, with large-scale hydro electricity production and wind generation contributing the majority of the remainder as Chart 7.1 shows. For the first time in 2007 wind (with a 9 per cent share) contributed more than large scale hydro in primary input terms. Only 2 per cent of renewable energy comes from renewable sources other than biomass, wind and large-scale hydro. These include solar, small-scale hydro and geothermal aquifers.

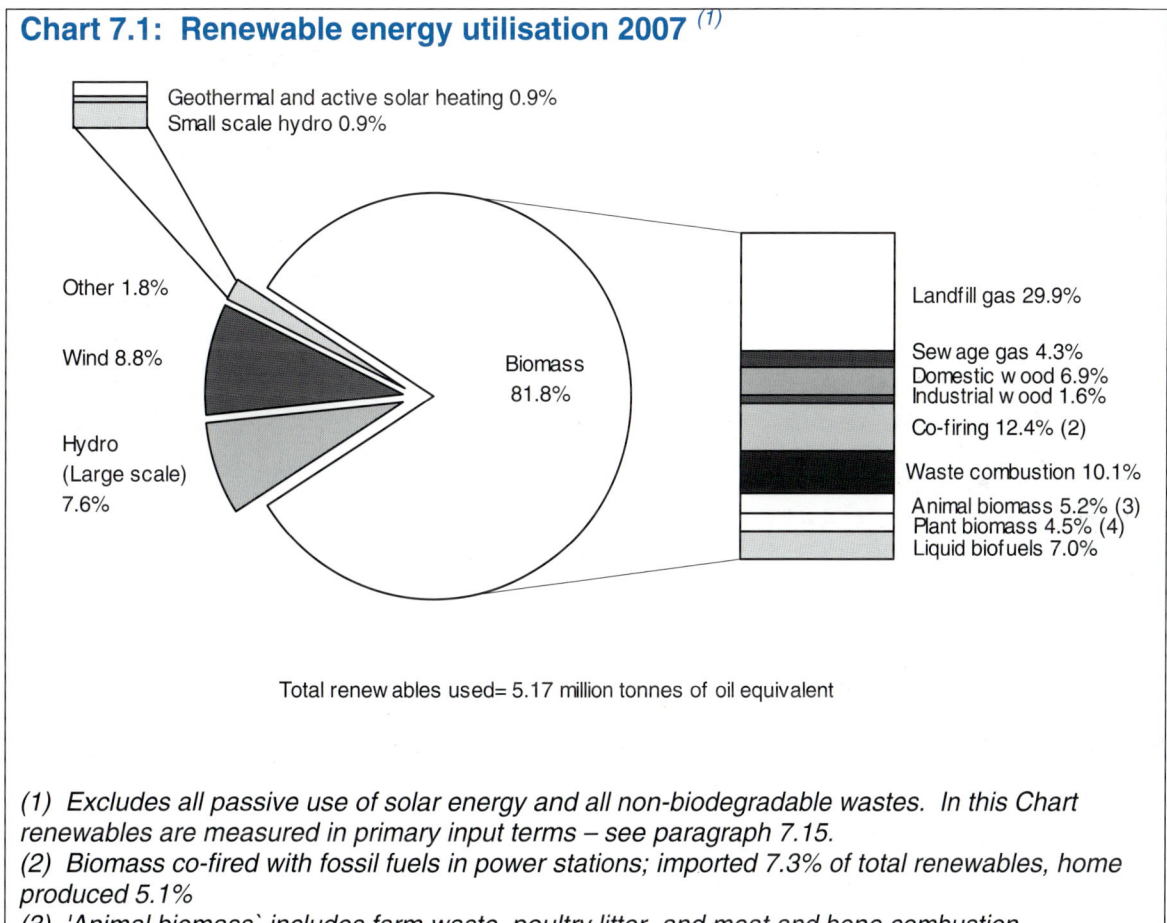

Chart 7.1: Renewable energy utilisation 2007 [1]

Geothermal and active solar heating 0.9%
Small scale hydro 0.9%

Other 1.8%

Wind 8.8%

Hydro (Large scale) 7.6%

Biomass 81.8%

Landfill gas 29.9%

Sewage gas 4.3%
Domestic wood 6.9%
Industrial wood 1.6%

Co-firing 12.4% (2)

Waste combustion 10.1%

Animal biomass 5.2% (3)
Plant biomass 4.5% (4)
Liquid biofuels 7.0%

Total renewables used= 5.17 million tonnes of oil equivalent

(1) Excludes all passive use of solar energy and all non-biodegradable wastes. In this Chart renewables are measured in primary input terms – see paragraph 7.15.
(2) Biomass co-fired with fossil fuels in power stations; imported 7.3% of total renewables, home produced 5.1%
(3) 'Animal biomass` includes farm waste, poultry litter, and meat and bone combustion.
(4) 'Plant biomass' includes straw and energy crops.

7.16   79 per cent of the renewable energy produced in 2007 was transformed into electricity. This is a decrease from 83 per cent in 2006 and 85 per cent in 2005, because the use of biofuels for transport has grown faster than the use of renewables for electricity. While biomass appears to dominate the picture when fuel inputs are being measured, hydro electricity and wind power together provide a larger contribution when the output of electricity is being measured as Table 7.4 shows. This is because on an energy supplied basis (see Chapter 5, paragraph 5.27) hydro (and also wind, wave and solar) inputs are assumed to be equal to the electricity produced. For landfill gas, sewage sludge, municipal solid waste and other renewables a substantial proportion of the energy content of the input is lost in the process of conversion to electricity as the flow chart (page 174, illustrates).

Capacity of, and electricity generated from, renewable sources (Table 7.4)

7.17   Table 7.4 shows the capacity of, and the amounts of electricity generated from, each renewable source. Total electricity generation from renewables in 2007 amounted to 19,664 GWh, an increase of 1,548 GWh (+8.5 per cent) on 2006. The main contributors to this substantial increase were 917 GWh from onshore wind (+26 per cent), 439 GWh from large scale hydro (+11 per cent), 253 GWh (+6 per cent) from landfill gas, 132 GWh (+20 per cent) from offshore wind, and 94 GWh (+9 per cent) from municipal solid waste combustion. There was a 572 GWh decrease in co-firing of biomass with fossil fuels (-23 per cent). The increase from large scale hydro was greater than the decrease recorded in 2006, which was attributable to drier weather and took hydro generation to a new record level. Even so, generation from wind (both onshore and offshore) overtook hydro to become the largest renewables technology in output terms, with both closely followed by landfill gas. Co-firing of biomass was the next most prominent. In 2007, 27 per cent of the electricity generated from renewables was from wind, 26 per cent was from hydro sources, 24 per cent from landfill gas, 10 per cent from co-firing, and 13 per cent from other biofuels.

7.18   As a result, all renewable sources provided 4.98 per cent of the electricity generated in the United Kingdom in 2007, 0.43 percentage points higher than in 2006. Chart 7.2 shows the growth in the proportion of electricity produced from renewable sources. It includes the progress towards the renewables targets set under the RO and RD (see paragraphs 7.10 to 7.14 above and 7.24 below).

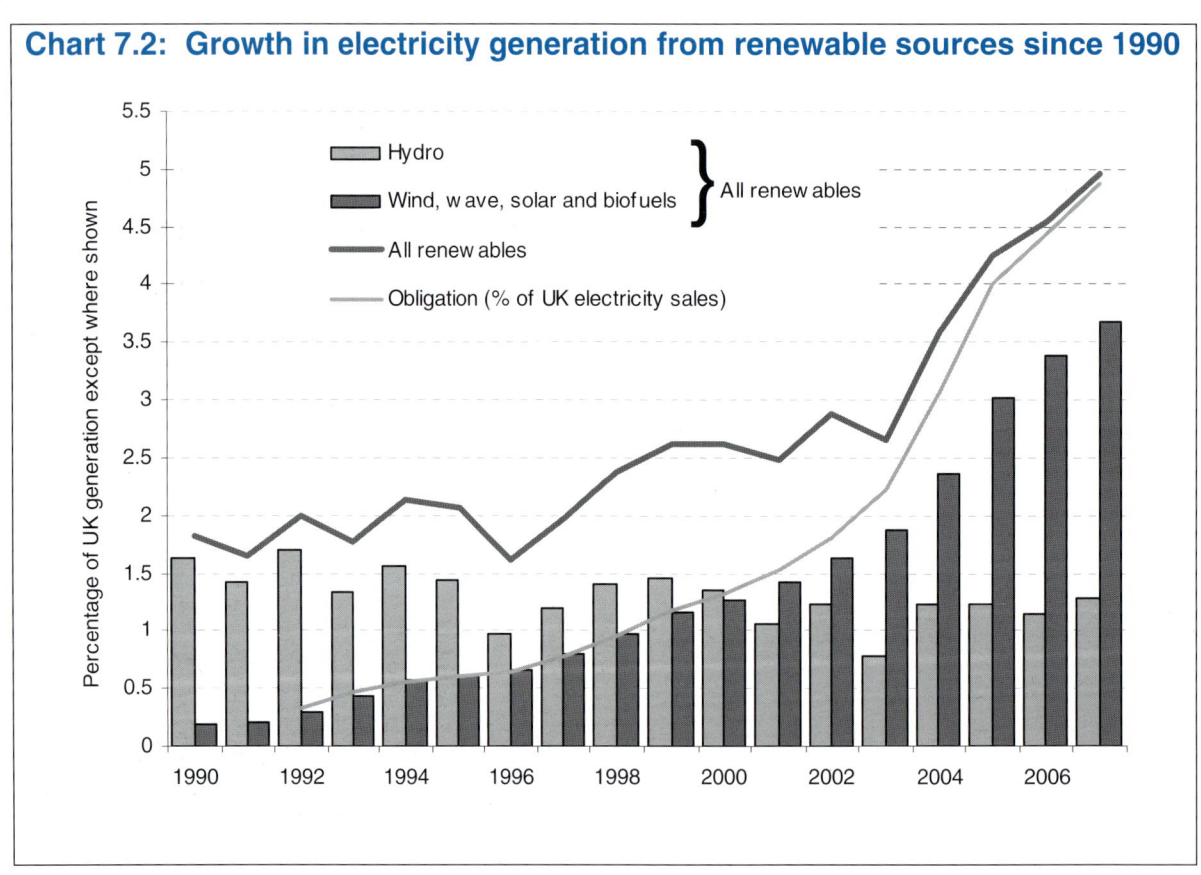

Chart 7.2: Growth in electricity generation from renewable sources since 1990

7.19   There was a 13 per cent increase (+665 MWe) in the installed generating capacity of renewable sources in 2007, mainly as a result of a 26 per cent increase (+433 MWe) in onshore wind capacity and a 30 per cent increase (+90 MWe) in offshore wind capacity. There was also a 5 per increase (+44 MWe) in the capacity fuelled by landfill gas and a 37 per cent increase (+83 MWe) in capacity fuelled by animal or plant biomass.

7.20   Chart 7.3 (which covers all renewables capacity except large scale hydro) illustrates the continuing increase in the electricity generation capacity from all significant renewable sources. This upward trend in the capacity of renewable sources will continue as recently consented onshore and offshore windfarms and other projects come on stream.

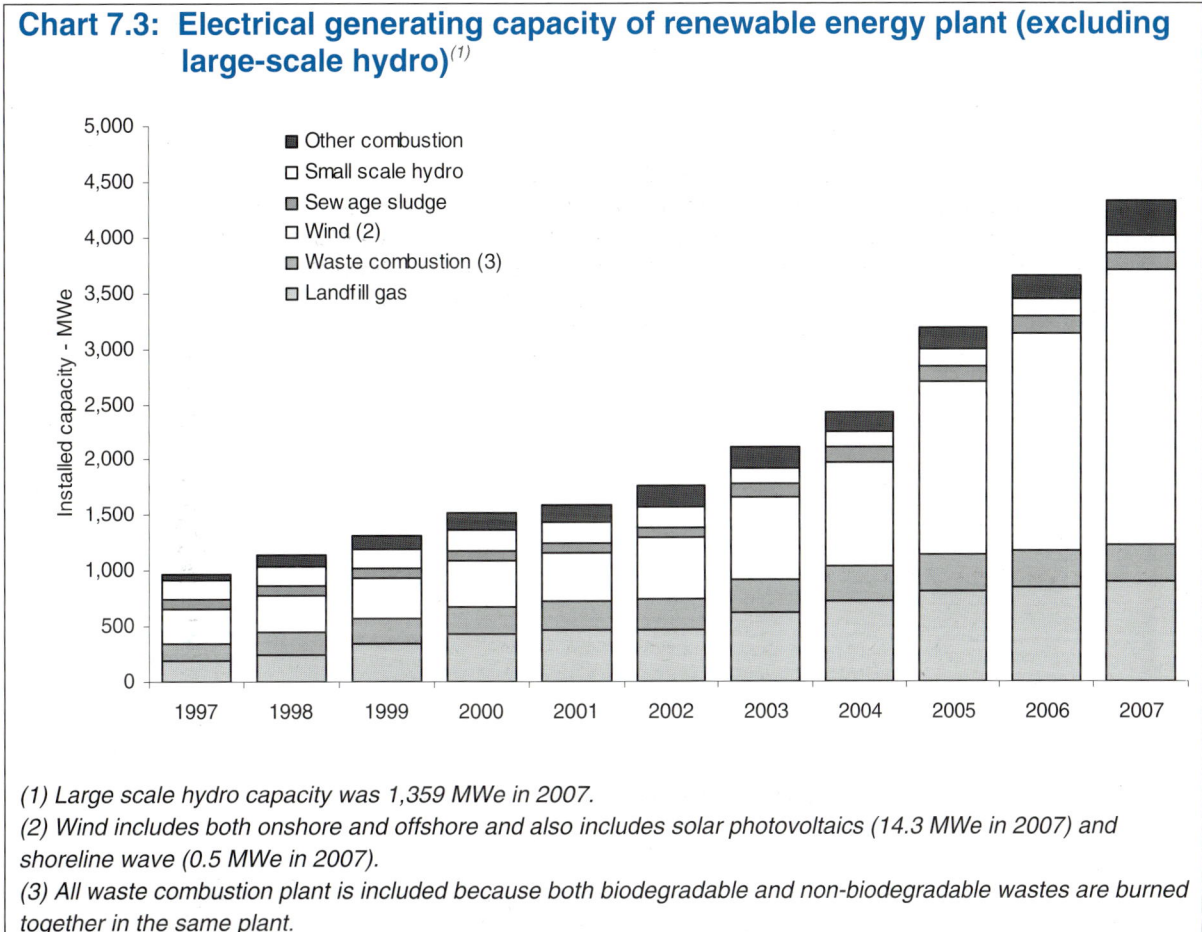

Chart 7.3: Electrical generating capacity of renewable energy plant (excluding large-scale hydro)[1]

(1) Large scale hydro capacity was 1,359 MWe in 2007.

(2) Wind includes both onshore and offshore and also includes solar photovoltaics (14.3 MWe in 2007) and shoreline wave (0.5 MWe in 2007).

(3) All waste combustion plant is included because both biodegradable and non-biodegradable wastes are burned together in the same plant.

7.21   In 2007, (excluding large-scale hydro which largely pre-date the introduction of NFFO) 27 per cent of electricity from renewables was generated under NFFO contracts. If ex-NFFO sites (NFFO 1 and 2 in England and Wales) are included the proportion increases to 35 per cent. Table 7.4, however, includes both electricity generated outside of these contracts and electricity from large-scale hydro schemes, and thus reports on total electricity generation from renewables. All electricity generated from renewables is also reported within the tables of Chapter 5 of this Digest (eg Table 5.6).

7.22   Plant load factors in Table 7.4 have been calculated in terms of installed capacity and express the average hourly quantity of electricity generated as a percentage of the average capacity at the beginning and end of the year. In the past the overall figure has been heavily influenced by the availability of hydro capacity during the year, which in turn has been influenced by the amount of rainfall during the preceding period. Low rainfall in the winter of 2002/2003 led to 2003 having particularly low hydro load factors. The dry weather in 2006 had a lesser effect. Two factors contributed to the lower load factor for wind in 2003. Firstly 110 MWe was installed late in the year and had little opportunity to contribute to generation. Secondly, the long hot summer of 2003 was not as windy as previous years. The load factors for biomass in 2006 and 2007 were lower than in 2005 because of a combination of factors. Firstly, some sites did not generate, although their capacity is still available, and secondly, one large new site did not begin to operate until late in 2006 and another until late in 2007. As a result the overall load factor for renewables and wastes was again below the record level seen on 2004. Plant load factors for all generating plant in the UK are shown in Chapter 5, Table 5.10.

7.23   To overcome the biasing of load factors for wind caused by new turbines coming on stream either early or late in a calendar year, Table 7.4 also contains a second statistic to describe the load factor of wind turbines. This statistic is calculated in the same way as the other load factors but includes only those wind farms that have operated throughout the calendar year with an unchanged configuration. See paragraphs 7.76 and 7.77 for the full definitions. In 2007, this "unchanged configuration" load factor for onshore wind farms is slightly lower than the all-onshore wind factor because of more new turbines coming on stream early in the calendar year compared with the latter half of the year. For offshore wind farms the reverse is true with the "unchanged configuration" load factor; higher than the all-offshore load factor. This is because one new offshore wind farm came on stream in the latter half of the year, and also because in some months some other new turbines had generation lower than their potential.

Electricity generated from renewable sources: Renewables Obligation and Renewables Directive bases (Table 7.5)

7.24   Electricity generated in the UK from renewable sources eligible under the RO in 2007 was 9.3 per cent greater than in 2006. This compares with growth of 10.5 per cent in 2006, 33 per cent in 2005 and 38 per cent in 2004. Electricity generated in the UK from renewable sources eligible under the Renewables Directive in 2007 was 8.5 per cent greater than in 2006. This compares with growth of 6.9 per cent in 2006, 20 per cent in 2005 and 33 per cent in 2004. Chart 7.3 shows the growth in the proportion of electricity produced from renewable sources under the Renewables Obligation and international definitions.

Renewable sources used to generate electricity and heat (Table 7.6)

7.25   Between 2006 and 2007 there was an increase of 3.7 per cent in the **input** of renewable sources into electricity generation. Wind grew by 25 per cent, hydro by 11 per cent, but biomass by only 0.4 per cent.

7.26   Compared to 2001, total inputs to electricity generation have doubled, aided by a doubling of the use of biomass and a more than quadrupling of the use of wind.

7.27   Table 7.6 also shows the contribution from renewables to heat generation. Renewables used to generate heat are now shown to have declined to a low point in 2000 but since picked up to be less than 20 per cent lower in 2007 than the level 10 years earlier. The decline was mainly due to tighter emissions controls discouraging on-site burning of biomass, especially wood waste by industry. Domestic use of wood provides the main contribution to renewables used for heat, but the use of animal and plant biomass has shown strong growth in recent years. In addition, the use of active solar heating has almost tripled in the last five years.

7.28   This year a re-assessment of the use of wood by households has taken place, resulting in the estimates being increased compared with earlier years. Further details are given at paragraph 7.51.

Biofuels for transport

7.29   AEA has researched the UK production and consumption of biofuels for transport in 2007 and combined this with the information they had previously obtained for 2006. Their findings are as follows:

7.30   Around 485 million litres of biodiesel were produced in the UK in 2007 (up from 291 million litres in 2006). Biodiesel consumption figures can be obtained from figures published by HM Revenue and Customs (HMRC) derived from road fuel taxation statistics. The most usual way for biodiesel to be sold is for it to be blended with ultra-low sulphur diesel fuel and thus it is reported as part of the road transport use of diesel in Chapter 3. The duty payable on biodiesel is just over half the duty payable on road diesel and in blended fuels the duty payable is proportionate to the duty payable on the constituent fuels. These HMRC figures show that 347 million litres of biodiesel were consumed in 2007, up from 169 million litres in 2006 and 33 million litres in 2005. This implies that around 138 million litres of biodiesel were exported in 2007. The total annual capacity for biodiesel production in the UK could reach 1,600 million litres per year in 2010 if all the planned plant become operational and the existing plant operate at full capacity. This production level would be equivalent to just around 6 per cent of the UK's diesel consumption in 2007. This reduced capacity, compared to that reported as planned in last year's Digest, is due to a number of plants having closed or are planning to close citing that it is currently uneconomic to produce biodiesel in the UK because of subsidised US imports, the price of biodiesel being too low, a shortage of feedstock, and high feedstock prices.

7.31   HMRC data show that 153 million litres of bioethanol was consumed in the UK in 2007, up from 95 million litres in 2006, and 85 million litres in 2005. Only one UK plant was in production in 2007, and so the majority of the bioethanol was imported. If all planned plants became operational on current planned timescales, their combined capacity would be around 600 million litres by 2011, equivalent to 2.4 per cent of the UK's petrol consumption in 2007. Some of the capacity reported as planned in last year's Digest will not now go ahead due to the uncertain future market.

7.32   The HMRC data have been converted from litres to tonnes of oil equivalent and the data are now shown as additional rows in Table 7.6. In 2007, 7 per cent of the renewable sources used in the UK in primary input terms were liquid biofuels for transport, up from 4 per cent in 2006 and less than half a per cent in 2003.

7.1 Commodity balances 2007

Renewables and waste

Thousand tonnes of oil equivalent

	Wood waste	Wood	Poultry litter, meat and bone, and farm waste	Straw, SRC, and other plant-based biomass (3)	Sewage gas	Landfill gas
Supply						
Production	81	358	270	495	221	1,547
Other sources	-	-	-	-	-	-
Imports	-	-	-	378	-	-
Exports	-	-	-	-	-	-
Marine bunkers	-	-	-	-	-	-
Stock change (1)	-	-	-	-	-	-
Transfers	-	-	-	-	-	-
Total supply	**81**	**358**	**270**	**873**	**221**	**1,547**
Statistical difference (2)	-	-	-	-	-	-
Total demand	**81**	**358**	**270**	**873**	**221**	**1,547**
Transformation	-	-	**223**	**775**	**170**	**1,534**
Electricity generation	-	-	223	775	170	1,534
Major power producers	-	-	203	422	-	-
Autogenerators	-	-	19	354	170	1,534
Heat generation	-	-	-	-	-	-
Petroleum refineries	-	-	-	-	-	-
Coke manufacture	-	-	-	-	-	-
Blast furnaces	-	-	-	-	-	-
Patent fuel manufacture	-	-	-	-	-	-
Other	-	-	-	-	-	-
Energy industry use	-	-	-	-	-	-
Electricity generation	-	-	-	-	-	-
Oil and gas extraction	-	-	-	-	-	-
Petroleum refineries	-	-	-	-	-	-
Coal extraction	-	-	-	-	-	-
Coke manufacture	-	-	-	-	-	-
Blast furnaces	-	-	-	-	-	-
Patent fuel manufacture	-	-	-	-	-	-
Pumped storage	-	-	-	-	-	-
Other	-	-	-	-	-	-
Losses	-	-	-	-	-	-
Final consumption	**81**	**358**	**48**	**97**	**52**	**14**
Industry	**81**	-	**46**	**25**	-	**14**
Unclassified	81	-	46	25	-	14
Iron and steel	-	-	-	-	-	-
Non-ferrous metals	-	-	-	-	-	-
Mineral products	-	-	-	-	-	-
Chemicals	-	-	-	-	-	-
Mechanical engineering, etc	-	-	-	-	-	-
Electrical engineering, etc	-	-	-	-	-	-
Vehicles	-	-	-	-	-	-
Food, beverages, etc	-	-	-	-	-	-
Textiles, leather, etc	-	-	-	-	-	-
Paper, printing, etc	-	-	-	-	-	-
Other industries	-	-	-	-	-	-
Construction	-	-	-	-	-	-
Transport	-	-	-	-	-	-
Air	-	-	-	-	-	-
Rail	-	-	-	-	-	-
Road	-	-	-	-	-	-
National navigation	-	-	-	-	-	-
Pipelines	-	-	-	-	-	-
Other	-	**358**	**2**	**72**	**52**	-
Domestic	-	358	-	-	-	-
Public administration	-	-	-	-	52	-
Commercial	-	-	-	-	-	-
Agriculture	-	-	2	72	-	-
Miscellaneous	-	-	-	-	-	-
Non energy use	-	-	-	-	-	-

(1) Stock fall (+), stock rise (-).
(2) Total supply minus total demand.
(3) SRC is short rotation coppice.

(4) Municipal solid waste, general industrial waste and hospital waste.
(5) The amount of shoreline wave included is less than 0.05 ktoe.

7.1 Commodity balances 2007 (continued)
Renewables and waste

Thousand tonnes of oil equivalent

Waste(4) and tyres	Geothermal and active solar heat	Hydro	Wind and wave (5)	Liquid biofuels for transport	Total renewables	
						Supply
956	47	438	453	396	5,262	Production
-	-	-	-	-	-	Other sources
-	-	-	-	-	378	Imports
-	-	-	-	-34	-34	Exports
-	-	-	-	-	-	Marine bunkers
-	-	-	-	-	-	Stock change (1)
-	-	-	-	-	-	Transfers
956	47	438	453	362	5,606	**Total supply**
-	-	-	-	-	-	**Statistical difference** (2)
956	47	438	453	362	5,606	**Total demand**
785	1	438	453	-	4,378	**Transformation**
785	1	438	453	-	4,378	Electricity generation
58	-	356	-	-	1,039	Major power producers
727	1	81	453	-	3,339	Autogenerators
-	-	-	-	-	-	Heat generation
-	-	-	-	-	-	Petroleum refineries
-	-	-	-	-	-	Coke manufacture
-	-	-	-	-	-	Blast furnaces
-	-	-	-	-	-	Patent fuel manufacture
-	-	-	-	-	-	Other
-	-	-	-	-	-	**Energy industry use**
-	-	-	-	-	-	Electricity generation
-	-	-	-	-	-	Oil and gas extraction
-	-	-	-	-	-	Petroleum refineries
-	-	-	-	-	-	Coal extraction
-	-	-	-	-	-	Coke manufacture
-	-	-	-	-	-	Blast furnaces
-	-	-	-	-	-	Patent fuel manufacture
-	-	-	-	-	-	Pumped storage
-	-	-	-	-	-	Other
-	-	-	-	-	-	**Losses**
171	46	-	-	362	1,228	**Final consumption**
90	-	-	-	-	256	**Industry**
90	-	-	-	-	256	Unclassified
-	-	-	-	-	-	Iron and steel
-	-	-	-	-	-	Non-ferrous metals
-	-	-	-	-	-	Mineral products
-	-	-	-	-	-	Chemicals
-	-	-	-	-	-	Mechanical engineering, etc
-	-	-	-	-	-	Electrical engineering, etc
-	-	-	-	-	-	Vehicles
-	-	-	-	-	-	Food, beverages, etc
-	-	-	-	-	-	Textiles, leather, etc
-	-	-	-	-	-	Paper, printing, etc
-	-	-	-	-	-	Other industries
-	-	-	-	-	-	Construction
-	-	-	-	362	362	**Transport**
-	-	-	-	-	-	Air
-	-	-	-	-	-	Rail
-	-	-	-	362	362	Road
-	-	-	-	-	-	National navigation
-	-	-	-	-	-	Pipelines
81	46	-	-	-	610	**Other**
23	46	-	-	-	427	Domestic
39	-	-	-	-	90	Public administration
10	-	-	-	-	10	Commercial
-	-	-	-	-	74	Agriculture
9	-	-	-	-	9	Miscellaneous
-	-	-	-	-	-	**Non energy use**

191

7.2 Commodity balances 2006

Renewables and waste

	Wood waste	Wood	Poultry litter, meat and bone, and farm waste	Straw, SRC, and other plant-based biomass (3)	Sewage gas	Landfill gas
Supply						
Production	81	322r	174	538	195r	1,465
Other sources	-	-	-	-	-	-
Imports	-	-	-	497	-	-
Exports	-	-	-	-	-	-
Marine bunkers	-	-	-	-	-	-
Stock change (1)	-	-	-	-	-	-
Transfers	-	-	-	-	-	-
Total supply	81	322r	174	1,035	195r	1,465
Statistical difference (2)	-	-	-	-	-	-
Total demand	81	322r	174	1,035	195r	1,465
Transformation	-	-	149	948	149r	1,451
Electricity generation	-	-	149	948	149r	1,451
Major power producers	-	-	129	543	-	-
Autogenerators	-	-	19	405	149r	1,451
Heat generation	-	-	-	-	-	-
Petroleum refineries	-	-	-	-	-	-
Coke manufacture	-	-	-	-	-	-
Blast furnaces	-	-	-	-	-	-
Patent fuel manufacture	-	-	-	-	-	-
Other	-	-	-	-	-	-
Energy industry use	-	-	-	-	-	-
Electricity generation	-	-	-	-	-	-
Oil and gas extraction	-	-	-	-	-	-
Petroleum refineries	-	-	-	-	-	-
Coal extraction	-	-	-	-	-	-
Coke manufacture	-	-	-	-	-	-
Blast furnaces	-	-	-	-	-	-
Patent fuel manufacture	-	-	-	-	-	-
Pumped storage	-	-	-	-	-	-
Other	-	-	-	-	-	-
Losses	-	-	-	-	-	-
Final consumption	81	322r	25	87	46r	14
Industry	81	-	23	16	-	14
Unclassified	81	-	23	16	-	14
Iron and steel	-	-	-	-	-	-
Non-ferrous metals	-	-	-	-	-	-
Mineral products	-	-	-	-	-	-
Chemicals	-	-	-	-	-	-
Mechanical engineering, etc	-	-	-	-	-	-
Electrical engineering, etc	-	-	-	-	-	-
Vehicles	-	-	-	-	-	-
Food, beverages, etc	-	-	-	-	-	-
Textiles, leather, etc	-	-	-	-	-	-
Paper, printing, etc	-	-	-	-	-	-
Other industries	-	-	-	-	-	-
Construction	-	-	-	-	-	-
Transport	-	-	-	-	-	-
Air	-	-	-	-	-	-
Rail	-	-	-	-	-	-
Road	-	-	-	-	-	-
National navigation	-	-	-	-	-	-
Pipelines	-	-	-	-	-	-
Other	-	322r	2	72	46r	-
Domestic	-	322r	-	-	-	-
Public administration	-	-	-	-	46r	-
Commercial	-	-	-	-	-	-
Agriculture	-	-	2	72	-	-
Miscellaneous	-	-	-	-	-	-
Non energy use	-	-	-	-	-	-

(1) Stock fall (+), stock rise (-).
(2) Total supply minus total demand.
(3) SRC is short rotation coppice.
(4) Municipal solid waste, general industrial waste and hospital waste.
(5) The amount of shoreline wave included is less than 0.05 ktoe.

7.2 Commodity balances 2006 (continued)

Renewables and waste

Thousand tonnes of oil equivalent

Waste(4) and tyres	Geothermal and active solar heat	Hydro	Wind and wave (5)	Liquid biofuels for transport	Total renewables	
						Supply
918r	38	395	363	232	4,721r	Production
-	-	-	-	-	-	Other sources
-	-	-	-	-	497r	Imports
-	-	-	-	-44	-44r	Exports
-	-	-	-	-	-	Marine bunkers
-	-	-	-	-	-	Stock change (1)
-	-	-	-	-	-	Transfers
918r	38	395	363	188	5,174r	**Total supply**
-	-	-	-	-	-	**Statistical difference (2)**
918r	38	395	363	188	5,174r	**Total demand**
773r	1	395	363	-	4,229r	**Transformation**
773r	1	395	363	-	4,229r	Electricity generation
59	-	318	-	-	1,049r	Major power producers
714r	1	77	363	-	3,180r	Autogenerators
-	-	-	-	-	-	Heat generation
-	-	-	-	-	-	Petroleum refineries
-	-	-	-	-	-	Coke manufacture
-	-	-	-	-	-	Blast furnaces
-	-	-	-	-	-	Patent fuel manufacture
-	-	-	-	-	-	Other
-	-	-	-	-	-	**Energy industry use**
-	-	-	-	-	-	Electricity generation
-	-	-	-	-	-	Oil and gas extraction
-	-	-	-	-	-	Petroleum refineries
-	-	-	-	-	-	Coal extraction
-	-	-	-	-	-	Coke manufacture
-	-	-	-	-	-	Blast furnaces
-	-	-	-	-	-	Patent fuel manufacture
-	-	-	-	-	-	Pumped storage
-	-	-	-	-	-	Other
-	-	-	-	-	-	**Losses**
145r	37	-	-	188	945r	**Final consumption**
65	-	-	-	-	198r	**Industry**
65	-	-	-	-	198r	Unclassified
-	-	-	-	-	-	Iron and steel
-	-	-	-	-	-	Non-ferrous metals
-	-	-	-	-	-	Mineral products
-	-	-	-	-	-	Chemicals
-	-	-	-	-	-	Mechanical engineering, etc
-	-	-	-	-	-	Electrical engineering, etc
-	-	-	-	-	-	Vehicles
-	-	-	-	-	-	Food, beverages, etc
-	-	-	-	-	-	Textiles, leather, etc
-	-	-	-	-	-	Paper, printing, etc
-	-	-	-	-	-	Other industries
-	-	-	-	-	-	Construction
-	-	-	-	188	188r	**Transport**
-	-	-	-	-	-	Air
-	-	-	-	-	-	Rail
-	-	-	-	188	188r	Road
-	-	-	-	-	-	National navigation
-	-	-	-	-	-	Pipelines
81	37	-	-	-	560r	**Other**
23	37	-	-	-	382r	Domestic
39	-	-	-	-	84r	Public administration
10	-	-	-	-	10	Commercial
-r	-	-	-	-	74	Agriculture
9	-	-	-	-	9	Miscellaneous
-	-	-	-	-	-	**Non energy use**

7.3 Commodity balances 2005

Renewables and waste

Thousand tonnes of oil equivalent

	Wood waste	Wood	Poultry litter, meat and bone, and farm waste	Straw, SRC, and other plant-based biomass (3)	Sewage gas	Landfill gas
Supply						
Production	81	287r	176	621	208r	1,421
Other sources	-	-	-	-	-	-
Imports	-	-	-	421	-	-
Exports	-	-	-	-	-	-
Marine bunkers	-	-	-	-	-	-
Stock change (1)	-	-	-	-	-	-
Transfers	-	-	-	-	-	-
Total supply	81	287r	176	1,042	208r	1,421
Statistical difference (2)	-	-	-	-	-	-
Total demand	81	287r	176	1,042	208r	1,421
Transformation	-	-	161	956	154r	1,407
Electricity generation	-	-	161	956	154r	1,407
Major power producers	-	-	159	582	-	-
Autogenerators	-	-	2	373	154r	1,407
Heat generation	-	-	-	-	-	-
Petroleum refineries	-	-	-	-	-	-
Coke manufacture	-	-	-	-	-	-
Blast furnaces	-	-	-	-	-	-
Patent fuel manufacture	-	-	-	-	-	-
Other	-	-	-	-	-	-
Energy industry use	-	-	-	-	-	-
Electricity generation	-	-	-	-	-	-
Oil and gas extraction	-	-	-	-	-	-
Petroleum refineries	-	-	-	-	-	-
Coal extraction	-	-	-	-	-	-
Coke manufacture	-	-	-	-	-	-
Blast furnaces	-	-	-	-	-	-
Patent fuel manufacture	-	-	-	-	-	-
Pumped storage	-	-	-	-	-	-
Other	-	-	-	-	-	-
Losses	-	-	-	-	-	-
Final consumption	81	287r	15	86	54r	14
Industry	81	-	12	14	-	14
Unclassified	81	-	12	14	-	14
Iron and steel	-	-	-	-	-	-
Non-ferrous metals	▪	-	-	-	-	-
Mineral products	-	-	-	-	-	-
Chemicals	-	-	-	-	-	-
Mechanical engineering, etc	-	-	-	-	-	-
Electrical engineering, etc	-	-	-	-	-	-
Vehicles	-	-	-	-	-	-
Food, beverages, etc	-	-	-	-	-	-
Textiles, leather, etc	-	-	-	-	-	-
Paper, printing, etc	-	-	-	-	-	-
Other industries	-	-	-	-	-	-
Construction	-	-	-	-	-	-
Transport	-	-	-	-	-	-
Air	-	-	-	-	-	-
Rail	-	-	-	-	-	-
Road	-	-	-	-	-	-
National navigation	-	-	-	-	-	-
Pipelines	-	-	-	-	-	-
Other	-	287r	2	72	54r	-
Domestic	-	287r	-	-	-	-
Public administration	-	-	-	-	54r	-
Commercial	-	-	-	-	-	-
Agriculture	-	-	2	72	-	-
Miscellaneous	-	-	-	-	-	-
Non energy use	-	-	-	-	-	-

(1) Stock fall (+), stock rise (-).
(2) Total supply minus total demand.
(3) SRC is short rotation coppice.
(4) Municipal solid waste, general industrial waste and hospital waste.
(5) The amount of shoreline wave included is less than 0.05 ktoe.

7.3 Commodity balances 2005 (continued)

Renewables and waste

Thousand tonnes of oil equivalent

Waste(4) and tyres	Geothermal and active solar heat	Hydro	Wind and wave (5)	Liquid biofuels for transport	Total renewables	
849	31r	423	250	8	4,354	**Supply**
-	-	-	-	-	-	Other sources
-	-	-	-	66	487	Imports
-	-	-	-	-	-	Exports
-	-	-	-	-	-	Marine bunkers
-	-	-	-	-	-	Stock change (1)
-	-	-	-	-	-	Transfers
849	31r	423	250	74	4,841r	**Total supply**
-	-	-	-	-	-	Statistical difference (2)
849	31r	423	250	74	4,841r	**Total demand**
688	1	423	250	-	4,041r	**Transformation**
688	1	423	250	-	4,041	Electricity generation
89	-	329r	-	-	1,160	Major power producers
599	1	94r	250	-	2,881r	Autogenerators
-	-	-	-	-	-	Heat generation
-	-	-	-	-	-	Petroleum refineries
-	-	-	-	-	-	Coke manufacture
-	-	-	-	-	-	Blast furnaces
-	-	-	-	-	-	Patent fuel manufacture
-	-	-	-	-	-	Other
-	-	-	-	-	-	**Energy industry use**
-	-	-	-	-	-	Electricity generation
-	-	-	-	-	-	Oil and gas extraction
-	-	-	-	-	-	Petroleum refineries
-	-	-	-	-	-	Coal extraction
-	-	-	-	-	-	Coke manufacture
-	-	-	-	-	-	Blast furnaces
-	-	-	-	-	-	Patent fuel manufacture
-	-	-	-	-	-	Pumped storage
-	-	-	-	-	-	Other
-	-	-	-	-	-	**Losses**
161	30r	-	-	74	800r	**Final consumption**
68	-	-	-	-	189r	**Industry**
68	-	-	-	-	189r	Unclassified
-	-	-	-	-	-	Iron and steel
-	-	-	-	-	-	Non-ferrous metals
-	-	-	-	-	-	Mineral products
-	-	-	-	-	-	Chemicals
-	-	-	-	-	-	Mechanical engineering, etc
-	-	-	-	-	-	Electrical engineering, etc
-	-	-	-	-	-	Vehicles
-	-	-	-	-	-	Food, beverages, etc
-	-	-	-	-	-	Textiles, leather, etc
-	-	-	-	-	-	Paper, printing, etc
-	-	-	-	-	-	Other industries
-	-	-	-	-	-	Construction
-	-	-	-	74	74	**Transport**
-	-	-	-	-	-	Air
-	-	-	-	-	-	Rail
-	-	-	-	74	74	Road
-	-	-	-	-	-	National navigation
-	-	-	-	-	-	Pipelines
93	30r	-	-	-	538r	**Other**
23	30r	-	-	-	340r	Domestic
51	-	-	-	-	105r	Public administration
10	-	-	-	-	10	Commercial
-	-	-	-	-	74	Agriculture
9	-	-	-	-	9	Miscellaneous
-	-	-	-	-	-	**Non energy use**

7.4 Capacity of, and electricity generated from, renewable sources

	2003	2004	2005	2006	2007
Installed Capacity (MWe) *(1)*					
Wind:					
Onshore	678.4	809.4	1,351.2	1,650.7	2,083.4
Offshore *(2)*	63.8	123.8	213.8	303.8	393.8
Shoreline wave	0.5	0.5	0.5	0.5	0.5
Solar photovoltaics	6.0	8.2	10.9	14.3r	14.3
Hydro:					
Small scale	130.0	142.9	157.9	153.4r	166.2
Large scale *(3)*	1,354.5	1,355.9	1,343.2	1,361.4r	1,358.7
Biomass:					
Landfill gas	619.1	722.2	817.8	856.2	900.6
Sewage sludge digestion	123.7r	131.9r	139.6r	146.4r	151.7
Municipal solid waste combustion	298.8	307.4	321.4	326.5	326.4
Animal Biomass *(4)*	94.4	86.5	86.6	88.9	114.4
Plant Biomass *(5)*	89.5	89.8	99.5	132.4	189.5
Total biomass and wastes	1,225.5r	1,337.8r	1,464.9r	1,550.4r	1,682.6
Total	**3,458.7r**	**3,778.4r**	**4,542.4r**	**5,034.4r**	**5,699.5**
Co-firing (6)	92.4	146.2	308.8	310.2	247.6
Generation (GWh)					
Wind:					
Onshore *(7)*	1,276	1,736	2,501	3,574	4,491
Offshore *(8)*	10	199	403	651	783
Solar photovoltaics	3	4	8	11r	11
Hydro:					
Small scale *(9)*	150	283	444	478r	534
Large scale *(3)*	2,987	4,561	4,478	4,115r	4,554
Biomass:					
Landfill gas	3,276	4,004	4,290	4,424	4,677
Sewage sludge digestion	394r	440r	470r	456r	517
Municipal solid waste combustion *(9)*	965	971	964	1,083	1,177
Co-firing with fossil fuels	602	1,022	2,533	2,528	1,956
Animal Biomass *(10)*	535	565	468	434	555
Plant Biomass *(11)*	402	362	382	363	409
Total biomass	6,174r	7,364r	9,107r	9,288r	9,291
Total generation	**10,600r**	**14,147r**	**16,940r**	**18,116r**	**19,664**
Non-biodegradable wastes (12)	579	583	578	651	707
Load factors (per cent) *(13)*					
Onshore wind	24.1	26.6	26.4	27.2r	27.5
Offshore wind (from 2004 only)	..	24.2	27.2	28.7r	25.6
Hydro	23.3	37.1	37.5	34.9r	38.2
Biomass (excluding co-firing)	62.4r	61.7r	58.3r	56.1r	56.8
Total (including wastes)	36.5	43.2r	41.1r	38.7r	39.2
Load factors on an unchanged configuration basis (per cent) *(14)*					
Onshore wind	26.2	29.2	28.1	26.7	27.3
Offshore wind (from 2006 only)	..	..	..	26.7r	28.3

(1) Capacity on a DNC basis is shown in Long Term Trends Table 7.1.1 available on the BERR web site - see paragraph 7.74.
(2) In 2007 excludes Beatrice (10 MW) which was only supplying an offshore oil platform.
(3) Excluding pumped storage stations. Capacities are as at the end of December.
(4) Includes the use of farm waste digestion, poultry litter and meat and bone.
(5) Includes the use of waste tyres, straw combustion, short rotation coppice and hospital waste.
(6) This is the proportion of fossil fuelled capacity used for co-firing of renewables based on the proportion of generation accounted for by the renewable source.
(7) Actual generation figures are given where available, but otherwise are estimated using a typical load factor or the design load factor, where known.
(8) Latest years include electricity from shoreline wave but this amounts to less than 0.05 GWh. Generation by Beatrice excluded (see note 2).
(9) Biodegradable part only.
(10) Includes the use of farm waste digestion, poultry litter combustion and meat and bone combustion.
(11) Includes the use of straw and energy crops.
(12) Non-biodegradable part of municipal solid waste plus waste tyres, hosptal waste and general industrial waste.
(13) Load factors are calculated based on installed capacity at the beginning and the end of the year - see paragraph 7.75.
(14) For a definition see paragraphs 7.76 and 7.77.

7.5 Electricity generated from renewable sources - Renewables Obligation basis and Renewables Directive basis

GWh

	2003	2004	2005	2006	2007
Generation : Renewables Obligation basis					
Wind:					
Onshore (1)	1,276	1,736	2,501	3,574	4,491
Offshore (2)	10	199	403	651	783
Solar photovoltaics	3	4	8	11r	11
Hydro:					
Small scale (1)	150	283	444	478r	534
Other hydro including refurbished large scale	560	1,353	1,710	1,672r	2,020
Biomass:					
Landfill gas	3,276	4,004	4,290	4,424	4,677
Sewage sludge digestion	394r	440r	470r	456r	517
Co-firing with fossil fuels	602	1,022	2,533	2,528	1,956
Animal Biomass (3)	535	565	468	434	555
Plant Biomass (4)	402	362	382	363	409
Total biomass	5,209r	6,393r	8,143r	8,204r	8,114
Total renewables generation on an obligation basis (5)	**7,207r**	**9,968r**	**13,209r**	**14,590r**	**15,953**
Generation : Renewables Directive basis					
Wind:					
Onshore (1)	1,276	1,736	2,501	3,574	4,491
Offshore (2)	10	199	403	651	783
Solar photovoltaics	3	4	8	11r	11
Hydro:					
Small scale (1)	150	283	444	478r	534
Large scale (6)	2,987	4,561	4,478	4,115r	4,554
Biomass:					
Landfill gas	3,276	4,004	4,290	4,424	4,677
Sewage sludge digestion	394r	440r	470r	456r	517
Municipal solid waste combustion (7)	965	971	964	1,083	1,177
Co-firing with fossil fuels	602	1,022	2,533	2,528	1,956
Animal Biomass (3)	535	565	468	434	555
Plant Biomass (4)	402	362	382	363	409
Total biomass	6,174r	7,364r	9,107r	9,288r	9,291
Total renewables generation on a directive basis (5)	**10,600r**	**14,147r**	**16,940r**	**18,116r**	**19,664**
Imports of electricity certified as CCL exempt (8)	2,865	3,522	3,522	3,475	..

(1) Actual generation figures are given where available, but otherwise are estimated using a typical load factor or the design load factor, where known.

(2) Includes electricity from shoreline wave but this amounts to less than 0.05 GWh.

(3) Includes the use of farm waste digestion, poultry litter combustion and meat and bone combustion.

(4) Includes the use of straw and energy crops.

(5) See paragraphs 7.7 and 7.8 for definitions.

(6) Excluding pumped storage stations.

(7) Biodegradable part only.

(8) Mainly hydro electricity exported to England from France. In the 2005 Digest these figures were included within the Renewables Directive basis but, following clarification by the European Commission, they were removed and are included in this table for information. Ofgem were not able to provide the 2007 figure at the time of publication.

7.6 Renewable sources used to generate electricity and heat and for transport fuels[(1)(2)]

				Thousand tonnes of oil equivalent	
	2003	2004	2005	2006	2007
Used to generate electricity *(3)*					
Wind:					
Onshore	109.7	149.3	215.1	307.3	386.2
Offshore *(4)*	0.8	17.1	34.6	56.0	67.3
Solar photovoltaics	0.3	0.3	0.7	0.9r	0.9
Hydro:					
Small scale	12.9	24.3	38.2	41.0	46.0
Large scale *(5)*	256.9	392.2	385.0	353.9r	391.6
Biomass:					
Landfill gas	1,074.5	1,313.1	1,407.2	1,451.1	1,533.9
Sewage sludge digestion	129.3r	144.3r	154.3r	149.4r	169.5
Municipal solid waste combustion *(6)*	445.8	429.5	426.3	479.0	486.8
Co-firing with fossil fuels	197.3	335.1	830.7	829.0	641.4
Animal Biomass *(7)*	172.4	182.3	161.5	148.5	222.5
Plant Biomass *(8)*	131.8	118.8	125.2	119.0	134.1
Total biomass	2,151.3r	2,523.1r	3,105.1r	3,176.0r	3,188.2
Total	**2,531.8r**	**3,106.3r**	**3,778.7r**	**3,935.2r**	**4,080.1**
Non-biodegradable wastes *(9)*	273.8	263.9	262.0	293.7	298.3
Used to generate heat					
Active solar heating	19.8	24.6	29.4	36.3	44.9
Biomass:					
Landfill gas	13.6	13.6	13.6	13.6	13.6
Sewage sludge digestion	52.4r	54.8r	53.6r	45.7r	51.6
Wood combustion - domestic	207.8r	250.8r	286.6r	322.4r	358.3
Wood combustion - industrial	195.6	195.6	80.9	80.9	80.9
Animal Biomass *(10)*	0.3	2.0	14.4	24.9	47.8
Plant Biomass *(11)*	71.9	71.9	85.7	87.3	97.3
Municipal solid waste combustion *(6)*	33.7	33.7	33.7	33.7	33.7
Total biomass	575.3r	622.4r	568.5r	608.7r	683.2
Geothermal aquifers	0.8	0.8	0.8	0.8	0.8
Total	**595.9r**	**647.8r**	**598.7r**	**645.8r**	**728.9**
Non-biodegradable wastes *(9)*	117.1	115.7	127.5	111.6	137.3
Renewable sources used as transport fuels					
as Bioethanol	-	-	47.9	53.4	85.8
as Biodiesel	15.1	16.7	26.1	134.4	276.0
Total	**15.1**	**16.7**	**74.1r**	**187.8**	**361.8**
Total use of renewable sources and wastes					
Solar heating and photovoltaics	20.0	24.9	30.1	37.2r	45.8
Onshore and offshore wind *(4)*	110.5	166.4	249.7	363.3	453.5
Hydro	269.8	416.5	423.2	394.9r	437.6
Biomass	2,726.5r	3,145.4r	3,673.6r	3,784.7r	3,871.4
Geothermal aquifers	0.8	0.8	0.8	0.8	0.8
Transport fuels	15.1	16.7	74.1	187.8	361.8
Total	**3,142.8r**	**3,770.8r**	**4,451.4r**	**4,768.7r**	**5,170.8**
Non-biodegradable wastes *(9)*	390.9	379.6	389.5	405.3	435.6
All renewables and wastes *(12)*	**3,533.7r**	**4,150.4r**	**4,840.9r**	**5,174.0r**	**5,606.5**

(1) Includes some waste of fossil fuel origin.
(2) See paragraphs 7.33 to 7.77 for technical notes and definitions of the categories used in this table
(3) For wind, solar PV and hydro, the figures represent the energy content of the electricity supplied but for biomass the figures represent the energy content of the fuel used.
(4) Latest years includes electricity from shoreline wave but this is less than 0.05 ktoe.
(5) Excluding pumped storage stations.
(6) Biodegradable part only.
(7) Includes electricity from farm waste digestion, poultry litter combustion and meat and bone combustion.
(8) Includes electricity from straw and energy crops.
(9) Non-biodegradable part of municipal solid waste plus waste tyres, hospital waste, and general industrial waste.
(10) Includes heat from farm waste digestion, meat and bone combustion and sewage sludge combustion.
(11) Includes heat from straw, energy crops, paper and packaging.
(12) The figures in this row correspond to the total demand and total supply figures in Tables 7.1, 7.2 and 7.3.

Digest of United Kingdom Energy Statistics 2008

Annexes

Annex A: Energy and commodity balances, conversion factors and calorific values

Annex B: Glossary and acronyms

Annex C: Further sources

Annex D: Major events in the Energy Industry, 2006-2008

Department for Business, Enterprise and Regulatory Reform

Annex A
Energy and commodity balances, conversion factors and calorific values

Balance principles

A.1   This Annex outlines the principles behind the balance presentation of energy statistics. It covers these in general terms. Fuel specific details are given in the appropriate chapters of this publication.

A.2   Balances are divided into two types, each of which performs a different function.

a) *commodity balance* – a balance for each energy commodity that uses the units usually associated with that commodity. By using a single column of figures, it shows the flow of the commodity from its sources of supply through to its final use. Commodity balances are presented in the individual fuel chapters of this publication.

b) *energy balance* - presents the commodity balances in a common unit and places them alongside one another in a manner that shows the dependence of the supply of one commodity on another. This is useful as some commodities are manufactured from others. The layout of the energy balance also differs slightly from the commodity balance. The energy balance format is used in Chapter 1.

A.3   Energy commodities can be either primary or secondary. Primary energy commodities are drawn (extracted or captured) from natural reserves or flows, whereas secondary commodities are produced from primary energy commodities. Crude oil and coal are examples of primary commodities, whilst petrol and coke are secondary commodities manufactured from them. For balance purposes, electricity may be considered to be both primary electricity (for example, hydro, wind) or secondary (produced from steam turbines using steam from the combustion of fuels).

A.4   Both commodity and energy balances show the flow of the commodity from its production, extraction or import through to its final use.

A.5   A simplified model of the commodity flow underlying the balance structure is given in Chart A.1. It illustrates how primary commodities may be used directly and/or be transformed into secondary commodities. The secondary fuels then enter final consumption or may also be transformed into another energy commodity (for example, electricity produced from fuel oil). To keep the diagram simple these "second generation" flows have not been shown.

A.6   The arrows at the top of the chart represent flows to and from the "pools" of primary and secondary commodities, from imports and exports and, in the case of the primary pool, extraction from reserves (eg the production of coal, gas and crude oil).

Commodity balances (Tables 2.1 to 2.6, 3.1 to 3.4, 4.1, 5.1, and 7.1 to 7.3)

A.7   A commodity balance comprises a supply section and a demand section. The supply section gives available sources of supply (ie exports are subtracted). The demand section is divided into a transformation section, a section showing uses in the energy industries (other than for transformation) and a section covering uses by final consumers for energy or non-energy purposes. Final consumption for energy purposes is divided into use by sector of economic activity. The section breakdowns are described below.

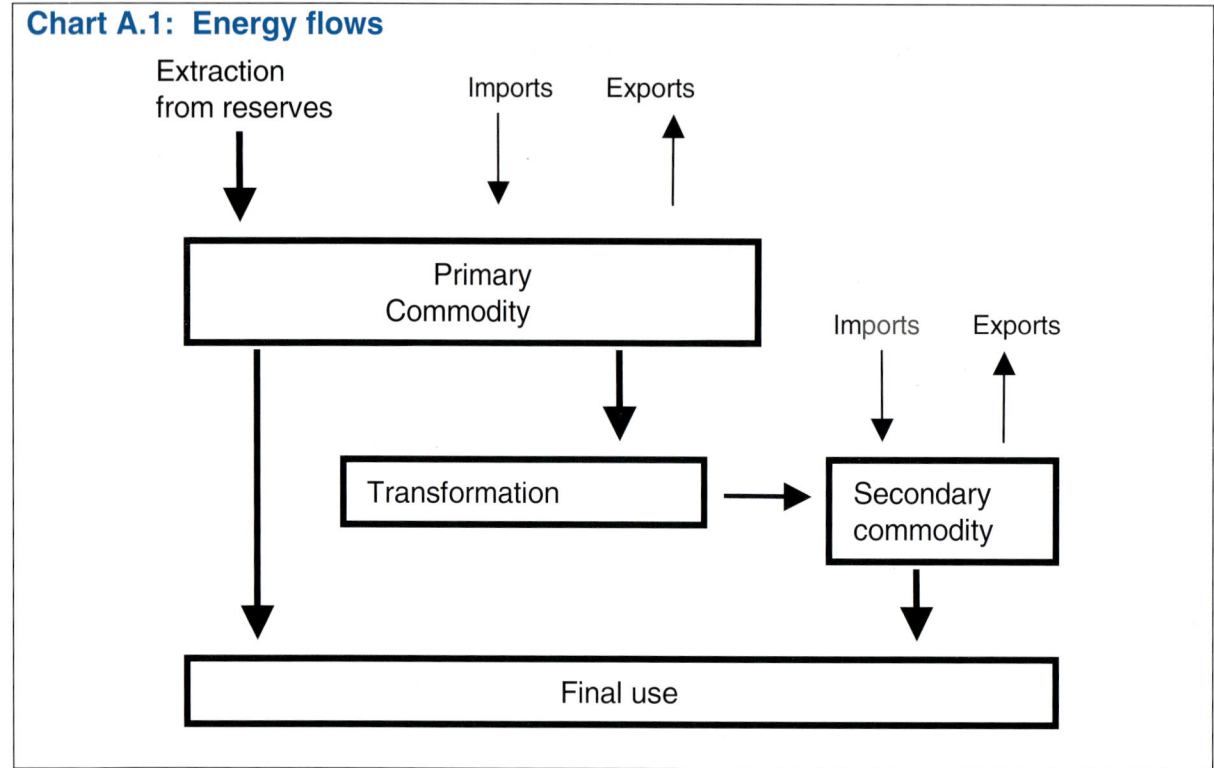

Chart A.1: Energy flows

Supply

Production

A.8   Production, within the commodity balance, covers indigenous production (extraction or capture of primary commodities) and generation or manufacture of secondary commodities. Production is always gross, that is, it includes the quantities used during the extraction or manufacturing process.

Other sources

A.9   Production from other sources covers sources of supply that do not represent "new" supply. These may be recycled products, recovered fuels (slurry or waste coal), or electricity from pumped storage plants. The production of these quantities will have been reported in an earlier accounting period or have already been reported in the current period of account. Exceptionally, the *Other sources* row in the commodity balances for ethane, propane and butane is used to receive transfers of these hydrocarbons from gas stabilisation plants at North Sea terminals. In this manner, the supplies of primary ethane, propane and butane from the North Sea are combined with the production of these gases in refineries, so that the disposals may be presented together in the balances.

Imports and exports

A.10   The figures for imports and exports relate to energy commodities moving into or out of the United Kingdom as part of transactions involving United Kingdom companies. Exported commodities are produced in the United Kingdom and imported commodities are for use within the United Kingdom (although some may be re-exported before or after transformation). The figures thus exclude commodities either exported from or imported into HM Revenue and Customs bonded areas or warehouses. These areas, although part of the United Kingdom, are regarded as being outside of the normal United Kingdom's customs boundary, and so goods entering into or leaving them are not counted as part of the statistics on trade used in the balances.

A.11   Similarly, commodities that only pass through the United Kingdom on their way to a final destination in another country are also excluded. However, for gas these transit flows are included because it is difficult to identify this quantity separately, without detailed knowledge of the contract information covering the trade. This means that for gas, there is some over statement of the level of imports and exports, but the net flows are correct.

A.12   The convention in these balances is that exports are shown with a negative sign.

Marine bunkers
A.13   These are deliveries of fuels (usually fuel oil or gas oil) to ships of any flag (including the United Kingdom) for consumption during the voyage to other countries. Marine bunkers are treated rather like exports and shown with a negative sign.

Stock changes
A.14   Additions to (- sign) and withdrawals from stocks (+ sign) held by producers and transformation industries correspond to withdrawals from and additions to supply, respectively.

Transfers
A.15   There are several reasons why quantities may be transferred from one commodity balance to another:
- a commodity may no longer meet the original specification and be reclassified;
- the name of the commodity may change through a change in use;
- to show quantities returned to supply from consumers. These may be by-products of the use of commodities as raw materials rather than fuels.

A.16   A quantity transferred from a balance is shown with a negative sign to represent a withdrawal from supply and with a positive sign in the receiving commodity balance representing an addition to its supply.

Total supply
A.17   The total supply available for national use is obtained by summing the flows above this entry in the balance.

Total demand
A.18   The various figures for the disposals and/or consumption of the commodities are summed to provide a measure of the demand for them. The main categories or sectors of demand are described in paragraphs A.32 to A.42.

Statistical difference
A.19   Any excess of supply over demand is shown as a statistical difference. A negative figure indicates that demand exceeds supply. Statistical differences arise when figures are gathered from a variety of independent sources and reflect differences in timing, in definition of coverage of the activity, or in commodity definition. Differences also arise for methodological reasons in the measurement of the flow of the commodity eg if there are differences between the volumes recorded by the gas producing companies and the gas transporting companies. A non-zero statistical difference is normal and, provided that it is not too large, is preferable to a statistical difference of zero as this suggests that a data provider has adjusted a figure to balance the account.

Transformation
A.20   The transformation sector of the balance covers those processes and activities that transform the original primary (and sometimes secondary) commodity into a form which is better suited for specific uses than the original form. Most of the transformation activities correspond to particular energy industries whose main business is to manufacture the product associated with them. Certain activities involving transformation take place to make products that are only partly used for energy needs (coke oven coke) or are by-products of other manufacturing processes (coke oven and blast furnace gases). However, as these products and by-products are then used, at least in part, for their energy content they are included in the balance system.

A.21 The figures given under the activity headings of this sector represent the quantities used for transformation. The production of the secondary commodities will be shown in the *Production* row of the corresponding commodity balances.

Electricity generation

A.22 The quantities of fuels burned for the generation of electricity are shown in their commodity balances under this heading. The activity is divided into two parts, covering the major power producers (for whom the main business is the generation of electricity for sale) and autogenerators (whose main business is not electricity generation but who produce electricity for their own needs and may also sell surplus quantities). The amounts of fuels shown in the balance represent the quantities consumed for the gross generation of electricity. Where a generator uses combined heat and power plant, the figures include only the part of the fuel use corresponding to the electricity generated.

A.23 In relation to autogenerators' data, the figures for quantities of fuel used for electricity generation appear under the appropriate fuel headings in the *Transformation* sector heading for *Autogenerators,* whilst the electricity generated appears in the *Electricity* column under *Production.* A breakdown of the information according to the branch of industry in which the generation occurs is not shown in the balance but is given in Chapter 1, Table 1.9. The figures for energy commodities consumed by the industry branches shown under final consumption include all use of electricity, but exclude the fuels combusted by the industry branches to generate the electricity.

Heat generation

A.24 The quantities of fuel burned to generate heat that is sold under the provision of a contract to a third party are shown in their commodity balances under this heading. It includes heat that is generated and sold by combined heat and power plants and by community heating schemes (also called district heating).

Petroleum refineries

A.25 Crude oil, natural gas liquids and other oils needed by refineries for the manufacture of finished petroleum products are shown under this heading.

Coke manufacture and blast furnaces

A.26 Quantities of coal for coke ovens and all fuels used within blast furnaces are shown under this heading. The consumption of fuels for heating coke ovens and the blast air for blast furnaces are shown under *Energy industry use.*

Patent fuel manufacture

A.27 The coals and other solid fuels used for the manufacture of solid patent fuels are reported under this heading.

Other

A.28 Any minor transformation activities not specified elsewhere are captured under this heading.

Energy industry use

A.29 Consumption by both extraction and transformation industries to support the transformation process (but not for transformation itself) are included here according to the energy industry concerned. Typical examples are the consumption of electricity in power plants (eg for lighting, compressors and cooling systems) and the use of extracted gases on oil and gas platforms for compressors, pumps and other uses. The headings in this sector are identical to those used in the transformation sector with the exception of *Pumped storage.* In this case, the electricity used to pump the water to the reservoir is reported.

Losses

A.30 This heading covers the intrinsic losses that occur during the transmission and distribution of electricity and gas (including manufactured gases). Other metering and accounting differences for gas and electricity are within the statistical difference, as are undeclared losses in other commodities.

Final consumption

A.31 *Final consumption* covers both final energy consumption (by different consuming sectors) and the use of energy commodities for non-energy purposes, that is *Non energy use.* Final consumption occurs when the commodities used are not for transformation into secondary commodities. The energy concerned disappears from the account after use. Any fuel used for electricity generation by final consumers is identified and reported separately within the transformation sector. When an enterprise generates electricity, the figure for final consumption of the industrial sector to which the enterprise belongs includes its use of the electricity it generates itself (as well as supplies of electricity it purchases from others) but does not include the fuel used to generate that electricity.

A.32   The classification of consumers according to their main business follows, as far as practicable, the *Standard Industrial Classification (SIC2003)*. The qualifications to, and constraints on, the classification are described in the technical notes to Chapter 1. Table 1E in Chapter 1 shows the breakdown of final consumers used, and how this corresponds to the SIC2003.

Industry
A.33   Two sectors of industry (iron and steel and chemicals) require special mention because the activities they undertake fall across the transformation, final consumption and non-energy classifications used for the balances. Also, the data permitting an accurate allocation of fuel use within each of these major divisions are not readily available.

Iron and steel
A.34   The iron and steel industry is a heavy energy user for transformation and final consumption activities. Figures shown under final consumption for this industry branch reflect the amounts that remain after quantities used for transformation and energy sector own use have been subtracted from the industry's total energy requirements. Use of fuels for transformation by the industry may be identified within the transformation sector of the commodity balances.

A.35   The amounts of coal used for coke manufacture by the iron and steel industry are in the transformation sector of the coal balance. Included in this figure is the amount of coal used for coke manufacture by the companies outside of the iron and steel industry, ie solid fuel manufacturers. The corresponding production of coke and coke oven gas may be found in the commodity balances for these products. The use of coke in blast furnaces is shown in the commodity balance for coke, and the gases produced from blast furnaces and the associated basic oxygen steel furnaces are shown in the production row of the commodity balance for blast furnace gas.

A.36   Fuels used for electricity generation by the industry are included in the figures for electricity generation by autogenerators and are not distinguishable as being used by the iron and steel sector in the balances. Electricity generation and fuel used for this by broad industry group are given in Table 1.9.

A.37   Fuels used to support coke manufacture and blast furnace gas production are included in the quantities shown under *Energy industry use.* These gases and other fuels do not enter coke ovens or blast furnaces, but are used to heat the ovens and the blast air supplied to furnaces.

Chemicals
A.38   The petro-chemical industry uses hydrocarbon fuels (mostly oil products and gases) as feedstock for the manufacture of its products. Distinguishing the energy use of delivered fuels from their non-energy use is complicated by the absence of detailed information. The procedures adopted to estimate the use are described in paragraphs A.41 and A.42 under *Non energy use.*

Transport
A.39   Figures under this heading are almost entirely quantities used strictly for transport purposes. However, the figures recorded against road transport usually include some fuel that is actually consumed in some "off-road" activities. Similarly, figures for railway fuels include some amounts of burning oil not used directly for transport purposes. Transport sector use of electricity includes all electricity used in industries classified to SIC2003 Groups 60 to 63. Fuels supplied to cargo and passenger ships undertaking international voyages are reported as *Marine bunkers* (see paragraph A.13). Supplies to fishing vessels are included under "agriculture".

Other sectors
A.40   The classification of all consumers groups under this heading, except *domestic*, follows *SIC2003* and is described in Table 1E in Chapter 1. The consistency of the classification across

different commodities cannot be guaranteed because the figures reported are dependent on what the data suppliers can provide.

Non energy use

A.41   The non energy use of fuels may be divided into two types. They may be used directly for their physical properties e.g. lubricants or bitumen used for road surfaces, or by the petro-chemical industry as raw materials for the manufacture of goods such as plastics. In their use by the petro-chemical industry, relatively little combustion of the fuels takes place and the carbon and/or hydrogen they contain are largely transferred into the finished product. However, in some cases heat from the manufacturing process or from combustion of by-products may be used. Data for this energy use are rarely available. Depending on the feedstock, non energy consumption is either estimated or taken to be the deliveries to the chemicals sector.

A.42   Both types of non energy use are shown under the *Non energy use* heading at the foot of the balances.

The energy balance (Tables 1.1 to 1.3)

Principles

A.43   The energy balance conveniently presents:

- an overall view of the United Kingdom's energy supplies;
- the relative importance of each energy commodity;
- dependence on imports;
- the contribution of our own fossil and renewable resources;
- the interdependence of commodities on one another.

A.44   The energy balance is constructed directly from the commodity balances by expressing the data in a common unit, placing them beside one another and adding appropriate totals. Heat sold is also included as a fuel. However, some rearrangement of the commodity balance format is required to show transformation of primary into secondary commodities in an easily understood manner.

A.45   Energy units are widely used as the common unit, and the current practice for the United Kingdom and the international organisations which prepare balances is to use the tonne of oil equivalent or a larger multiple of this unit, commonly thousands. One tonne of oil equivalent is defined as 10^7 kilocalories (41.868 gigajoules). The tonne of oil equivalent is another unit of energy like the gigajoule, kilocalorie or kilowatt hour, rather than a physical quantity. It has been chosen as it is easier to visualise than the other units. Due to the natural variations in heating value of primary fuels such as crude oil, it is rare that one tonne of oil has an energy content equivalent to one tonne of oil equivalent, however it is generally within a few per cent of the heating value of a tonne of oil equivalent. The energy figures are calculated from the natural units of the commodity balances by multiplying by factors representing the calorific (heating) value of the fuel. The gross calorific values of fuels are used for this purpose. When the natural unit of the commodity is already an energy unit (electricity in kilowatt hours, for example) the factors are just constants, converting one energy unit to another.

A.46   Most of the underlying definitions and ideas of commodity balances can be taken directly over into the energy balance. However, production of secondary commodities and, in particular, electricity are treated differently and need some explanation. The components of the energy balance are described below, drawing out the differences of treatment compared with the commodity balances.

Primary supply

A.47   Within the energy balance, the production row covers only extraction of primary fuels and the generation of primary energy (hydro, nuclear, wind). Note the change of row heading from *Production* in the commodity balances to *Indigenous production* in the energy balance. Production of secondary fuels and secondary electricity are shown in the transformation sector and not in the indigenous production row at the top of the balance.

A.48   For fossil fuels, indigenous production represents the marketable quantity extracted from the reserves. Indigenous production of *Primary electricity* comprises hydro-electricity, wind and nuclear energy. The energy value for hydro-electricity is taken to be the energy content of the electricity produced from the hydro power plant and not the energy available in the water driving the turbines. A similar approach is adopted for electricity from wind generators. The electricity is regarded as the primary energy form because there are currently no other uses of the energy resource "upstream" of the generation. The energy value attached to nuclear electricity is discussed in paragraph A.52.

A.49   The other elements of the supply part of the balance are identical to those in the commodity balances. In particular, the sign convention is identical, so that figures for exports and international marine bunkers carry negative signs. A stock build carries a negative sign to denote it as a withdrawal from supply whilst a stock draw carries a positive sign to show it as an addition to supply.

A.50   The *Primary supply* is the sum of the figures above it in the table, taking account of the signs, and expresses the national requirement for primary energy commodities from all sources and foreign supplies of secondary commodities. It is an indicator of the use of indigenous resources and external energy supplies. Both the amount and mixture of fuels in final consumption of energy commodities in the United Kingdom will differ from the primary supply. The "mix" of commodities in final consumption will be much more dependent on the manufacture of secondary commodities, in particular electricity.

Transformation

A.51   Within an energy balance the presentation of the inputs to and outputs from transformation activities requires special mention, as it is carried out using a compact format. The transformation sector also plays a key role in moving primary electricity from its own column in the balance into the electricity column, so that it can be combined with electricity from fossil fuelled power stations and the total disposals shown.

A.52   Indigenous production of primary electricity comprises nuclear electricity, hydro electricity and electricity from wind generation. Nuclear electricity is obtained by passing steam from nuclear reactors through conventional steam turbine sets. The heat in the steam is considered to be the primary energy available and its value is calculated from the electricity generated using the average thermal efficiency of nuclear stations, currently 38.6 in the United Kingdom. The electrical energy from hydro and wind is transferred from the *Primary electricity* column to the *Electricity* column using the *transfers* row because electricity is the form of primary energy and no transformation takes place. However, because the form of the nuclear energy is the steam from the nuclear reactors, the energy it contains is shown entering electricity generation and the corresponding electricity produced is included with all electricity generation in the figure, in the same row, under the *Electricity* column.

A.53   Quantities of fuels entering transformation activities (fuels into electricity generation and heat generation, crude oil into petroleum product manufacture (refineries), or coal into coke ovens) are shown with a negative sign to represent the input and the resulting production is shown as a positive number.

A.54   For electricity generated by Major power producers, the inputs are shown in the *Major power producers* row of the *coal, manufactured fuel, primary oils, petroleum products, gas, renewables* and *primary electricity* columns. The total energy input to electricity generation is the sum of the values in these first seven columns. The *Electricity* column shows total electricity generated from these inputs and the transformation loss is the sum of these two figures, given in the *Total* column.

A.55   Within the transformation sector, the negative figures in the *Total* column represent the losses in the various transformation activities. This is a convenient consequence of the sign convention chosen for the inputs and outputs from transformation. Any positive figures represent a transformation gain and, as such, are an indication of incorrect data.

A.56   In the energy balance, the columns containing the input commodities for electricity generation, heat generation and oil refining are separate from the columns for the outputs. However, for the transformation activities involving solid fuels this is only partly the case. Coal used for the manufacture of coke is shown in the coke manufacture row of the transformation section in the coal column, but the related coke and coke oven gas production are shown combined in the *Manufactured fuels* column. Similarly, the input of coke to blast furnaces and the resulting production of blast furnace gas are not identifiable and have been combined in the *Manufactured fuels* column in the *Blast furnace* row. As a result, only the net loss from blast furnace transformation activity appears in the column.

A.57   The share of each commodity or commodity group in primary supply can be calculated from the table. This table also shows the demand for primary as well as foreign supplies. Shares of primary supplies may be taken from the *Primary supply* row of the balance. Shares of fuels in final consumption may be calculated from the final consumption row.

Energy industry use and final consumption
A.58   The figures for final consumption and energy industry use follow, in general, the principles and definitions described under commodity balances in paragraphs A.29 to A.42.

Standard conversion factors

1 tonne of oil equivalent (toe)	= 10^7 kilocalories
	= 396.83 therms
	= 41.868 GJ
	= 11,630 kWh
100,000 British thermal units (Btu)	= 1 therm

This Digest follows UK statistical practice and uses the term "billion" to refer to one thousand million or 10^9

The following prefixes are used for multiples of joules, watts and watt hours:

kilo (k)	= 1,000	or 10^3
mega (M)	= 1,000,000	or 10^6
giga (G)	= 1,000,000,000	or 10^9
tera (T)	= 1,000,000,000,000	or 10^{12}
peta (P)	= 1,000,000,000,000,000	or 10^{15}

WEIGHT

1 kilogramme (kg)	= 2.2046 pounds (lb)
1 pound (lb)	= 0.4536 kg
1 tonne (t)	= 1,000kg
	= 0.9842 long ton
	= 1.102 short ton (sh tn)
1 Statute or long ton	= 2,240 lb
	= 1.016 t
	= 1.102 sh tn

VOLUME

1 cubic metre (cu m)	= 35.31 cu ft
1 cubic foot (cu ft)	= 0.02832 cu m
1 litre	= 0.22 Imperial gallons (UK gal)
1 UK gallon	= 8 UK pints
	= 1.201 US gallons (US gal)
	= 4.54609 litres
1 barrel	= 159.0 litres
	= 34.97 UK gal
	= 42 US gal

LENGTH

1 mile	= 1.6093 kilometres
1 kilometre (km)	= 0.62137 miles

TEMPERATURE

1 scale degree Celsius (C)	= 1.8 scale degrees Fahrenheit (F)

For conversion of temperatures: $^\circ C = 5/9 \ (^\circ F -32)$; $^\circ F = 9/5 \ ^\circ C +32$

Average conversion factors for petroleum

	Imperial gallons per tonne	Litres per tonne		Imperial gallons per tonne	Litres per tonne
Crude oil:			Gas/diesel oil:		
Indigenous	264	1,199	Gas oil	254	1,155
Imported	260	1,181	Marine diesel oil	254	1,155
Average of refining throughput	262	1,192			
			Fuel oil:		
Ethane	601	2,730	All grades	223	1,014
Propane	426	1,937	Light fuel oil:		
Butane	381	1,731	1% or less sulphur	233	1,059
Naphtha (l.d.f.)	319	1,450			
			Medium fuel oil:		
Aviation gasoline	309	1,405	1% or less sulphur	230	1,047
Motor spirit:			Heavy fuel oil:		
All grades	299	1,361	1% or less sulphur	222	1,011
Unleaded Super	297	1,351			
Ultra low sulphur petrol	299	1,361			
Lead replacement petrol	299	1,360	Lubricating oils:		
			White	244	1,111
Middle distillate feedstock	230	1,043	Greases	240	1,090
Kerosene:			Other	248	1,127
Aviation turbine fuel	274	1,246	Bitumen	215	975
Burning oil	274	1,244	Petroleum coke	186	843
			Petroleum waxes	258	1,173
DERV fuel:			Industrial spirit	274	1,247
0.005% or less sulphur	264	1,199	White spirit	280	1,273

Note: The above conversion factors, which for refined products have been compiled by the UK Petroleum Industry Association, apply to the year 2007, and are only approximate for other years.

Fuel conversion factors for converting fossil fuels to carbon dioxide, 2006

	kg CO_2 per tonne	kg CO_2 per kWh	kg CO_2 per litre
Gases			
Natural Gas		0.185	
Liquid fuels			
LPG		0.214	1.495
Gas oil	3190	0.252	2.674
Fuel oil	3223	0.268	3.169*
Burning oil	3150	0.245	2.518
Naptha	3131	0.237	2.203*
Petrol	3135	0.240	2.315
Diesel	3164	0.250	2.630
Aviation spirit	3128	0.238	2.233
Aviation turbine fuel	3150	0.245	2.518
Solid fuels			
Industrial coal	2457	0.330	
Domestic coal	2523	0.298	
Coking coal	2810	0.332	

All emission factors are based on a Gross Calorific Value basis

*BERR estimates

The information above is based on Defra's greenhouse gas (GHG) conversion factors for company reporting, available at: www.defra.gov.uk/environment/business/envrp/conversion-factors.htm
The figures are derived for Defra by AEA, based on the National Atmospheric Emissions Inventory, more information on the Inventory is available at: www.naei.org.uk/reports.php

A.1 Estimated average calorific values of fuels 2007

	GJ per tonne net	GJ per tonne gross		GJ per tonne net	GJ per tonne gross
Coal:			Renewable sources:		
All consumers (weighted average) (1)	25.6	26.9	Domestic wood (2)	12.3	13.9
Power stations (1)	24.9	26.2	Industrial wood (3)	12.1	13.7
Coke ovens (1)	29.0	30.5	Straw	12.8	15.0
Low temperature carbonisation plants			Poultry litter	7.4	8.8
and manufactured fuel plants	27.9	29.4	Meat and bone	15.6	18.6
Collieries	28.3	29.8	General industrial waste	15.2	16.0
Agriculture	26.6	28.0	Hospital waste	13.3	14.0
Iron and steel	28.9	30.4	Municipal solid waste (4)	6.7	9.5
Other industries (weighted average)	25.8	27.2	Refuse derived waste (4)	13.0	18.5
Non-ferrous metals	24.1	25.4	Short rotation coppice (5)	9.3	11.1
Food, beverages and tobacco	28.9	30.4	Tyres	..	32.0
Chemicals	25.4	26.7			
Textiles, clothing, leather etc.	28.0	29.5	Petroleum:		
Pulp, paper, printing etc.	27.9	29.4	Crude oil (weighted average)	43.4	45.7
Mineral products	26.2	27.6	Petroleum products (weighted average)	43.6	45.8
Engineering (mechanical and			Ethane	48.1	50.7
electrical engineering and					
vehicles)	28.0	29.5	Butane and propane (LPG)	47.0	49.5
Other industries	27.0	28.4	Light distillate feedstock for gasworks	45.1	47.5
			Aviation spirit and wide cut gasoline	45.0	47.4
			Aviation turbine fuel	43.9	46.2
Domestic			Motor spirit	44.7	47.1
House coal	29.0	30.5	Burning oil	43.9	46.2
Anthracite and dry steam coal	32.1	33.8	Gas/diesel oil (DERV)	43.3	45.5
Other consumers	27.8	29.3	Fuel oil	41.5	43.6
Imported coal (weighted average)	25.7	27.1	Power station oil	41.5	43.6
Exports (weighted average)	30.9	32.5	Non-fuel products (notional value)	41.1	43.2

	GJ per tonne net	GJ per tonne gross		MJ per cubic metre net	MJ per cubic metre gross
Coke (including low temperature			Natural gas produced (6)	35.7	39.7
carbonisation cokes)	29.8	29.8	Natural gas consumed (7)	35.5	39.4
Coke breeze	24.8	24.8	Coke oven gas	16.2	18.0
Other manufactured solid fuel	30.9	32.5	Blast furnace gas	3.0	3.0
			Landfill gas (8)	19-23	21-25
			Sewage gas (8)	19-23	21-25

(1) Applicable to UK consumption - based on calorific value for home produced coal plus imports and, for "All consumers" net of exports.

(2) On an "as received" basis; seasoned logs at 25% moisture content. On a "dry" basis 18.6 GJ per tonne.

(3) Average figure covering a range of possible feedstock; at 25% moisture content. On a "dry" basis 18.6 GJ per tonne.

(4) Average figure based on survey returns.

(5) On an "as received" basis; at 40% moisture content. On a "dry" basis 18.6 GJ per tonne.

(6) The gross calorific value of natural gas can also be expressed as 11.026 kWh per cubic metre. This value represents the average calorific value seen for gas when extracted. At this point it contains not just methane, but also some other hydrocarbon gases (ethane, butane, propane). These gases are removed before the gas enters the National Transmission System for sale to final consumers.

(7) Home produced and imported gas. This weighted average of calorific values will approximate the average for the year that readers will see quoted on their gas bills. It can also be expressed as 10.948 kWh per cubic metre.

(8) Calorific value varies depending on the methane content of the gas

Note: The above estimated average calorific values apply only to the year 2007. For calorific values of fuels in earlier years see Tables A.2 and A.3 and previous issues of this Digest. See the notes in Chapter 1, paragraph 1.52 regarding net calorific values. The calorific values for coal other than imported coal are based on estimates provided by the main coal producers, but with some exceptions as noted on Table A.2. The calorific values for petroleum products have been calculated using the method described in Chapter 1, paragraph 1.29. The calorific values for coke oven gas, blast furnace gas, coke and coke breeze are currently being reviewed jointly by BERR and the Iron and Steel Statistics Bureau (ISSB).

Data reported in this Digest in 'thousand tonnes of oil equivalent' have been prepared on the basis of 1 tonne of oil equivalent having an energy content of 41.868 gigajoules (GJ), (1 GJ = 9.478 therms) - see notes in Chapter 1, paragraphs 1.26 to 1.27.

A.2 Estimated average gross calorific values of fuels 1980, 1990, 2000 and 2004 to 2007

GJ per tonne (gross)

	1980	1990	2000	2004	2005	2006	2007
Coal							
All consumers *(1)(2)*	25.6	25.5	26.2	26.1	25.8	25.7	25.7
All consumers - home produced plus imports minus exports *(1)*	..	..	27.0	26.7	26.9	26.8	26.9
Power stations *(2)*	23.8	24.8	25.6	25.4	25.0	25.0	25.4
Power stations - home produced plus imports *(1)*	..	..	26.0	26.1	26.1	26.2	26.2
Coke ovens *(2)*	30.5	30.2	31.2	31.6	32.5	32.3	32.8
Coke ovens - home produced plus imports *(1)*	..	..	30.4	30.5	30.5	30.5	30.5
Low temperature carbonisation plants and manufactured fuel plants	19.1	29.2	30.3	30.5	29.6	29.5	29.4
Collieries	27.0	28.6	29.6	29.9	29.8	30.0	29.8
Agriculture	30.1	28.9	29.2	28.0	28.0	28.0	28.0
Iron and steel industry *(3)*	29.1	28.9	30.7	30.4	30.4	30.4	30.4
Other industries *(1)*	27.1	27.8	26.7	26.6	26.6	26.6	27.2
Non-ferrous metals		23.1	25.1	24.8	24.5	25.0	25.4
Food, beverages and tobacco	28.6	28.1	29.5	29.4	29.8	29.0	30.4
Chemicals	25.8	27.3	28.7	26.6	26.6	26.7	26.7
Textiles, clothing, leather and footwear	27.5	27.7	30.4	29.5	29.5	29.6	29.5
Pulp, paper, printing, etc.	26.5	27.9	28.7	28.7	28.9	29.4	29.4
Mineral products *(4)*	..	28.2	27.0	27.9	27.6	27.6	27.6
Engineering *(5)*	27.7	28.3	29.3	30.6	30.7	30.4	29.5
Other industry *(6)*	28.4	28.5	30.2	27.8	25.9	25.5	28.4
Domestic							
House coal	30.1	30.2	30.9	30.9	30.7	30.5	30.5
Anthracite and dry steam coal	33.3	33.6	33.6	33.8	33.8	33.9	33.8
Other consumers	27.5	27.5	29.2	29.8	29.1	29.6	29.3
Imported coal *(1)*	..	28.3	28.0	27.1	27.3	27.2	27.1
of which Steam coal	..	..	26.6	26.5	26.6	26.5	26.5
Coking coal	..	..	30.4	30.4	30.4	30.4	30.4
Anthracite	..	..	31.2	30.4	30.4	31.8	32.6
Exports *(1)*	..	29.0	32.0	32.3	32.7	32.5	32.5
of which Steam coal	..	..	31.0	29.9	32.9	32.2	32.2
Anthracite	..	..	32.6	32.5	32.6	32.5	32.6
Coke *(7)*	28.1	28.1	29.8	29.8	29.8	29.8	29.8
Coke breeze	24.4	24.8	24.8	24.8	24.8	24.8	24.8
Other manufactured solid fuels *(1)*	27.6	27.6	30.8	31.8	32.5	32.5	32.5
Petroleum							
Crude oil *(1)*	45.2	45.6	45.7	45.7	45.7	45.7	45.7
Liquified petroleum gas	49.6	49.4	49.4	49.4	49.5	49.5	49.5
Ethane	52.3	50.6	50.7	50.7	50.7	50.7	50.7
LDF for gasworks/Naphtha	47.8	47.9	47.7	47.5	47.6	47.5	47.7
Aviation spirit and wide-cut gasoline (AVGAS and AVTAG)	47.2	47.3	47.3	47.5	47.4	47.4	47.4
Aviation turbine fuel (AVTUR)	46.4	46.2	46.2	46.2	46.2	46.2	46.2
Motor spirit	47.0	47.0	47.0	47.1	47.0	47.1	47.1
Burning oil	46.5	46.2	46.2	46.2	46.2	46.2	46.2
Vaporising oil	45.9	45.9	..				..
Gas/diesel oil (including DERV)	45.5	45.4	45.6	45.6	45.7	45.6	45.5
Fuel oil	42.8	43.2	43.1	43.5	43.5	43.3	43.6
Power station oil	42.8	43.2	43.1	43.5	43.5	43.3	43.6
Non-fuel products (notional value)	42.2	43.2	43.8	43.4	42.9	43.1	43.2
Petroleum coke	..	39.5	35.8	35.8	35.8	35.8	35.7
Natural Gas *(8)*	..	38.4	39.4	39.6	39.6	39.8	39.7

(1) Weighted averages.
(2) Home produced coal only.
(3) From 2001 onwards almost entirely sourced from imports.
(4) Based on information provided by the British Cement Industry Association; almost all coal used by this sector in the latest 4 years was imported.

(5) Mechanical engineering and metal products, electrical and instrument engineering and vehicle manufacture.
(6) Includes construction.
(7) Since 1995 the source of these figures has been the ISSB.
(8) Natural Gas figures are shown in MJ per cubic metre

A.3 Estimated average net calorific values of fuels 1980, 1990, 2000 and 2004 to 2007

GJ per tonne (net)

	1980	1990	2000	2004	2005	2006	2007
Coal							
All consumers (1)(2)	24.3	24.2	24.9	24.8	24.5	24.4	24.4
All consumers - home produced plus imports minus exports (1)	..	..	25.7	25.4	25.6	25.5	25.6
Power stations (2)	22.6	23.6	24.3	24.1	23.8	23.8	24.1
Power stations - home produced plus imports (1)	..	..	24.7	24.8	24.8	24.9	24.9
Coke ovens (2)	29.0	28.7	29.6	30.0	30.9	30.7	31.2
Coke ovens - home produced plus imports (1)	..	..	28.9	29.0	29.0	29.0	29.0
Low temperature carbonisation plants and manufactured fuel plants	18.1	27.7	28.8	29.0	28.1	28.0	27.9
Collieries	25.7	27.2	28.1	28.4	28.3	28.5	28.3
Agriculture	28.6	27.5	27.7	26.6	26.6	26.6	26.6
Iron and steel industry (3)	27.6	27.5	29.2	28.9	28.9	28.9	28.9
Other industries (1)	25.7	26.4	25.4	25.3	25.3	25.3	25.8
Non-ferrous metals	..	21.9	23.8	23.6	23.3	23.8	24.1
Food, beverages and tobacco	27.2	26.7	28.0	27.9	28.3	27.6	28.9
Chemicals	24.5	25.9	27.3	25.3	25.3	25.4	25.4
Textiles, clothing, leather and footwear	26.1	26.3	28.9	28.0	28.0	28.1	28.0
Pulp, paper, printing, etc.	25.2	26.5	27.3	27.3	27.5	27.9	27.9
Mineral products (4)	..	26.8	25.7	26.5	26.2	26.2	26.2
Engineering (5)	26.3	26.9	27.8	29.1	29.2	28.9	28.0
Other industry (6)	27.0	27.1	28.7	26.4	24.6	24.2	27.0
Domestic							
House coal	28.6	28.7	29.4	29.4	29.2	29.0	29.0
Anthracite and dry steam coal	31.6	31.9	31.9	32.1	32.1	32.2	32.1
Other consumers	26.1	26.1	27.7	28.3	27.6	28.1	27.8
Imported coal (1)	..	26.9	26.6	25.7	25.9	25.8	25.7
of which Steam coal	..	..	25.3	25.2	25.3	25.2	25.2
Coking coal	..	..	28.9	28.9	28.9	28.9	28.9
Anthracite	..	..	29.6	28.9	28.9	30.2	31.0
Exports (1)	..	27.6	30.4	30.7	31.1	30.9	30.9
of which Steam coal	..	..	29.5	28.4	31.3	30.6	30.6
Anthracite	..	..	31.0	30.9	31.0	30.9	31.0
Coke (7)	28.1	28.1	29.8	29.8	29.8	29.8	29.8
Coke breeze	24.4	24.8	24.8	24.8	24.8	24.8	24.8
Other manufactured solid fuels (1)	26.2	26.2	29.3	30.2	30.9	30.9	30.9
Petroleum							
Crude oil (1)	42.9	43.3	43.4	43.4	43.4	43.4	43.4
Liquified petroleum gas	47.1	46.9	46.9	46.9	47.1	47.0	47.0
Ethane	49.7	48.1	48.2	48.1	48.1	48.1	48.1
LDF for gasworks/Naphtha	45.4	45.5	45.3	45.1	45.2	45.1	45.3
Aviation spirit and wide-cut gasoline (AVGAS and AVTAG)	44.8	44.9	44.9	45.1	45.0	45.0	45.0
Aviation turbine fuel (AVTUR)	44.1	43.9	43.9	43.9	43.9	43.9	43.9
Motor spirit	44.7	44.7	44.7	44.8	44.7	44.7	44.7
Burning oil	44.2	43.9	43.9	43.9	43.9	43.9	43.9
Vaporising oil	43.6	43.6	..	..	..	..	..
Gas/diesel oil (including DERV)	43.2	43.1	43.3	43.3	43.4	43.4	43.3
Fuel oil	40.7	41.0	40.9	41.3	41.3	41.2	41.5
Power station oil	40.7	41.0	40.9	41.3	41.3	41.2	41.5
Non-fuel products (notional value)	40.1	41.0	41.6	41.2	40.8	40.9	41.1
Petroleum coke	..	37.5	34.0	34.0	34.0	34.0	33.9
Natural Gas (8)	-	34.6	35.5	35.6	35.6	35.8	35.7

For footnotes see table A.2
The net calorific values of natural gas and coke oven gas are the gross calorific values x 0.9.

Annex B
Glossary and Acronyms

Advanced gas-cooled reactor (AGR) A type of nuclear reactor cooled by carbon dioxide gas.

AEA Energy & Environment Part of the AEA Group, comprising the former Future Energy Solutions and NETCEN.

AES Association of Electricity Supplies

Anthracite Within this publication, anthracite is coal classified as such by UK coal producers and importers of coal. Typically it has a high heat content making it particularly suitable for certain industrial processes and for use as a domestic fuel.

Anthropogenic Produced by human activities.

Associated Gas Natural gas found in association with crude oil in a reservoir, either dissolved in the oil or as a cap above the oil.

Autogeneration Generation of electricity by companies whose main business is not electricity generation, the electricity being produced mainly for that company's own use.

Aviation spirit A light hydrocarbon oil product used to power piston-engined aircraft power units.

Aviation turbine fuel The main aviation fuel used for powering aviation gas-turbine power units (jet aircraft engine).

BE British Energy

Benzole A colourless liquid, flammable, aromatic hydrocarbon by-product of the iron and steel making process. It is used as a solvent in the manufacture of styrenes and phenols but is also used as a motor fuel.

BERR Department for Business, Enterprise and Regulatory Reform

BETTA British Electricity Trading and Transmission Arrangements (BETTA) refer to changes to electricity generation, distribution and supply licences. On 1 April 2005, the England and Wales trading arrangements were extended to Scotland by the British Electricity Trading and Transmission Arrangements creating a single GB market for trading of wholesale electricity, with common arrangements for access to and use of GB transmission system. From 1 April 2005, NGC has become the System Operator for the whole of GB. BETTA replaced NETA (see page 222) on 4 April 2005.

Biodiesel (FAME - biodiesel produced to BS EN 14214). Produced from vegetable oils or animal fats by mixing them with ethanol or methanol to break them down.

Bioethanol Created from crops rich in starch or sugar by fermentation, distillation and finally dehydration.

Biogas	Energy produced from the anaerobic digestion of sewage and industrial waste.
Bitumen	The residue left after the production of lubricating oil distillates and vacuum gas oil for upgrading plant feedstock. Used mainly for road making and construction purposes.
Blast furnace gas	Mainly produced and consumed within the iron and steel industry. Obtained as a by-product of iron making in a blast furnace, it is recovered on leaving the furnace and used partly within the plant and partly in other steel industry processes or in power plants equipped to burn it. A similar gas is obtained when steel is made in basic oxygen steel converters; this gas is recovered and used in the same way.
Breeze	Breeze can generally be described as coke screened below 19 mm (¾ inch) with no fines removed but the screen size may vary in different areas and to meet the requirements of particular markets.
BG	British Gas
BOS	Basic Oxygen Steel furnace gas
BNFL	British Nuclear Fuels plc.
BRE	Building Research Establishment
Burning oil	A refined petroleum product, with a volatility in between that of motor spirit and gas diesel oil primarily used for heating and lighting.
Butane	Hydrocarbon (C_4H_{10}), gaseous at normal temperature but generally stored and transported as a liquid. Used as a component in Motor Spirit to improve combustion, and for cooking and heating (see LPG).
Calorific values (CVs)	The energy content of a fuel can be measured as the heat released on complete combustion. The SI (Système International - see page 224) derived unit of energy and heat is the Joule. This is the energy per unit volume of the fuel and is often measured in GJ per tonne. The energy content can be expressed as an upper (or gross) value and a lower (or net) value. The difference between the two values is due to the release of energy from the condensation of water in the products of combustion. Gross calorific values are used throughout this publication.
CCA	Climate Change Agreement. Climate Change Agreements allow energy intensive business users to receive an 80 per cent discount from the Climate Change Levy (CCL), in return for meeting energy efficiency or carbon saving targets. The CCL is a tax on the use of energy in industry, commerce and the public sector. The aim of the levy is to encourage users to improve energy efficiency and reduce emissions of greenhouse gases.
CCL	Climate Change Levy. The Climate Change Levy is a tax on the use of energy in industry, commerce and the public sector, with offsetting cuts in employers' National Insurance Contributions and additional support for energy efficiency schemes and renewable sources of energy. The aim of the levy is to encourage users to improve energy efficiency and reduce emissions of greenhouse gases.

CO$_2$	Carbon dioxide. Carbon dioxide contributes about 60 per cent of the potential global warming effect of man-made emissions of greenhouse gases. Although this gas is naturally emitted by living organisms, these emissions are offset by the uptake of carbon dioxide by plants during photosynthesis; they therefore tend to have no net effect on atmospheric concentrations. The burning of fossil fuels, however, releases carbon dioxide fixed by plants many millions of years ago, and thus increases its concentration in the atmosphere.
Co-firing	The burning of biomass products in fossil fuel power stations
Coke oven coke	The solid product obtained from carbonisation of coal, principally coking coal, at high temperature, it is low in moisture and volatile matter. Used mainly in iron and steel industry.
Coke oven gas	Gas produced as a by-product of solid fuel carbonisation and gasification at coke ovens, but not from low temperature carbonisation plants. Synthetic coke oven gas is mainly natural gas which is mixed with smaller amounts of blast furnace and basic oxygen steel furnace gas to produce a gas with almost the same quantities as coke oven gas.
Coking coal	Within this publication, coking coal is coal sold by producers for use in coke ovens and similar carbonising processes. The definition is not therefore determined by the calorific value or caking qualities of each batch of coal sold, although calorific values tend to be higher than for steam coal. Not all coals form cokes. For a coal to coke it must exhibit softening and agglomeration properties, ie the end product must be a coherent solid.
Colliery methane	Methane released from coal seams in deep mines which is piped to the surface and consumed at the colliery or transmitted by pipeline to consumers.
Combined cycle gas Turbine (CCGT)	Combined cycle gas turbine power stations combine gas turbines and steam turbines which are connected to one or more electrical generators in the same plant. The gas turbine (usually fuelled by natural gas or oil) produces mechanical power (to drive the generator) and heat in the form of hot exhaust gases. These gases are fed to a boiler, where steam is raised at pressure to drive a conventional steam turbine, which is also connected, to an electrical generator.
Combined Heat and Power (CHP)	CHP is the simultaneous generation of usable heat and power (usually electricity) in a single process. The term CHP is synonymous with cogeneration and total energy, which are terms often used in the United States or other Member States of the European Community. The basic elements of a CHP plant comprise one or more prime movers driving electrical generators, where the steam or hot water generated in the process is utilised via suitable heat recovery equipment for use either in industrial processes, or in community heating and space heating. For further information see Chapter 6 paragraph 6.34.
CHPQA	Combined Heat and Power Quality Assurance Scheme
Conventional thermal power stations	These are stations which generate electricity by burning fossil fuels to produce heat to convert water into steam, which then powers steam turbines.

Cracking/conversion   A refining process using combinations of temperature, pressure and in some cases a catalyst to produce petroleum products by changing the composition of a fraction of petroleum, either by splitting existing longer carbon chain or combining shorter carbon chain components of crude oil or other refinery feedstock's. Cracking allows refiners to selectively increase the yield of specific fractions from any given input petroleum mix depending on their requirements in terms of output products.

CRC   Carbon Reduction Commitment. The Carbon Reduction Commitment is a proposed mandatory cap and trade scheme that will apply to large non energy-intensive organisations in the public and private sectors. It is anticipated that the scheme will have cut carbon emissions by 1.2 million tonnes of carbon per year by 2020.

Crude oil   A mineral oil consisting of a mixture of hydrocarbons of natural origins, yellow to black in colour, of variable density and viscosity.

DEFRA   Department for Environment, Food and Rural Affairs

DERV   Diesel engined road vehicle fuel used in internal combustion engines that are compression-ignited (see gas diesel oil on page 220).

DFT   Department for Transport

Distillation   A process of separation of the various components of crude oil and refinery feedstocks using the different temperatures of evaporation and condensation of the different components of the mix received at the refineries.

DNC   Declared net capacity and capability are used to measure the maximum power available from generating stations at a point in time. See Chapter 5 paragraphs 5.55 and 5.56 and Chapter 7 paragraph 7.68 for a fuller definition.

DNO   Distribution Network Operator

Downstream   Used in oil and gas processes to cover the part of the industry after the production of the oil and gas. For example, it covers refining, supply and trading, marketing and exporting.

DTI   Department of Trade and Industry

DUKES   Digest of United Kingdom Energy Statistics, the Digest provides essential information for everyone, from economists to environmentalists and from energy suppliers to energy users.

ECA   Enhanced Capital Allowances

EHCS   English House Condition Survey

Embedded Generation   Embedded generation is electricity generation by plant which has been connected to the distribution networks of the public electricity distributors rather than directly to the National Grid Company's transmission systems. Typically they are either smaller stations located on industrial sites, or combined heat and power plant, or renewable energy plant such as wind farms, or refuse burner generators. The category also includes some domestic generators such as those with electric solar panels. For a description of the current structure of the electricity industry in the UK see Chapter 5 paragraphs 5.3 to 5.9.

Energy use	Energy use of fuel mainly comprises use for lighting, heating or cooling, motive power and power for appliances. See also non-energy use on page 222.
ESA	European System of National and Regional Accounts. An integrated system of economic accounts which is the European version of the System of National Accounts (SNA).
EESs	The Energy Efficiency Commitment (formerly known as Energy Efficiency Standards of Performance) is an obligation placed on all energy suppliers to offer help and advice to their customers to improve the energy efficiency of their homes.
Ethane	A light hydrocarbon gas (C_2H_6) in natural gas and refinery gas streams (see LPG).
EU-ETS	European Union Emissions Trading Scheme. This began on 1st January 2005 and involves the trading of emissions allowances as means of reducing emissions by a fixed amount.
EUROSTAT	Statistical Office of the European Communities (SOEC).
Exports	For some parts of the energy industry, statistics on trade in energy related products can be derived from two separate sources. Firstly, figures can be reported by companies as part of systems for collecting data on specific parts of the energy industry (eg as part of the system for recording the production and disposals of oil from the UK continental shelf). Secondly, figures are also available from the general systems that exist for monitoring trade in all types of products operated by HM Revenue and Customs.
FES	Future Energy Solutions, now known as AEA Energy & Environment, part of the AEA Group.
Feedstock	In the refining industry, a product or a combination of products derived from crude oil, destined for further processing other than blending. It is distinguished from use as a chemical feedstock etc. See non-energy use on page 222.
Final energy consumption	Energy consumption by final user – ie which is not being used for transformation into other forms of energy.
Fossil fuels	Coal, natural gas and fuels derived from crude oil (for example petrol and diesel) are called fossil fuels because they have been formed over long periods of time from ancient organic matter.
Fuel oils	The heavy oils from the refining process; used as fuel in furnaces and boilers of power stations, industry, in domestic and industrial heating, ships, locomotives, metallurgic operations, and industrial power plants etc.
Fuel oil - Light	Fuel oil made up of heavier straight-run or cracked distillates and used in commercial or industrial burner installations not equipped with pre-heating facilities.
Fuel oil - Medium	Other fuel oils, sometimes referred to as bunker fuels, which generally require pre-heating before being burned, but in certain climatic conditions do not require pre-heating.

Fuel oil - Heavy

Other heavier grade fuel oils which in all situations require some form of pre-heating before being burned.

Fuel poverty

The common definition of a fuel poor household is one needing to spend in excess of 10 per cent of household income to achieve a satisfactory heating regime (21°C in the living room and 18°C in the other occupied rooms).

Gas Diesel Oil

The medium oil from the refinery process; used as a fuel in diesel engines (ie internal combustion engines that are compression-ignited), burned in central heating systems and used as a feedstock for the chemical industry.

GDP

Gross Domestic Product.

GDP deflator

An index of the ratio of GDP at current prices to GDP at constant prices. It provides a measure of general price inflation within the whole economy.

Gigajoule (GJ)

A unit of energy equal to 10^9 joules (see note on joules on page 221).

Gigawatt (GW)

A unit of electrical power, equal to 10^9 watts.

Heat sold

Heat (or steam) that is produced and sold under the provision of a contract. Heat sold is derived from heat generated by Combined Heat and Power (CHP) plants and from community heating schemes without CHP plants.

HMRC

HM Revenue and Customs.

Imports

See the first paragraph of the entry for exports on page 219. Before the 1997 edition of the Digest, the term "arrivals" was used to distinguish figures derived from the former source from those import figures derived from the systems operated by HM Revenue and Customs. To make it clearer for users, a single term is now being used for both these sources of figures (the term imports) as this more clearly states what the figures relate to, which is goods entering the UK.

International Energy Agency (IEA)

The IEA is an autonomous body located in Paris which was established in November 1974 within the framework of the Organisation for Economic Co-operation and Development (OECD) to implement an international energy programme.

Indigenous production

For oil this includes production from the UK Continental Shelf both onshore and offshore.

Industrial spirit

Refined petroleum fractions with boiling ranges up to 200°C dependent on the use to which they are put – eg seed extraction, rubber solvents, perfume etc.

ISSB

Iron and Steel Statistics Bureau

ITF

Industry Technology Facilitator

Joules	A joule is a generic unit of energy in the conventional SI system (see note on page 224). It is equal to the energy dissipated by an electrical current of 1 ampere driven by 1 volt for 1 second; it is also equal to twice the energy of motion in a mass of 1 kilogram moving at 1 metre per second.
Kilowatt (kW)	1,000 watts
Landfill gas	The methane-rich biogas formed from the decomposition of organic material in landfill.
LDF	Light distillate feedstock
LDZ	Local distribution zone
Liquefied natural Gas (LNG)	Natural gas that has been converted to liquid form for ease of storage or transport.
Liquefied petroleum Gas (LPG)	Gas usually propane or butane, derived from oil and put under pressure so that it is in liquid form. Often used to power portable cooking stoves or heaters and to fuel some types of vehicle, eg some specially adapted road vehicles, forklift trucks.
Lead Replacement Petrol (LRP)	An alternative to Leaded Petrol containing a different additive to lead (in the UK usually potassium based) to perform the lubrication functions of lead additives in reducing engine wear.
Lubricating oils	Refined heavy distillates obtained from the vacuum distillation of petroleum residues. Includes liquid and solid hydrocarbons sold by the lubricating oil trade, either alone or blended with fixed oils, metallic soaps and other organic and/or inorganic bodies.
Magnox	A type of gas-cooled nuclear fission reactor developed in the UK, so called because of the magnesium alloy used to clad the uranium fuel.
Major power producers	Companies whose prime purpose is the generation of electricity (paragraph 5.51 of Chapter 5 gives a full list of major power producers).
Megawatt (MW)	1,000 kilowatts. MWe is used to emphasise when electricity is being measured. MWt is used when heat ("thermal") is being measured.
Micro CHP	Micro CHP is a new technology that is expected to make a significant contribution to domestic energy efficiency in the future.
MMC	Monopolies and Mergers Commission
Motor spirit	Blended light petroleum product used as a fuel in spark-ignition internal combustion engines (other than aircraft engines).
NAEI	National Atmospheric Emissions Inventory
National Allocation Plan (NAP)	Under the EU Emissions Trading Scheme (EU-ETS) Directive each EU country must have a National Allocation Plan which lays down the overall contribution of the EU-ETS participants (the "cap") for the country and the allowances that each sector and each individual installation covered under the Directive is allocate, effectively stating how much that sector can emit over the trading period of the scheme

Naphtha	(Light distillate feedstock) – Petroleum distillate boiling predominantly below 200°C.
Natural gas	Natural gas is a mixture of naturally occurring gases found either in isolation, or associated with crude oil, in underground reservoirs. The main component is methane; ethane, propane, butane, hydrogen sulphide and carbon dioxide may also be present, but these are mostly removed at or near the well head in gas processing plants.
Natural gas - compressed	Natural gas that has been compressed to reduce the volume it occupies to make it easier to transport other than in pipelines. Whilst other petroleum gases can be compressed such that they move into liquid form, the volatility of natural gas is such that liquefaction cannot be achieved without very high pressures and low temperatures being used. As such, the compressed form is usually used as a "half-way house".
Natural gas liquids (NGLs)	A mixture of liquids derived from natural gas and crude oil during the production process, including propane, butane, ethane and gasoline components (pentanes plus).
NDA	Nuclear Decommissioning Authority
NETA	New Electricity Trading Arrangements - In England and Wales these arrangements replaced "the pool" from 27 March 2001. The arrangements are based on bi-lateral trading between generators, suppliers, traders and customers and are designed to be more efficient, and provide more market choice.
NETCEN	National Environment Technology Centre, now known as AEA Energy & Environment, part of the AEA Group.
NIE	Northern Ireland Electricity
NI NFFO	Northern Ireland Non Fossil Fuel Obligation
Non-energy use	Includes fuel used for chemical feedstock, solvents, lubricants, and road making material.
NFFO	Non Fossil Fuel Obligation. The 1989 Electricity Act empowers the Secretary of State to make orders requiring the Regional Electricity Companies in England and Wales to secure specified amounts of electricity from renewable sources.
NFPA	Non Fossil Purchasing Agency
NO$_x$	Nitrogen oxides. A number of nitrogen compounds including nitrogen dioxide are formed in combustion processes when nitrogen in the air or the fuel combines with oxygen. These compounds can add to the natural acidity of rainfall.
NSCP	National Statistics Code of Practice
NUTS	Nonmenclature of Units for Territorial Statistics
OFGEM	The regulatory office for gas and electricity markets
OFT	Office of Fair Trading

Orimulsion	An emulsion of bitumen in water that was used as a fuel in some power stations until 1997.
ONS	Office for National Statistics
OTS	Overseas Trade Statistics of the United Kingdom
OXERA	Oxford Economic Research Association Ltd
Patent fuel	A composition fuel manufactured from coal fines by shaping with the addition of a binding agent (typically pitch). The term manufactured solid fuel is also used.
Petrochemical feedstock	All petroleum products intended for use in the manufacture of petroleum chemicals. This includes middle distillate feedstock of which there are several grades depending on viscosity. The boiling point ranges between 200°C and 400°C.
Petroleum cokes	Carbonaceous material derived from hydrocarbon oils, uses for which include metallurgical electrode manufacture and in the manufacture of cement.
PILOT	Phase 2 (PILOT) is the successor body to the Oil & Gas Industry Task Force (OGITF) and was established on 1 January 2000, to secure the long-term future of the oil and gas industry in the UK. A forum that brings together Government and industry to address the challenges facing the oil and gas industry. One outcome of PILOT's work is the published Code of Practice on Supply Chain Relationships.
Photovoltaics	The direct conversion of solar radiation into electricity by the interaction of light with the electrons in a semiconductor device or cell.
Plant capacity	The maximum power available from a power station at a point in time (see also Chapter 5 paragraph 5.55).
Plant loads, demands and efficiency	Measures of how intensively and efficiently power stations are being used. These terms are defined in Chapter 5 paragraphs 5.57 and 5.58
PPRS	Petroleum production reporting system. Licensees operating in the UK Continental Shelf are required to make monthly returns on their production of hydrocarbons (oil and gas) to BERR. This information is recorded in the PPRS, which is used to report flows, stocks and uses of hydrocarbon from the well-head through to final disposal from a pipeline or terminal (see DUKES internet annex F on BERR's energy statistics web site for further information).
Process oils	Partially processed feedstocks which require further processing before being classified as a finished product suitable for sale. They can also be used as a reaction medium in the production process.
Primary fuels	Fuels obtained directly from natural sources, eg coal, oil and natural gas.
Primary electricity	Electricity obtained other than from fossil fuel sources, eg nuclear, hydro and other non-thermal renewables. Imports of electricity are also included.

Propane	Hydrocarbon containing three carbon atoms (C_3H_8), gaseous at normal temperature, but generally stored and transported under pressure as a liquid.
PWR	Pressurised water reactor. A nuclear fission reactor cooled by ordinary water kept from boiling by containment under high pressure.
Reforming	Processes by which the molecular structure of different fractions of petroleum can be modified. It usually involves some form of catalyst, most often platinum, and allows the conversion of lower grades of petroleum product into higher grades, improving their octane rating. It is a generic term for processes such as cracking, cyclization, dehydrogenation and isomerisation. These processes generally led to the production of hydrogen as a by-product, which can be used in the refineries in some desulphurization procedures.
Refinery fuel	Petroleum products produced by the refining process that are used as fuel at refineries.
Renewable energy sources	Renewable energy includes solar power, wind, wave and tide, and hydroelectricity. Solid renewable energy sources consist of wood, straw, short rotation coppice, other biomass and the biodegradable fraction of wastes. Gaseous renewables consist of landfill gas and sewage gas. Non-biodegradable wastes are not counted as a renewables source but appear in the Renewable sources of energy chapter of this Digest for completeness.
Reserves	With oil and gas these relate to the quantities identified as being present in underground cavities. The actual amounts that can be recovered depend on the level of technology available and existing economic situations. These continually change; hence the level of the UK's reserves can change quite independently of whether or not new reserves have been identified.
RD	Renewables Directive – this proposes that EU Member States adopt national targets that are consistent with the overall EU target of 12 per cent of energy (22.1 per cent of electricity) from renewables by 2010.
RESTATS	The Renewable Energy Statistics System
RO	Renewables Obligation – this is an obligation on all electricity suppliers to supply a specific proportion of electricity from eligible renewable sources.
ROCs	Renewables Obligation Certificates
SEPN	Sustainable Energy Policy Network represents the body of people responsible for delivering the white paper directly or indirectly through having links to business and other organisations nationally and regionally.
SI (Système International)	Refers to the agreed conventions for the measurement of physical quantities.

SIC	Standard Industrial Classification in the UK. Last revised in 2003 and known as SIC(2003), replaced previous classifications SIC(92), SIC(80) and SIC(68). SIC(92) was compatible with European Union classification NACE Rev1 (Nomenclature générale des activités économiques dans les Communautés européennes as revised in October 1990) and similarly SIC(2003) is consistent with NACE Rev1.1 which came into effect in January 2003. Classification systems need to be periodically revised because over time new products, processes and industries emerge.
Secondary fuels	Fuels derived from natural primary sources of energy. For example electricity generated from burning coal, gas or oil is a secondary fuel, as are coke and coke oven gas.
Steam coal	Within this publication, steam coal is coal classified as such by UK coal producers and by importers of coal. It tends to be coal having lower calorific values; the type of coal that is typically used for steam raising.
SO_2	Sulphur Dioxide. Sulphur dioxide is a gas produced by the combustion of sulphur-containing fuels such as coal and oil.
SOEC	Statistical Office of the European Communities
SRO	Scottish Renewable Orders
Synthetic coke oven gas	Mainly a natural gas, which is mixed with smaller amounts of blast furnace, and BOS (basic oxygen steel furnace) gas to produce a gas with almost the same quantities as coke oven gas.
Temperature correction	The temperature corrected series of total inland fuel consumption indicates what annual consumption might have been if the average temperature during the year had been the same as the average for the years 1961 to 1990.
Terawatt (TW)	1,000 gigawatts
TWh	Terawatt Hour
Thermal Sources of Electricity	These include coal, oil, natural gas, nuclear, landfill gas, sewage gas, municipal solid waste, farm waste, tyres, poultry litter, short rotation coppice, straw, coke oven gas, blast furnace gas, and waste products from chemical processes.
Tonne of oil equivalent (toe)	A common unit of measurement which enables different fuels to be compared and aggregated. (See Chapter 1 paragraphs 1.26 to 1.27 for further information and Annex A page 209 for conversion factors).
Tars	Viscous materials usually derived from the destructive distillation of coal which are by-products of the coke and iron making processes.
Therm	A common unit of measurement similar to a tonne of oil equivalent which enables different fuels to be compared and aggregated. (see Annex A).

Thermal efficiency

The thermal efficiency of a power station is the efficiency with which heat energy contained in fuel is converted into electrical energy. It is calculated for fossil fuel burning stations by expressing electricity generated as a percentage of the total energy content of the fuel consumed (based on average gross calorific values). For nuclear stations it is calculated using the quantity of heat released as a result of fission of the nuclear fuel inside the reactor.

UKCS

United Kingdom Continental Shelf

UKOOA

United Kingdom Offshore Operators Association

UKPIA

UK Petroleum Industry Association. The trade association for the UK petroleum industry.

UKSA

UK Statistics Authority

Ultra low sulphur Diesel (ULSD)

A grade of diesel fuel which has a much lower sulphur content (less than 0.005 per cent or 50 parts per million) and of a slightly higher volatility than ordinary diesel fuels. As a result it produces fewer emissions when burned. As such it enjoys a lower rate of excise duty in the UK than ordinary diesel (by 3 pence per litre) to promote its use. Virtually 100 per cent of sales of DERV fuel in the UK are ULSD.

Ultra low sulphur Petrol (ULSP)

A grade of motor spirit with a similar level of sulphur to ULSD (less than 0.005 per cent or 50 parts per million). In the March 2000 Budget it was announced that a lower rate of excise duty than ordinary petrol for this fuel would be introduced during 2000, which was increased to 3 pence per litre in the March 2001 Budget. It has quickly replaced ordinary premium grade unleaded petrol in the UK market place.

Upstream

A term to cover the activities related to the exploration, production and delivery to a terminal or other facility of oil or gas for export or onward shipment within the UK.

USBS

United States Bureau of Standards refers to legislation that sets minimum safety standards in the coal market and mining industry.

VAT

Value added tax

Watt (W)

The conventional unit to measure a rate of flow of energy. One watt amounts to 1 joule per second.

White spirit

A highly refined distillate with a boiling range of about $150°C$ to $200°C$ used as a paint solvent and for dry cleaning purposes etc.

Annex C
Further sources of United Kingdom energy publications

Some of the publications listed below give shorter term statistics, some provide further information about energy production and consumption in the United Kingdom and in other countries, and others provide more detail on a country or fuel industry basis. The list also covers recent publications on energy issues and policy, including statistical information, produced or commissioned by BERR. The list is not exhaustive and the titles of publications and publishers may alter. Unless otherwise stated, all titles are available from

BERR Publications Orderline
Web: www.berr.gov.uk/publications
Phone: 0845 015 0010
Address: ADMAIL 528, London SW1W 0YT
Email: publications@berr.gsi.gov.uk

and can also be found on the BERR Web site at www.berr.gov.uk/energy/.

Department for Business, Enterprise and Regulatory Reform publications on energy

Energy Statistics
Monthly, quarterly and annual statistics on production and consumption of overall energy and individual fuels in the United Kingdom together with energy prices is available in MS Excel format on the Internet at: www.berr.gov.uk/energy/statistics/source/index.html.

Energy Trends
A quarterly publication covering all major aspects of energy. It provides a comprehensive picture of energy production and use and contains analysis of data and articles covering energy issues. Available on subscription, with Quarterly Energy Prices (see below). Annual subscriptions run from June to March and are available at £40 to UK subscribers from Amey Plc, 7th Floor, Clarence House, Clarence Place, Newport, Wales NP19 7AA, Tel. 01633 224712. A subscription form is available at: www.berr.gov.uk/energy/statistics/publications/trends/index.html An electronic version of the latest eight editions can be found at the same address. Single copies are available from the BERR Publications Orderline priced at £6.

Quarterly Energy Prices
A quarterly publication containing tables, charts and commentary covering energy prices to domestic and industrial consumers for all the major fuels as well as presenting comparisons of fuel prices in the European Union and G7 countries. Available on subscription, with Energy Trends, (details given above). An electronic version of the latest eight editions can be found at www.berr.gov.uk/energy/statistics/publications/prices/index.html Single copies are available from the BERR Publications Orderline priced at £8.

UK Energy in Brief 2008
An annual publication summarising the latest statistics on energy production, consumption and prices in the United Kingdom. The figures are taken from "Digest of UK Energy Statistics". Available free from BERR, EMU/ESD, Bay 299, 1 Victoria Street, London, SW1H 0ET, tel. 020-7215 2697/2698 and from the BERR Publications Orderline. It is also available on the web site at: www.berr.gov.uk/energy/statistics/publications/in-brief/page17222.html

Energy Flow Chart

A triennial publication illustrating the flow of primary fuels from home production and imports to their eventual final uses. They are shown in their original state and after being converted into different kinds of energy by the secondary fuel producers. The 2008 edition of the chart shows the flows for 2007. Available free from BERR, EMU/ESD, Bay 299, 1 Victoria Street, London, SW1H 0ET, tel. 020-7215 2697/2698 and from the BERR Publications Orderline. It is also available on the web site at: www.berr.gov.uk/energy/statistics/publications/flowchart/page37716.html

UK Energy Sector Indicators 2008

An annual publication designed to show in headline form the progress that has been made in implementing the four key energy policy goals as set out in the 2003 Energy White Paper, and reiterated in the 2007 Energy White Paper. Available free from BERR, EMU/ESD, Bay 299, 1 Victoria Street, London, SW1H 0ET, tel. 020-7215 2697/2698 and from the BERR Publications Orderline. It is also available on the web site at:
www.berr.gov.uk/energy/statistics/publications/indicators/page46000.html. A further set of background indicators (charts and tables) will be available on the BERR website (address as above) in October 2008.

Energy Consumption in the United Kingdom

Energy consumption in the United Kingdom brings together statistics from a variety of sources to produce a comprehensive review of energy consumption in the UK since the 1970s. The data describes the key trends in energy consumption in the UK since 1970 with a particular focus on trends since 1990. The information is presented in five sections covering overall energy consumption and energy consumption in the transport, domestic, industrial and service sectors. It includes an analysis of the factors driving the changes in energy consumption, the impact of increasing activity, increased efficiency, and structural change in the economy. It is also available on the web site at:
www.berr.gov.uk/energy/statistics/publications/ecuk/page17658.html

Energy White Paper, 'Meeting the Energy Challenge'

The Government's Energy White Paper, 'Meeting the Energy Challenge' was published by the Secretary of State for Trade and Industry on 23 May 2007, and sets out the Government's international and domestic energy strategy. It shows how the Government are implementing the measures set out in the Energy Review Report in 2006, as well as those announced since, including in the Pre-Budget Report in 2006 and the Budget in 2007. The White Paper is available on the web site at: www.berr.gov.uk/energy/whitepaper/page39534.html and in hard copy from The Stationery Office.

Fifth Annual Report on progress towards the 2003 Energy White Paper goals

The Government's 2003 Energy White Paper, "Our energy future - creating a low carbon economy", was published by the Secretary of State for Trade and Industry on 24 February 2003. This report describes progress made towards: cutting the United Kingdom's carbon emissions; maintaining the reliability of the UK's energy supplies; promoting competitive energy markets in the UK; and reducing the number of people living in fuel poverty in the UK. The 2003 Energy White Paper is available on the BERR web site at: www.berr.gov.uk/energy/policy-strategy/energy-white-paper-2003/page21223.html and in hard copy from The Stationery Office. The 5th Annual Progress Report was published on 17 July 2008 and is available at: www.berr.gov.uk/files/file46645.pdf

Energy Markets Outlook

The Energy Markets Outlook report provides energy market information on security of supply, looking forward over a fifteen-year time span. The intention is to help develop a shared understanding of the longer-term outlook for energy supply and demand, and to help understand emerging risks that could affect security of supply. Available free from BERR (020 7215 6566). It is also available on the web site at: www.berr.gov.uk/energy/energymarketsoutlook/page41839.html

Energy Bill
The Energy Bill was published by the Secretary of State for Business, Enterprise and Regulatory Reform on 10 January 2008 in tandem with the Nuclear White Paper. The Bill, alongside the Climate Change and Planning Bills, contains the legislative provisions required to implement UK energy policy following the publication of the Energy Review 2006 and the Energy White Paper 2007. This policy is driven by the two long-term energy challenges faced by the UK: tackling climate change by reducing carbon dioxide emissions, and ensuring secure, clean and affordable energy. The Bill has been published on the internet at: http://services.parliament.uk/bills/2007-08/energy.html

Nuclear White Paper, 'Meeting the energy challenge: a White Paper on nuclear power'
The Government's Nuclear White Paper, 'Meeting the energy challenge: a White Paper on nuclear power'' was published by the Secretary of State for Business, Enterprise and Regulatory Reform on 10 January 2008. The White Paper is the response to the Government's public consultation "The Future of Nuclear Power" on whether it is in the public interest to allow energy companies to invest in new nuclear power stations. The White Paper is available on the BERR website at: www.berr.gov.uk/energy/nuclear-whitepaper/page42765.html and in hard copy from The Stationery Office.

Climate Change Bill
The Climate Change Bill was published by the Secretary of State for Environment, Food and Rural Affairs on 14 November 2007. The Bill will create a new approach to managing and responding to climate change in the UK through: setting ambitious targets, taking powers to help achieve them, strengthening the institutional framework, enhancing the UK's ability to adapt to the impact of climate change and establishing clear and regular accountability to the UK, Parliament and devolved legislatures. The Bill contains provisions that will set a legally binding target for reducing UK carbon dioxide emission by at least 26 per cent by 2020 and at least 60 per cent by 2050, compared to 1990 levels. The Bill has been published on the internet at: http://services.parliament.uk/bills/2007-08/climatechangehl.html

Planning Bill
The Planning Bill was published by the Secretary of State for Communities and Local Government on 27 November 2007. The Bill introduces a new system for approving major infrastructure of national importance, such as energy developments like nuclear power, and replaces current regimes under several pieces of legislation. The objective is to streamline these decisions and avoid long public inquiries. The Bill has been published on the internet at: http://services.parliament.uk/bills/2007-08/planning.html

UK Energy and CO2 emissions projection: updated projections to 2020
This report provides key information on updated energy and emission projections. The latest projections have been updated for the Energy White Paper, and show the impact of the Energy White Paper measures and the EU ETS on energy demand, energy mix and carbon emissions between now and 2020. The report is to be found at: www.berr.gov.uk/files/file39580.pdf. Previous projections are available at: www.berr.gov.uk/files/file26363.pdf

The UK Fuel Poverty Strategy: November 2001
Produced by BERR and Defra. The strategy sets out the Government's objectives, policies and targets for alleviating fuel poverty in the UK over the next 10 years. Available free from the BERR Publications Orderline and on the web site at: www.berr.gov.uk/files/file16495.pdf

The UK Fuel Poverty Strategy, 5th Annual Progress Report 2007
Produced by Defra and BERR in association with the Devolved Administrations. This report sets out the progress that has been made on tackling fuel poverty and is available on the website at www.berr.gov.uk/files/file42720.pdf. It is accompanied by detailed annexes also on the website at: www.berr.gov.uk/energy/fuel-poverty/strategy/index.html. The fifth annual report is also available free from BERR's Publications Orderline.

Other publications including energy information

General
Digest of Welsh Statistics (annual); *Welsh Assembly Government*
Eurostat Yearbook (annual); *Statistical Office of the European Communities - Eurostat*
Eurostatistics - Data for Short Term Analysis; *Statistical Office of the European Communities – Eurostat*
High Level Summary of Statistics: Key Trends for Scotland; *Scottish Government*
Monthly Digest of Statistics; *Office for National Statistics*
Northern Ireland Annual Abstract of Statistics; *Northern Ireland Statistics and Research Agency*
Overseas Trade Statistics of the United Kingdom; *H.M. Revenue and Customs*
      - Business Monitor OTS1 (monthly) (trade with countries outside the EC)
      - Business Monitor OTS2 (monthly) (trade with the EC and the world)
      - Business Monitor OTSQ (quarterly) (trade with the EC)
      - Business Monitor OTSA (annually) (trade with the EC and the world)
Regional Trends (annual); *Office for National Statistics*
Regional Yearbook (annual); *Statistical Office of the European Communities – Eurostat*
Statistics in Focus - energy and industry (ad hoc); *Statistical Office of the European Communities - Eurostat*
United Kingdom Minerals Yearbook; *British Geological Survey*

Energy
BP Statistical Review of World Energy (annual); *BP*
Energy - Monthly Statistics; *Statistical Office of the European Communities – Eurostat*
Energy - Yearly Statistics; *Statistical Office of the European Communities – Eurostat*
Energy Statistics and Balances of Non-OECD Countries (annual); *International Energy Agency*
Energy Statistics and Balances of OECD Countries (annual); *International Energy Agency*
UN Energy Statistical Yearbook (annual); *United Nations Statistical Office*

Coal
Annual Reports and Accounts of The Coal Authority and the private coal companies; (*apply to the Headquarters of the company concerned*)
Coal Information (annual); *International Energy Agency*
Coal Statistics (quarterly); *International Energy Agency*

Electricity
Annual Report of The Office of Gas and Electricity Markets; *OFGEM*
Annual Reports and Accounts of the Electricity Supply Companies, Distributed Companies and Generators; (*apply to the Headquarters of the company concerned*)
Electricity Information (annual); *International Energy Agency*
Electricity Statistics (quarterly); *International Energy Agency*
National Grid - Seven Year Statement - (annual); *National Grid*

Environment
e-Digest of Environmental Statistics; *Department for Environment, Food and Rural Affairs (Defra)*.
Sustainable development indicators in your pocket; *Department for Environment, Food and Rural Affairs (Defra)*

Oil and gas
Annual Reports and Accounts of National Grid, Centrica and other independent gas supply companies; (contact *the Headquarters of the company concerned directly*)
Oil and Gas Information (annual); *International Energy Agency*
Oil and Gas Statistics (quarterly); *International Energy Agency*
UK Petroleum Industry Statistics Consumption and Refinery Production (annual and quarterly); *Energy Institute*

Prices

Energy Prices and Taxes (quarterly); *International Energy Agency*

Gas and Electricity Market Statistics (annual); *Statistical Office of the European Communities - Eurostat (substitutes the previous publications "Energy prices", "Gas prices" and "Electricity prices")*

Renewables

Renewables Information (annual); *International Energy Agency*

Useful energy related web sites

The BERR web site can be found at www.berr.gov.uk, the energy information and statistics web site is at www.berr.gov.uk/energy/statistics/index.html

Other Government web sites

Central Office of Information	www.coi.gov.uk
Department for Communities and Local Government.	www.communities.gov.uk
Department for Environment, Food and Rural Affairs	www.defra.gov.uk
Department for Transport	www.dft.gov.uk
HM Government Online	www.direct.gov.uk
HM Revenue and Customs	www.hmrc.gov.uk
Northern Ireland Executive	www.northernireland.gov.uk
Ofgem (The Office of Gas and Electricity Markets)	www.ofgem.gov.uk
The Scottish Government	www.scotland.gov.uk
The Scottish Parliament	www.scottish.parliament.uk
UK Parliament	www.parliament.uk
UK Statistics Authority	www.statisticsauthority.gov.uk
Welsh Assembly Government	www.wales.gov.uk

Other useful energy related web sites

AEA Energy & Environment	www.aea-energy-and-environment.com
Association of Electricity Producers	www.aepuk.com
BP	www.bp.com
British Geological Survey	www.bgs.ac.uk
British Wind Energy Association	www.bwea.com
Building Research Establishment	www.bre.co.uk
Coal Authority	www.coal.gov.uk
Energywatch	www.energywatch.org.uk
Energy Institute	www.energyinst.org.uk
Energy Networks Association	www.energynetworks.org
Environmental Industries Sector Unit	www.eisu.org.uk
Europa (European Union Online)	http://europa.eu/
Eurostat	http://epp.eurostat.ec.europa.eu/
Interconnector (UK) Ltd	www.interconnector.com
International Energy Agency	www.iea.org
Iron and Steel Statistics Bureau	www.issb.co.uk
National Grid	www.nationalgrid.com
Oil & Gas UK	www.ukooa.co.uk
The Stationery Office	www.tso.co.uk
UK Air Quality Archive	www.airquality.co.uk
UK Petroleum Industry Association	www.ukpia.com
United Nations Statistics Division	http://unstats.un.org/unsd/default.htm
US Department of Energy	www.energy.gov
US Energy Information Administration	www.eia.doe.gov

Annex D
Major events in the Energy Industry

2008

Energy Bill

The Energy Bill was published in January 2008 in tandem with the Nuclear White Paper. The purpose of the Bill, alongside the Climate Change and Planning Bills, is to update and strengthen the legislative framework so that it is appropriate for today's energy market and fit for the challenges to be faced on climate change and security of supply. Key elements of the bill are:

- Create a regulatory framework to enable private sector investment in Carbon Capture and Storage projects while also protecting the environment;
- Ensure adequate funding provision be made by potential developers of new nuclear power stations to pay the full costs of decommissioning and their full share of waste management costs.
- Strengthen and simplify the regulatory framework to give investors more clarity and certainty, reducing costs and risks for private sector investment in offshore gas supply projects such as offshore storage and liquefied natural gas infrastructure.
- Strengthen the Renewables Obligation (RO) to drive greater and more rapid development of renewables in the UK. Proposals include amending the RO to give more support to new and emerging technologies such as offshore wind, wave and tidal by banding the Obligation.
- Measures to be brought forward for offshore renewables decommissioning, ensuring that companies have adequate decommissioning funds so that both the tax payer and the offshore environment is protected.

Carbon Capture and Storage

The Government published, in June 2008, a consultation on the legislative framework for Carbon Capture and Storage (CCS), including carbon capture readiness. The consultation sets out the Government's views on CCS as a 'high potential' carbon abatement technology and asks for views on what more can be done to promote, develop and deploy CCS in the UK, EU and globally.

Coal

Tower Colliery, the last deep mine in Wales, closed in January 2008, thirteen years after its workforce rescued it from the pit closure programme.

Electricity

The Government announced, in January 2008, its support for electricity generated from geopressure through the Renewables Obligation scheme, which provides companies using green energy sources with assistance in competing with fossil fuel generators.

Emissions Trading

The proposals announced by the European Commission in January 2008 for tackling climate change and delivering a low carbon economy in Europe put the EU Emissions Trading Scheme at the heart of EU climate policy, including establishing an EU wide central cap on emissions covered by the EU ETS to 2020 and beyond.

EU Energy Review

The European Commission's proposals for tackling climate change and delivering a low carbon economy in Europe were announced in January 2008. The proposals implement the decisions agreed by EU Heads of State and Government at the 2007 Spring European Council. For the UK, the Commission's proposals include:

- a reduction of 16 per cent in UK greenhouse gas emissions from sectors not covered by the EU ETS by 2020 from 2005 levels;
- for 15 per cent of the energy consumed in the UK to come from renewable sources by 2020;
- for 10 per cent of road transport fuels to come from renewable sources, subject to them being produced in a sustainable way.

Fuel Poverty

A raft of new measures was agreed at a Fuel Poverty Summit hosted by OFGEM in May 2008 to help vulnerable consumers access the best available tariffs.

In April 2008, the six largest energy suppliers individually agreed to spend an extra £225m over three years to help those squeezed by rising fuel bills, which could lift 100,000 households out of fuel poverty.

Heat

In January 2008, following a commitment in the 2007 Energy White Paper, BERR along with Defra and DCLG published the Heat Call for Evidence. The Call for Evidence will play an important part in developing a strategy for heat given that half of all UK's CO_2 emissions arise from the use of heat.

Nuclear

In June 2008 the Secretary of State for Business, Enterprise and Regulatory Reform announced the creation of a Nuclear Development Forum to provide regular discussions between Government and industry, and an Office of Nuclear Development to provide a single focus within Government on the development of new nuclear.

The Government's response to its nuclear consultation, in the form of a White Paper, was published alongside the Energy Bill in January 2008. Following the consultation, the Government has now decided that it is in the public interest to allow private sector energy companies to invest in new nuclear power stations. Building of new nuclear power stations is expected to commence in 2013-2014 with operation commencing in 2017-2020.

Oil and Gas

Brent crude oil prices topped $100 a barrel for the first time in March 2008, and rose as high as $141 a barrel in June 2008.

Industrial action by oil workers at the Grangemouth refinery in April 2008 led to the temporary closure of the Forties oil pipeline, which provides 30% of the UK's daily oil output from the North Sea.

A record breaking 2,297 blocks or part blocks in UK waters were offered, in February 2008, for exploration in the 25th Offshore Oil and Gas Licensing Round.

Renewables

The Government published, in June 2008, a consultation on the UK Renewable Energy Strategy. The consultation puts forward a package of measures to drive up the use and deployment of renewable energy as part of the UK goal to tackle climate change and ensure security of supply and to enable the UK to meet its EU 2020 target.

2008 (continued) The Crown Estate launched, in June 2008, its round 3 leasing programme for the delivery of up to 25 GW (gigawatts) of new offshore windfarm sites by 2020. The announcement was made at the BWEA (British Wind Energy Association) conference in central London.

The UK's offshore renewable industry will benefit from measures, jointly announced by the Government and Ofgem in January 2008, to connect at least £2 billion of investment to the national grid. The investment will support the delivery of the necessary infrastructure to connect 8 gigawatts of planned offshore wind generation.

2007 ### Energy Policy
The Planning Reform Bill was published in November 2007. The Bill will make the planning system quicker, more transparent and easier for the public to become involved in, and will reform the planning system for major infrastructure projects including climate change and energy security. At a local level, the Bill and other reforms published at the same time will allow people to install small-scale renewable power sources such as solar panels and wind turbines without planning permission if they do not affect their neighbours.

The Energy Markets Outlook report, published in October 2007, provides energy market information on security of supply, looking forward over a fifteen-year time span. The report is intended to help develop a shared understanding of the longer-term outlook for energy supply and demand, and to help understand emerging risks that could affect security of supply.

Machinery of Government changes announced in June 2007 resulted in Energy Policy being transferred from the former Department of Trade and Industry to the new Department for Business, Enterprise and Regulatory Reform.

Energy White paper
The 2007 Energy White Paper (EWP), 'Meeting the Energy Challenge' was published in May 2007. The paper reiterated the Government's commitment to the four key energy policy goals. The EWP, in response to the twin challenges of climate change and security of supply announced a strategy to deliver energy security and accelerate the transition to a low carbon economy. The key elements of the strategy are:
- Establish an international framework to tackle climate change;
- Provide legally binding carbon targets for the whole UK economy, progressively reducing emissions;
- Make further progress in achieving fully competitive and transparent international markets;
- Encourage more energy saving through better information, incentives and regulations;
- Provide more support for low carbon technologies; and
- Ensure the right conditions for investment, including improvements to the planning system.

Carbon Capture and Storage
The launch of the competition to build one of the world's first commercial-scale carbon capture and storage (CCS) plants was among the measures set out in a major speech on climate change given by the Prime Minister at a WWF event in November 2007.

The 2007 Energy White Paper provided further details of the competition to develop the UK's first commercial-scale demonstration of a Carbon Capture and Storage (CCS) Plant. The CCS plant will be due to be operational early in the next decade. This competition was announced in the 2007 Budget.

Carbon Reduction Commitment

The 2007 Energy White Paper outlined plans to introduce a mandatory cap and trade scheme, a Carbon Reduction Commitment, which will apply to the largest non-energy intensive public and private sector organisations.

Climate Change

The Climate Change Bill was published in November 2007. The Bill sets out a framework that will put Britain on the path to become a low-carbon economy, with clear, legally binding targets to reduce carbon dioxide emissions by at least 60% by 2050, and 26 to 32% by 2020, against 1990 levels. As well as setting clear targets, the Bill provides a pathway to achieve those reductions through a system of five-year carbon budgets set fifteen years ahead.

The Government published the Energy Measures Report, Addressing Climate Change and Fuel Poverty – energy measures information for Local Government, in September 2007. The report sets out the steps that local authorities can take to: improve energy efficiency; increase the levels of microgeneration and low carbon technologies; reduce greenhouse gas emissions; and reduce the number of households living in fuel poverty.

In September 2007 the location of the Energy Technologies Institute (ETI), backed by up to £550m of Government investment announced by the Chancellor in the 2006 Budget, was announced. The ETI, based in Loughborough, will bring more focus, ambition and collaboration to the UK's energy, science and engineering drive. It will have a potential budget of over £1bn.

The Government's blueprint for tackling climate change was published in March 2007. The draft bill, the first of its kind in any country, and accompanying strategy, set out a framework for moving the UK to a low-carbon economy.

Electricity

In May 2007 National Grid and TenneT Holding announced plans to construct a 260km, 1,000 MW electricity interconnector "BritNed" between the Netherlands and the UK. It is expected to be commissioned by late 2010.

Severe weather caused widespread damage to power lines in England and Wales, with supplies to over 1.2m customers being affected in January 2007. The Electricity industry performed well in restoring supply, with over 90% of interrupted supplies back on within 24 hours.

Emissions Trading

The Government published the National Allocation Plan (NAP) in March 2007; the plan includes a list of the allowances to be allocated to individual installations covered by the EU ETS for each year in the second phase.

EU Energy Review

The Council of the European Union at its Spring meeting in March 2007 called on Member States and EU institutions to pursue actions to develop a sustainable integrated European climate and energy policy. These include increasing security of supply; ensuring the competitiveness of European economies and the availability of affordable energy, and promoting environmental sustainability and combating climate change.

The European Commission published its Strategic Energy Review in January 2007, outlining proposals for the development of the internal energy market in the European Union. These include the greater unbundling of energy network businesses from other activities, more effective regulation and greater transparency.

Fuel Poverty

The Government published its UK Fuel Poverty Strategy Fifth Annual Progress Report in December 2007, reporting on the progress made since the last report, highlighting key areas for attention during the coming year, and setting out the fuel poverty figures for 2005. As part of the progress report, annexes were also produced on methodology, detailed analyses of the fuel poor, fuel poverty monitoring and company schemes and case studies.

Measures worth £2.3bn over three years to tackle fuel poverty and home energy efficiency, the Carbon Emissions Reduction Target (CERT), were laid in Parliament in December 2007. The Government also announced an £800m three year grant for the Warm Front Scheme, which could assist 400,000 of the poorest households in England.

The 2007 Energy White Paper outlined a range of new policies which will see a further 200,000 households being taken out of fuel poverty by 2010. These included: providing a benefit entitlement check to all households that require one; enabling the sharing of benefit information; putting in place a cross-Government communications campaign in time for next winter, and encouraging energy suppliers to do more.

Nuclear

In May 2007 alongside the Energy White Paper a consultation on the role of nuclear power in a low carbon UK economy was launched. The consultation seeks views on whether the private sector should be allowed to build new nuclear power stations.

In February 2007 the High Court in a judicial review decided that the consultation process on new nuclear power that preceded the Energy Review Report had not been adequate. The Government then decided that a new consultation on nuclear power was required.

Oil and Gas

In October 2007 the Langeled pipeline to Easington in Yorkshire from the Ormen Lange field in Norway, became connected directly to the British gas network.

The May 2007 Energy White Paper announced legislation to allow the storage of natural gas under the seabed and unloading of Liquefied Natural Gas at sea.

A milestone was reached in March 2007 when three new fields were approved, taking the total number of offshore oil and gas developments in the North Sea's 40 years history to 350.

The Excelerate project, a world first 'liquid gas' shipment on Teesside, with enough power in one 20,000 tonne shipment to fuel 60,000 homes began operation in February 2007.

The very large Buzzard field, one of the largest oil discoveries to be developed in British waters in more than a decade, started production in January 2007.

Renewables

Harnessing the vast potential of the UK's island status entered a new phase in December 2007 when the Government announced proposals to open up its seas to up to 33GW (gigawatts) of offshore wind energy.

A major expansion of energy from renewable sources was among the measures set out in a major speech on climate change given by the Prime Minister at a WWF event in November 2007. It was announced that tidal lagoons and barrages below one gigawatt capacity will receive extra support through the Renewables Obligation, potentially benefiting lagoons proposed for Rhyl, Swansea Bay and elsewhere.

A new report, Essential role of renewables generation in achieving zero carbon homes, published in November 2007, from the Renewables Advisory Board (RAB) which advises Government on renewable energy issues, provided the first in depth analysis of the role of on site energy generation in the delivery of the Government's policy of ensuring that all new homes are zero carbon from 2016. Amongst it findings is the conclusion that the policy could drive a market for onsite renewable worth £2.3 billion a year from 2016.

The May 2007 Energy White Paper announced legislation to band the Renewables Obligation to benefit offshore wind, wave, tidal and other emerging technologies.

The Government announced the development of the licensing regime for would be electricity transmission owners in the UK's pioneering offshore wind energy sector in March 2007. After a joint consultation the DTI decided in agreement with Ofgem that Britain's monopoly electricity transmission network owners will have the opportunity to compete against a wider range of transmission companies to build, own and maintain the links.

The opening of the Braes and Doune wind farm in February 2007 took the UK's wind generation capacity above 2 GW, making the UK one of only 8 countries in the world to have achieved this level.

Sustainable Energy Policy

The Fourth Annual Report on progress towards the 2003 Energy White Paper goals was published in May 2007, reviewing progress made over the last 12 months. Published as a supplement to the Fourth Annual Report, UK Energy Sector Indicators was also published in May 2007.

2006 **Sustainable Energy Policy**

On 16th February 2006 the Government issued a consultation paper on carbon dioxide emissions projections for industrial sectors covered by the EU Emissions Trading Scheme. These projections are being used as an input to the development of the UK National Allocation Plan for Phase II of the scheme, informing the allocation of carbon dioxide allowances to installations in the 2008-12 period. The consultation closes on April 13th.

The Third Annual Report to the Energy White Paper was published in May 2006, reviewing progress made over the last 12 months towards the targets and strategy for energy policy until 2050. Published as a supplement to the Third Annual Report, Energy Sector Indicators 2006 was also published in May 2006.

Energy Review

The Energy Review Report "The Energy Challenge" detailing what needs to be done to stay on track to meeting the goals in the 2003 Energy White Paper was published in July 2006. The Report outlined proposals designed to reduce demand, secure a mix of clean, low-carbon sources, and streamline the planning process for energy projects.

The 12 week consultation period for the energy review ended on the 12 April 2006 with over 5,300 responses from individuals, businesses, academics and NGOs. A summary and analysis of the responses is available on the BERR website at: www.berr.gov.uk/files/file31631.pdf

Climate Change

The Office of Climate Change (OCC) was established in October 2006. It is a shared resource across Government established with the aim of ensuring that analysis and policy work is consistent and supports the overall climate change strategy.

The Stern Review of economics of climate change was published in October 2006 and confirmed that climate change is real and is a problem that can only be solved by collective international action. The Review demonstrated that urgent action is needed to mitigate the effects of climate change and that the costs of global action to mitigate the most dangerous effects of climate change are significant but manageable, as long as action is taken multilaterally.

An ambitious programme to tackle climate change domestically and to secure agreement on action to reduce global greenhouse gas emissions was published by the Government on 28 March 2006. The Programme is expected to reduce the UK's emissions of greenhouse gases to 23-25 per cent below base year levels and reduce the UK's carbon dioxide emissions to 15-18 per cent below 1990 levels by 2010.

Fuel Poverty

The Government published its UK Fuel Poverty Strategy Fourth Annual Report in June 2006, reporting on the progress made since the last year's report, and setting out the fuel poverty figures for 2004. Also published were a series of annexes, setting out the detailed profile of the fuel poor, as well as outlining the many activities in which energy companies are involved to tackle fuel poverty and associated problems. In the report, we also responded to the recommendations made by The Fuel Poverty Advisory Group in its Third Annual Report, which was published in March 2006. The new Home Heat Helpline operated by the Energy Retail Association has been up and running since October 2005.

Emissions Trading

The UK's National Allocation Plan (NAP) for the second phase of the EU Emissions Trading Scheme (2008-2012) was accepted without change by the European Commission in December 2006.

On 21st March 2006 the Government published 15 (independently produced) reports reviewing and revising methodologies for calculating the New Entrants' benchmarks for sectors covered by Phase II of the EU – Emission Trading Scheme. The benchmarks will form the basis for the calculation of New Entrant allocations of carbon dioxide allowances to installations in the 2008-2012 period and will therefore form an important part of the development of the UK National Allocation Plan for Phase II of the EU – Emissions Trading Scheme

2006 (continued) The draft Phase II National Allocation Plan was published for consultation, alongside the Climate Change Programme Review, on 28 March. The number of CO2 allowances for Phase II will be set within a range representing a reduction of 3 to 8 MtC a year against projected business as usual emissions. The final decision on this should be made in June. The NAP also incorporate proposals on scope (some new sectors added in); allocation methodology; sector classification and new entrant policy.

Coal

British Gas owner Centrica announced in November 2006 that it wants to build the UK's first clean coal power plant, which could supply electricity to a million homes. The company has reached agreement with developer Progressive Energy to develop a station on Teesside, combining clean coal and carbon capture technology. Gasification technology will produce synthetic gas from coal, and the resulting carbon emissions would then be captured and stored. The Teesside plant would be the first coal-fired station since Drax was built in 1974. Centrica said it would have the fewest emissions of any fossil fuel power station in Britain.

UK Coal's Rossington colliery closed on 31 March 2006.

Oil and Gas

Market investment in new and enhanced UK gas infrastructure has continued with the completion of the Langeled pipeline in September 2006 and Balgzand-Bacton Line (BBL) pipeline in December 2006 allowing increased flows of Norwegian and Continental gas to the UK.

A milestone in the history of the UK's offshore oil and gas industry was reached with the start of drilling on the 10,000[th] well in the seabed around the British Isles in October 2006.

The UK is expected to be a net importer of oil and oil products in 2006, returning to being a net exporter in 2007 as a result of the very large Buzzard field that is due to commence production in the fourth quarter of 2006.

Electricity

Centrica announced in June 2006 that it would construct an 885 MW CCGT power station at Langage, Plymouth which is expected to be operational by 2008.

Renewables

Following the Energy Review, a two part consultation document was published in October 2006. Part 1 consulted on proposals to introduce changes to the Renewables Obligation and Part 2 was a statutory consultation on a small limited number of changes to the Renewables Obligation which came into effect from April 2007.

DTI announced development consent approval to two offshore windfarms, both in the Thames Estuary, in December 2006. London Array, at 1,000 megawatts, will be the largest offshore windfarm in the world when completed, while Thanet will be 300 megawatts.

The Renewables Obligation Order 2006 has been laid before Parliament. It will come into force on 1 April 2006.

A recent Carbon Trust report has concluded that wave and tidal could in time provide up to a fifth of UK's energy needs.

Winter Energy Supply
In July 2006, the Government established the Business Energy Forum to ensure that sound preparations were made for winter. This is a high level group, jointly chaired by BERR and CBI and bringing together Ofgem, National Grid, energy suppliers and users and other key players in the energy industry. BERR also created a dedicated page on its website, to provide information and signposting on winter energy supply issues.

The second annual report to Parliament on the security of gas and electricity supplies in Great Britain as required under section 172 of the Energy Act 2004 was published in July 2006.

For major events in earlier years see the BERR web site version of this annex at:
www.berr.gov.uk/energy/statistics/publications/dukes/page45537.html

Notes